Lecture Notes in Computer Science　　11312

Commenced Publication in 1973
Founding and Former Series Editors:
Gerhard Goos, Juris Hartmanis, and Jan van Leeuwen

More information about this series at http://www.springer.com/series/7407

Leah Epstein · Thomas Erlebach (Eds.)

Approximation and Online Algorithms

16th International Workshop, WAOA 2018
Helsinki, Finland, August 23–24, 2018
Revised Selected Papers

Editors
Leah Epstein
University of Haifa
Haifa, Israel

Thomas Erlebach
University of Leicester
Leicester, UK

ISSN 0302-9743   ISSN 1611-3349   (electronic)
Lecture Notes in Computer Science
ISBN 978-3-030-04692-7   ISBN 978-3-030-04693-4   (eBook)
https://doi.org/10.1007/978-3-030-04693-4

Library of Congress Control Number: 2018962355

LNCS Sublibrary: SL1 – Theoretical Computer Science and General Issues

This Springer imprint is published by the registered company Springer Nature Switzerland AG
The registered company address is: Gewerbestrasse 11, 6330 Cham, Switzerland

Preface

The 16th Workshop on Approximation and Online Algorithms (WAOA 2018) focused on the design and analysis of algorithms for online and computationally hard problems. Both kinds of problems have a large number of applications in a variety of fields. WAOA 2018 took place in Helsinki, Finland, during August 23–24, 2018 and was a success: It featured many interesting presentations and provided opportunity for stimulating discussions and interactions. WAOA 2018 was part of the ALGO 2018 event that also hosted ALGOCLOUD, ALGOSENSORS, ATMOS, ESA, IPEC, and WABI.

Topics of interest for WAOA 2018 were: graph algorithms, inapproximability results, network design, packing and covering, paradigms for the design and analysis of approximation and online algorithms, parameterized complexity, scheduling problems, algorithmic game theory, algorithmic trading, coloring and partitioning, competitive analysis, computational advertising, computational finance, cuts and connectivity, geometric problems, mechanism design, resource augmentation, and real-world applications. In response to the call for papers, we received 44 submissions. One submission was rejected as out of scope right away, and each of the remaining submissions was reviewed by at least three referees. The submissions were mainly judged on originality, technical quality, and relevance to the topics of the conference. Based on the reviews, the Program Committee selected 19 papers. This volume contains final revised versions of these papers as well as an invited contribution by our invited speaker Gerhard Woeginger. The EasyChair conference system was used to manage the electronic submissions, the review process, and the electronic Program Committee discussions. It made our task much easier.

We would like to thank all the authors who submitted papers to WAOA 2018 and all attendees of WAOA 2018, including the presenters of the accepted papers. A special thank you goes to the plenary invited speaker Gerhard Woeginger for accepting our invitation, giving a very nice talk, and contributing a paper to these proceedings. We would also like to thank the Program Committee members and the external reviewers for their diligent work in evaluating the submissions and their contributions to the electronic discussions. Furthermore, we are grateful to all the local organizers of ALGO 2018, especially the local co-chairs, Parinya Chalermsook, Petteri Kaski, and Jukka Suomela.

September 2018

Leah Epstein
Thomas Erlebach

Organization

Program Committee

Anna Adamaszek	University of Copenhagen, Denmark
János Balogh	University of Szeged, Hungary
Xujin Chen	Chinese Academy of Sciences, China
Leah Epstein (Co-chair)	University of Haifa, Israel
Thomas Erlebach (Co-chair)	University of Leicester, UK
Lene Monrad Favrholdt	University of Southern Denmark, Denmark
Kazuo Iwama	Kyoto University, Japan
Łukasz Jeż	University of Wrocław, Poland
Ralf Klasing	CNRS and University of Bordeaux, France
Kim-Manuel Klein	EPFL, Lausanne, Switzerland
Asaf Levin	The Technion, Israel
Minming Li	City University of Hong Kong, China
Friedhelm Meyer auf der Heide	Paderborn University, Germany
Gianpiero Monaco	University of L'Aquila, Italy
Sharath Raghvendra	Virginia Tech, Blacksburg, USA
Danny Segev	University of Haifa, Israel
Roberto Solis-Oba	The University of Western Ontario, Canada
Angelina Vidali	De Montfort University, UK
Alexander Wolff	Universität Würzburg, Germany

Additional Reviewers

Adamczyk, Marek
Berndt, Sebastian
Bilò, Davide
Bilò, Vittorio
Boyar, Joan
Bringmann, Karl
Byrka, Jarek
Böckenhauer, Hans-Joachim
Chau, Vincent
César San Felice, Mário
Dobrev, Stefan
Dosa, Gyorgy
Dürr, Christoph
Emek, Yuval

Engels, Christian
Fan, Chenglin
Feldkord, Björn
Feldotto, Matthias
Foucaud, Florent
Garncarek, Paweł
Gharibian, Sevag
Goldberg, Noam
Gouleakis, Themis
Grandoni, Fabrizio
Guo, Heng
Ikenmeyer, Christian
Kell, Nathaniel
Komm, Dennis

Kotrbcik, Michal
Kralovic, Rastislav
Kulkarni, Janardhan
Ladewig, Leon
Lahn, Nathaniel
Larsen, Kim S.
Levi, Reut
Lonc, Zbigniew
Malatyali, Manuel
Mazauric, Dorian
Melissourgos, Themistoklis
Mestre, Julian
Moscardelli, Luca
Mäcker, Alexander
Mömke, Tobias

Nikolidaki, Katerina
Page, Daniel R.
Panigrahi, Debmalya
Rawitz, Dror
Rohwedder, Lars
Rotenberg, Eva
Sidford, Aaron
Spoerhase, Joachim
Sun, Kevin
van Stee, Rob
Velaj, Yllka
Vinci, Cosimo
Wang, Changjun
Westphal, Stephan

Contents

Invited Contribution

Some Easy and Some Not so Easy Geometric Optimization Problems 3
  Gerhard J. Woeginger

Regular Papers

Deterministic Min-Cost Matching with Delays 21
  Yossi Azar and Amit Jacob Fanani

Sequential Metric Dimension 36
  Julien Bensmail, Dorian Mazauric, Fionn Mc Inerney, Nicolas Nisse,
  and Stéphane Pérennes

A Primal-Dual Online Deterministic Algorithm for Matching with Delays . . . 51
  Marcin Bienkowski, Artur Kraska, Hsiang-Hsuan Liu,
  and Paweł Schmidt

Advice Complexity of Priority Algorithms 69
  Allan Borodin, Joan Boyar, Kim S. Larsen, and Denis Pankratov

Approximating Node-Weighted k-MST on Planar Graphs. 87
  Jarosław Byrka, Mateusz Lewandowski, and Joachim Spoerhase

Exploring Sparse Graphs with Advice (Extended Abstract). 102
  Hans-Joachim Böckenhauer, Janosch Fuchs, and Walter Unger

Call Admission Problems on Grids with Advice (Extended Abstract). 118
  Hans-Joachim Böckenhauer, Dennis Komm, and Raphael Wegner

Improved Approximation Algorithms for Minimum Power
Covering Problems 134
  Gruia Calinescu, Guy Kortsarz, and Zeev Nutov

DISPATCH: An Optimally-Competitive Algorithm for Maximum Online
Perfect Bipartite Matching with i.i.d. Arrivals 149
  Minjun Chang, Dorit S. Hochbaum, Quico Spaen, and Mark Velednitsky

Strategic Contention Resolution in Multiple Channels 165
  George Christodoulou, Themistoklis Melissourgos, and Paul G. Spirakis

Sublinear Graph Augmentation for Fast Query Implementation. 181
  Artur Czumaj, Yishay Mansour, and Shai Vardi

Bin Packing Games with Weight Decision: How to Get a Small Value
for the Price of Anarchy   204
   Gyorgy Dosa, Hans Kellerer, and Zsolt Tuza

Probabilistic Embeddings of the Fréchet Distance   218
   Anne Driemel and Amer Krivošija

Algorithms for Dynamic NFV Workload   238
   Yaron Fairstein, Seffi (Joseph) Naor, and Danny Raz

Longest Increasing Subsequence Under Persistent Comparison Errors   259
   Barbara Geissmann

Cut Sparsifiers for Balanced Digraphs   277
   Motoki Ikeda and Shin-ichi Tanigawa

Reconfiguration of Graphs with Connectivity Constraints   295
   Nicolas Bousquet and Arnaud Mary

The Itinerant List Update Problem   310
   Neil Olver, Kirk Pruhs, Kevin Schewior, René Sitters, and Leen Stougie

The Price of Fixed Assignments in Stochastic Extensible Bin Packing   327
   Guillaume Sagnol, Daniel Schmidt genannt Waldschmidt,
   and Alexander Tesch

Author Index   349

Invited Contribution

Some Easy and Some Not so Easy Geometric Optimization Problems

Gerhard J. Woeginger[✉]

Department of Computer Science, RWTH Aachen, Aachen, Germany
woeginger@algo.rwth-aachen.de

Abstract. We survey complexity and approximability results for certain families of geometric optimization problems. We explain a generic approximation approach for maximization problems that is built around norms with polyhedral unit balls, and we pose a multitude of open problems.

Keywords: Combinatorial optimization · Approximation
Computational complexity · Geometry

1 Introduction

Let $\mathcal{G}$ be some fixed family of graphs. In this paper, we discuss geometric optimization problems of the following type.

Generic geometric optimization problem:
Instance: A finite set P of n points in the Cartesian space $\mathbb{R}^s$.
Goal: Choose an n-vertex graph from the family $\mathcal{G}$ and embed its vertices bijectively in P, so that the total length of the embedded edges is minimized/maximized.

If the family $\mathcal{G}$ consists of all n-vertex cycles C_n, then the resulting optimization problem is (a geometric special case of) the classic Travelling Salesman Problem [27]. If $\mathcal{G}$ consists of all perfect matchings, the resulting optimization problem is the geometric version of the classic matching problem [9]. And if $\mathcal{G}$ contains the graphs that are the disjoint union of 3-vertex cycles, then we get (a geometric special case of) the three-dimensional matching problem [9].

In the generic geometric optimization problem, the length of the embedded edges is measured according to some fixed norm. Recall that a norm is specified by its unit-ball $\mathcal{B} \subseteq \mathbb{R}^s$, which is a compact and convex set with non-empty interior, that is centrally symmetric with respect to the origin; see Fig. 1 for an illustration. The norm with unit ball $\mathcal{B}$ determines for any two points $x, y \in \mathbb{R}^s$ a distance $d(x, y)$ in the following way: First translate the underlying space so that point x coincides with the origin. Then determine the unique scaling factor $\lambda \geq 0$ by which one must rescale the unit ball $\mathcal{B}$ (shrinking for $\lambda < 1$, expanding

© Springer Nature Switzerland AG 2018
L. Epstein and T. Erlebach (Eds.): WAOA 2018, LNCS 11312, pp. 3–18, 2018.
https://doi.org/10.1007/978-3-030-04693-4_1

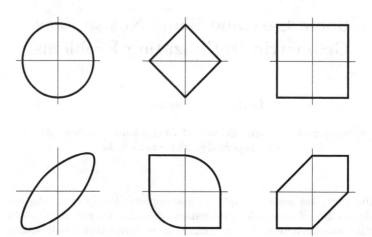

Fig. 1. Unit-balls for various norms in $\mathbb{R}^2$. The upper row contains the unit-balls of the Euclidean norm, the Manhattan norm, and the Maximum norm.

for $\lambda > 1$), such that point y lies on its boundary. The distance is then given by $d(x, y) = \lambda$. As $\mathcal{B}$ is centrally symmetric we have the symmetry $d(x, y) = d(y, x)$, and as $\mathcal{B}$ is convex we have the triangle inequality $d(x, y) + d(y, z) \geq d(x, z)$.

The most popular norms in $\mathbb{R}^s$ are the Euclidean norm, the Manhattan norm, and the Maximum norm. For two points $x = (x_1, \ldots, x_s)$ and $y = (y_1, \ldots, y_s)$ in s-dimensional space $\mathbb{R}^s$, the L_p distance for $1 \leq p < \infty$ is given by

$$d(x, y) = \left(\sum_{i=1}^{s} |x_i - y_i|^p \right)^{1/p}. \tag{1}$$

Equation (1) with $p = 1$ yields the Manhattan distance, and with $p = 2$ yields the Euclidean distance. The distance under the Maximum norm L_∞ is given by $d(x, y) = \max_{i=1}^{s} |x_i - y_i|$.

Now let us consider two unit-balls $\mathcal{B}_1$ and $\mathcal{B}_2$ that are *"very similar"* to each other, which in our context means that they satisfy for some small real number $\varepsilon > 0$ the relation

$$(1 - \varepsilon)\,\mathcal{B}_2 \subseteq \mathcal{B}_1 \subseteq (1 + \varepsilon)\,\mathcal{B}_2. \tag{2}$$

Then of course the distance $d_1(x, y)$ under the norm with unit-ball $\mathcal{B}_1$ and the distance $d_2(x, y)$ under the norm with unit-ball $\mathcal{B}_2$ satisfy the analogous relation

$$(1 - \varepsilon)d_2(x, y) \leq d_1(x, y) \leq (1 + \varepsilon)d_2(x, y). \tag{3}$$

See Fig. 2 for an illustration.

Crucial observation 1. *Let $\mathcal{G}$ be a family of graphs, and let $\mathcal{B}_1$ and $\mathcal{B}_2$ be two unit-balls that satisfy (2). Suppose that the geometric optimization problem for $\mathcal{G}$ under the norm with unit-ball $\mathcal{B}_2$ is polynomially solvable. Then the geometric optimization problem for $\mathcal{G}$ under the norm with unit-ball $\mathcal{B}_1$ has a polynomial time approximation algorithm with worst case guarantee $(1 - \varepsilon)/(1 + \varepsilon)$ in*

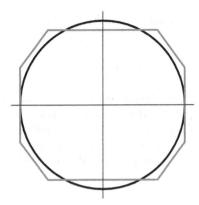

Fig. 2. Two unit-balls $\mathcal{B}_1$ and $\mathcal{B}_2$ that are "similar" to each other.

case of maximization and with worst case guarantee $(1 + \varepsilon)/(1 - \varepsilon)$ in case of minimization.

Here is the short proof of this observation for the case of minimization: Consider an optimal solution π_1 under unit-ball $\mathcal{B}_1$ with objective value v_1. Then the objective value of this (feasible) solution π_1 under the other unit-ball $\mathcal{B}_2$ is at most $v_1/(1 - \varepsilon)$. and hence the optimal solution π_2 under unit-ball $\mathcal{B}_2$ has objective value $v_2 \leq v_1/(1 - \varepsilon)$. Finally, note that solution π_2 can be computed in polynomial time and that its objective value under unit-ball $\mathcal{B}_1$ is at most $(1+\varepsilon)v_2$. The proof for maximization problems follows by symmetric arguments.

How useful is this crucial observation? For instance, what does it give us for the classic TSP, where one wants to find a Hamiltonian cycle of minimal total length? The answer to this question is simple and disappointing: For the minimization version of the TSP, the observation gives us nothing. Indeed, Itai, Papadimitriou and Swarcfiter [20] have shown that in $\mathbb{R}^2$ the TSP under the Euclidean norm is NP-hard. The construction in [20] also directly yields NP-hardness under the Manhattan norm, a rotation by 45 degrees yields NP-hardness under the Maximum norm, and some further minor modifications yield NP-hardness under an arbitrary norm. Hence, there are no unit-balls $\mathcal{B}_2$ to which the crucial observation could be applied.

Our next goal will be to show that the crucial observation yields strong approximation results for the maximization version of the TSP. For doing this, we first need to take a closer look at polyhedral norms.

2   Polyhedral Norms

A norm is called *polyhedral*, if its unit-ball is a polyhedron. Throughout the following sections, we will assume that the centrally symmetric polyhedral unit-ball $\mathcal{B}$ of such a polyhedral norm has $2f$ facets and is the intersection of the following halfspaces with normal vectors $h_1, \ldots, h_f$:

$$h_i \cdot x \leq 1$$
$$h_i \cdot x \geq -1$$

Note that we have scaled the normal vectors h_i so that all the right hand sides become ± 1. As an example, for the Manhattan norm in $\mathbb{R}^2$ the corresponding normal vectors are $h_1 = (1,1)$ and $h_2 = (1,-1)$, and for the Maximum norm in $\mathbb{R}^2$ the corresponding normal vectors are $h_1 = (1,0)$ and $h_2 = (0,1)$. Figure 3 shows the unit-ball of yet another polyhedral norm with $f = 3$.

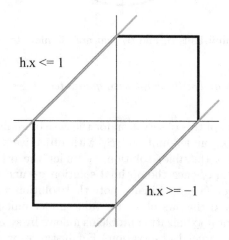

Fig. 3. The unit-ball of a polyhedral norm with $f = 3$.

The distance $d(x,y)$ between two points $x, y \in \mathbb{R}^s$ may then be written as

$$d(x,y) = \min \{\lambda : \ y \in x + \lambda \mathcal{B}\}$$
$$= \min \{\lambda : \ y - x \in \lambda \mathcal{B}\}$$
$$= \min \{\lambda : \ |h_i \cdot (y - x)| \leq \lambda \ \text{for } i = 1, \ldots, f\}$$
$$= \max \{|h_i \cdot (y - x)| : \ 1 \leq i \leq f\}$$
$$= \max \{h_i \cdot x - h_i \cdot y, \ -h_i \cdot x + h_i \cdot y : \ 1 \leq i \leq f\} \qquad (4)$$

Now let us consider a point set $P = \{p_1, \ldots, p_n\}$ that forms an instance of the generic geometric optimization problem. We create a corresponding edge-weighted bipartite multi-graph $G(P)$ with $n + f$ vertices in the following way:

– The vertices on one side of the bipartition are the points $p_1, \ldots, p_n$, and the vertices on the other side are the normal vectors $h_1, \ldots, h_f$.

– Every point p_j is connected to every normal vector h_i by two edges. There is a so-called red edge with weight $w^+(p_j, h_i) = p_j \cdot h_i$ and there is a so-called blue edge with weight $w^-(p_j, h_i) = -p_j \cdot h_i$.

Note that with this notation, the distance (4) between two points $x, y \in P$ can be rewritten as

$$d(x, y) = \max\{w^+(x, h_i) + w^-(y, h_i),$$
$$w^-(x, h_i) + w^+(y, h_i) : \ i = 1, \ldots, f\}$$

In the red-blue bipartite graph $G(P)$, there are f blue-red ways of going from x to y by first traversing a blue edge to some h_i and then traversing the red edge from h_i to y, and there are f red-blue ways that use a red edge followed by a blue edge. The distance between point x and point y is the largest weight of these $2f$ paths.

Crucial observation 2. *Let $\mathcal{G}$ be a family of graphs, and let $s \geq 2$ be some fixed integer. Suppose that for every polyhedral unit-ball $\mathcal{B}$ in $\mathbb{R}^s$, the geometric optimization problem for family $\mathcal{G}$ under the norm with unit-ball $\mathcal{B}$ is solvable in polynomial time. Then for every unit-ball $\mathcal{B}^*$ in $\mathbb{R}^s$, the geometric optimization problem for family $\mathcal{G}$ under the norm with unit-ball $\mathcal{B}^*$ possesses a polynomial time approximation scheme (PTAS).*

The correctness of this statement is straightforward: We simply approximate the unit-ball $\mathcal{B}^*$ by an appropriately chosen polyhedral unit-ball in the sense of (2). Observation 1 then yields an approximation algorithm with worst case guarantee $(1 - \varepsilon)/(1 + \varepsilon)$ for maximization and worst case guarantee $(1 + \varepsilon)/(1 - \varepsilon)$ for minimization problems. As ε tends to 0, these worst case guarantees tend to 1.

3 The Maximum TSP

Now let us turn to the maximization version of the Travelling Salesman Problem, Max-TSP for short.

Max-TSP
Instance: A set P of n points in Cartesian space $\mathbb{R}^s$; a unit-ball $\mathcal{B}$.
Goal: Find the longest round-trip through the points in P where distances are measured according to the norm with unit-ball $\mathcal{B}$.

In the general (non-geometric) Max-TSP, the distances between the points are specified explicitly as part of the input. This general version is of course NP-hard. APX-hardness of the general version follows from the arguments in Papadimitriou and Yannakakis [29]. The strongest known approximation algorithm for the general version has a worst case guarantee of 4/5 and is due to Dudycz, Marcinkowski, Paluch and Rybicki [17]. Kowalik and Mucha [24] reach an even better worst case guarantee of 7/8 for the cases where the distances satisfy the triangle inequality. The geometric versions of the Max-TSP behave in a more benevolent way. Serdyukov [32,33] and independently Barvinok [4] have shown that for all fixed dimensions and for all fixed norms, the Max-TSP has a PTAS. In this section we will discuss the PTAS of Barvinok, Fekete, Johnson, Tamir, Woeginger and Woodroofe [5] which is based on our Observations 1 and 2.

In the following paragraphs, we consider a fixed polyhedral unit-ball B with $2f$ facets in $\mathbb{R}^s$ and we determine the red-blue bipartite graph $G(P)$ for point set P as discussed in Sect. 2. A round-trip through the points in P is just a permutation π of the n points. We translate every permutation π into a corresponding subset $E(\pi)$ of the red and blue edges in $G(P)$: Whenever two points x and y are consecutive in the round-trip π, the edge set $E(\pi)$ contains the red-blue or blue-red edge pair that determines the distance $d(x, y)$ between x and y. Note that $E(\pi)$ contains n blue and n red edges, and note that the length of round-trip π equals the total weight of $E(\pi)$. As an example, Fig. 4 shows a situation with $n = 6$ points and a unit-ball with $f = 4$. The round-trip $\pi = \langle 1, 2, 5, 3, 4, 6 \rangle$ then translates into the depicted set $E(\pi)$ with six red and six blue edges.

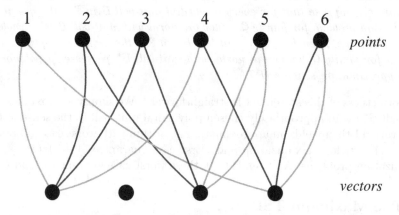

Fig. 4. The edge set $E(\pi)$ for the round-trip $\pi = \langle 1, 2, 5, 3, 4, 6 \rangle$. (Color figure online)

Now let us ignore the identities of the points from P in the subgraph of $G(P)$ that is induced by $E(\pi)$. Then the resulting subgraph has f main-vertices that correspond to the f normal vectors. These main-vertices are connected to each other by lots of paths of length 2, where the middle vertex of every such path is an anonymous vertex. This subgraph is called the *outline* of round-trip π. Every outline has the following two properties:

– The edges in the outline form a connected component.
– For every main-vertex, the number of incident red edges equals the number of incident blue edges.

There are only moderately many combinatorially different outlines: Every path of length 2 connects two of the f main-vertices, and its two edges have one of the four color combinations red-red, red-blue, blue-red, or blue-blue. Hence every such path carries one of $4f^2$ possible types. As an outline altogether contains n paths of length 2 that are partitioned into $4f^2$ types, there exist only $O(n^{4f^2})$ different outlines. It is easy to enumerate all outlines and to check that they satisfy the above two properties (on connectedness, and on red and blue degrees).

For every fixed outline, we find the best way of assigning the n points in P to the anonymized middle vertices on the n paths of length 2. Here the best way is of course the way that maximizes the total edge weight in the resulting subgraph of $G(P)$. If we assign a point to a middle vertex, this simply contributes the weight of the two incident red/blue edges. Hence the entire problem boils down to a bipartite matching problem which is solvable in polynomial time $O(n^3)$; see for instance Burkard, Dell'Amico and Martello [9]. The optimal assignment can be turned into a round-trip π with the same weight: Essentially, we need to find a special Eulerian cycle for the edge set $E(\pi)$ that changes edge color (red to blue, or blue to red) whenever it traverses one of the main-vertices. Such a Eulerian cycle exists in every outline by routine arguments and can be found by routine methods. To summarize: There are $O(n^{4f^2})$ outlines that each are handled in $O(n^3)$ time, and we arrive at the following result.

Theorem 1 *(Barvinok et al. [5]). Let $\mathcal{B}$ be a polyhedral unit-ball in $\mathbb{R}^s$. Then the Max-TSP with distances measured according to the norm with unit-ball $\mathcal{B}$ can be solved in polynomial time $O(n^{4f^2+3})$.*

Barvinok et al. [5] actually use a number of tricks (a more efficient type of outline; a better way of enumerating cases; faster optimization algorithms) and thereby get a better version of Theorem 1 with time complexity $O(n^{f-2} \log n)$. For the Manhattan norm and for the Maximum norm, Barvinok et al. [5] show how to solve the Max-TSP in $\mathbb{R}^2$ in linear time $O(n)$. At the ALGO conference in August 2018 in Helsinki, Bart Jansen asked whether further strong improvements on the running time are possible.

Open problem 1 *(Jansen [21]). Does there exist an FPT algorithm with parameter f for the Max-TSP under polyhedral norms with $2f$-facet unit-balls? In other words, do there exist a function $g : \mathbb{N} \to \mathbb{N}$ and some fixed integer c, such that the Max-TSP under polyhedral norms with $2f$-facet unit-balls is solvable in $O(n^c g(f))$ time?*

Intuitively, such a beautiful FPT algorithm should be just too good to be true, and perhaps one should better look for a W[1]-hardness argument. Finally, we may combine Theorem 1 with Observation 2 from Sect. 2 to get the following corollary:

Theorem 2 *(Barvinok et al. [5]). For any fixed norm in any fixed dimension, the Max-TSP has a PTAS.*

We close this section by listing some further results and some open problems on the Max-TSP. Barvinok et al. [5] show that the Max-TSP under the Maximum norm is APX-hard, if the dimension of the underlying Cartesian space is part of the input. The following problem was left open in [5] and does not look very difficult.

Open problem 2. *Establish the APX-hardness of the Max-TSP under every fixed L_p norm with $1 \leq p < \infty$, if the dimension of the underlying Cartesian space is part of the input.*

Furthermore, the Max-TSP under polyhedral norms is APX-hard in $\mathbb{R}^3$, if the number f of facets is part of the input [5]. The Euclidean Max-TSP is polynomially solvable in one-dimensional space [8, 22] and is NP-hard in three-dimensional space [5]. The two-dimensional case is wide open (though we expect it to be NP-hard).

Open problem 3. *Determine the complexity of the Euclidean Max-TSP in $\mathbb{R}^2$.*

Also the complexity of the Max-TSP in any fixed dimension $s \geq 2$ under any fixed non-polyhedral and non-Euclidean norm is open; polynomial time results for one of these problems would come as a big surprise. Some non-geometric tractable special cases of the Max-TSP are discussed by Burkard et al. [8] and Deineko and Woeginger [16].

4   The Maximum Three-Dimensional Matching Problem

Next, we want to discuss the maximization version of the three-dimensional matching problem, Max-3DM for short. We stress that in this context the word *"three-dimensional"* does not indicate an underlying three-dimensional geometric space, but simply means that a set of objects is to be partitioned into groups of size *three*.

Max-3DM

Instance: A set P of $3n$ points in Cartesian space $\mathbb{R}^s$; a unit-ball $\mathcal{B}$.
Goal: Partition the $3n$ points into n triangles, so that the sum of triangle perimeters is maximized (where the side lengths of the triangles are measured according to the norm with unit-ball $\mathcal{B}$).

In the general (non-geometric) Max-3DM, the distances between the points are specified explicitly as part of the input. The general Max-3DM is NP-hard, and its APX-hardness can be deduced from Kann [23]. The general Max-3DM allows a polynomial time approximation algorithm with constant worst case guarantee; for instance a worst case guarantee of $1/3$ can easily be extracted from the work of Bandelt, Crama and Spieksma [3]. Better worst case guarantees must be possible, but we are not aware of any serious work on the maximization version (we note that the literature does contain a number of approximation results on the minimization version). The geometric versions of the Max-3DM behave very similarly to the geometric versions of the Max-TSP as discussed in the preceding section: The cases with fixed polyhedral norms can be solved in polynomial time, and there is a PTAS for any fixed norm in any fixed dimension. We will now show how to recycle the approaches for the Max-TSP from the preceding section and how to carry them over to the Max-3DM.

Hence, let us once again fix some polyhedral unit-ball $\mathcal{B}$ with $2f$ facets in $\mathbb{R}^s$ and the red-blue bipartite graph $G(P)$ for point set P as discussed in Sect. 2. A partition π of P into triangles consists of n triangles with $3n$ sides. Similarly as in the preceding section, we translate such a partition π into a corresponding subset $E(\pi)$ of the red and blue edges in $G(P)$: Whenever two points x and y are in a common triangle, the edge set $E(\pi)$ contains the edge pair that determines the distance $d(x, y)$ between x and y. Every triangle $\Delta p_1 p_2 p_3$ in π is then translated into a six-cycle $C(\Delta)$ of edges that alternate between p_1, p_2, p_3 and three (not necessarily distinct) vertices that correspond to normal vectors; see Fig. 5. Note that $E(\pi)$ contains $3n$ blue and $3n$ red edges, and note that the sum of triangle perimeters in π equals the total weight of $E(\pi)$.

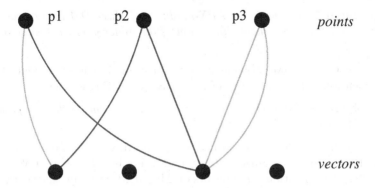

Fig. 5. The edges in $E(\pi)$ that correspond to the triangle $\Delta p_1 p_2 p_3$. (Color figure online)

Let us once again ignore the identities of the points in P. The resulting anonymized subgraph is called the outline of partition π; it has f main-vertices (that correspond to the f normal vectors) and many edges that arise from the six-cycles $C(\Delta)$. Note that an anonymized six-cycle $C(\Delta)$ is fully specified

- by the three main-vertices (in the even positions), and
- by the color (red/blue) of its first, third, fifth edge.

Consequently there exist at most $8f^3$ combinatorially different anonymized six-cycles. As an outline consists of the edges of n six-cycles that are partitioned into $8f^3$ types, there exist only $O(n^{8f^3})$ different outlines. For every fixed outline, we find the best way of assigning the n points in P to the anonymized vertices; this again boils down to a bipartite matching problem and hence is solvable in $O(n^3)$ time. The optimal assignment can easily be turned into a collection of triangles. We summarize:

Theorem 3 *(Custic, Klinz and Woeginger [14]). Let $\mathcal{B}$ be a polyhedral unit-ball in $\mathbb{R}^s$. Then the Max-3DM with distances measured according to the norm with unit-ball $\mathcal{B}$ can be solved in polynomial time $O(n^{8f^3+3})$.*

It would be nice to get a more civilized time complexity for Theorem 3, where the exponent of n grows linearly with f. And it would be nice to know whether there exists an FPT algorithm for this problem:

Open problem 4 *(Jansen [21]). Does there exist an FPT algorithm with parameter f for Max-3DM under polyhedral norms with $2f$-facet unit-balls?*

By combining Theorem 3 with Observation 2, we get the following corollary:

Theorem 4 *(Custic, Klinz and Woeginger [14]). For any fixed norm in any fixed dimension, Max-3DM has a PTAS.*

We mention some further results and some open problems on Max-3DM. Custic, Klinz and Woeginger [14] establish the NP-hardness of Max-3DM under any fixed L_p norm with $1 \leq p \leq \infty$, if the dimension of the underlying Cartesian space is part of the input. The approximability of these problems is unclear.

Open problem 5. *Establish the APX-hardness of Max-3DM under every fixed L_p norm with $1 \leq p \leq \infty$, if the dimension of the underlying Cartesian space is part of the input.*

The Euclidean Max-3DM problem in one-dimensional space is trivial [31]. In higher dimensions, the complexity of Euclidean Max-3DM is open.

Open problem 6. *For every fixed integer $s \geq 2$, determine the complexity of Euclidean Max-3DM in $\mathbb{R}^s$.*

Some non-geometric tractable special cases of Max-3DM are discussed by Burkard, Rudolf and Woeginger [10] and by Pferschy, Rudolf and Woeginger [30]. Finally, let us mention some results on Min-3DM, the minimization version of the three-dimensional matching problem. Crama and Spieksma [13] design a polynomial time approximation algorithm with worst case guarantee 4/3 for Min-3DM, if the distances satisfy the triangle inequality. Euclidean Min-3DM in $\mathbb{R}$ is trivial [30], and Euclidean Min-3DM in $\mathbb{R}^2$ is NP-hard [30]. The following open problem seems to be quite challenging.

Open problem 7. *Design a PTAS for Euclidean Min-3DM in $\mathbb{R}^2$.*

5   A More General View

The approaches in Sects. 3 and 4 for the Max-TSP and Max-3DM under polyhedral norms are based on the very same ideas: A feasible solution π is translated into a corresponding subset $E(\pi)$ of the red and blue edges in $G(P)$. The translation replaces every edge between two points x and y in π by the red-blue or blue-red edge pair that determines the distance $d(x,y)$ between x and y. The length of the feasible solution π equals the total weight of $E(\pi)$. Finally, a subset $E(\pi)$ of maximum weight in $G(P)$ can be computed in polynomial time.

- The approach only works for maximization problems. The translation step from π to $E(\pi)$ scrambles and merges the upper level maximization (find a feasible solution of maximum weight) and the lower level maximization (find the facet of $\mathcal{B}$ that determines the distance between x and y). Hence we may simply search for a subset $E(\pi)$ of maximum weight. In a minimization problem the upper level would minimize and the lower level would maximize, and the two levels would repel each other.

- The approach only works for max-sum type optimization problem, where the goal is to maximize the sum of the lengths of all edges. Another natural family are max-min type optimization problem, where the goal is to maximize the length of the shortest edge. When we move from the feasible solution π to the subset $E(\pi)$, we lose control over the individual edges. (But we stress that not everything is lost: As long as we are able to optimize over all polyhedral unit-balls for a max-min type problem, this will still imply a PTAS for arbitrary unit-balls.)
- The main ingredient is the polynomial time optimization algorithm for finding a maximum weight subset $E(\pi)$ in the bipartite graph $G(P)$. For this, we need a good understanding and a nice characterization of the feasible subsets $E(\pi)$. For making the anonymization of points/vertices work out, it seems to be necessary that the points fall into a small (constant) number of equivalence classes and that every point only interacts with a small (constant) number of other points.

For getting a better understanding of the approach, we suggest that the reader works through the following exercise. Which of the following six puzzle problems allows a polynomial time algorithm under polyhedral norms in fixed dimension (and hence also a PTAS)?

A. Given $5n$ points, partition them into n five-cycles of maximum total length.
B. Given $6n$ points, partition them into n six-cliques of maximum total length.
C. Given $2n$ points, partition them into two n-cliques of maximum total length.
D. Given $15n$ points, partition them into n seven-cliques and into n eight-cliques of maximum total length.
E. Given $\frac{1}{2}n(n+1)$ points, partition them into n cliques that respectively contain $1, 2, 3, \ldots, n$ points so that the total edge length is maximized.
F. Given n points together with a positive integer sequence $d_1, d_2, \ldots, d_n$ with $\sum d_i = 2n - 2$, find a spanning tree of maximum length for the points whose vertex degrees are $d_1, d_2, \ldots, d_n$.

The reader should have little difficulty in finding polynomial time algorithms for the three problems A, B, and D (and for similar problems of this flavor). We have no idea what to do with the other three problems C, E, and F. Section 6 lists some other (more natural) geometric maximization problems where the approach of Sects. 3 and 4 does not seem to apply.

Finally, we want to mention a slight extension of Observation 2: For getting the PTAS under arbitrary norms in some fixed dimension, it is sufficient to construct a PTAS under arbitrary polyhedral norms in this dimension (and it is not necessary to construct a polynomial time algorithm as imposed by Observation 2). However, we are not aware of a single application for which this slight extension would be useful.

6   Some Related Problems

In this concluding section we discuss three further geometric optimization problems that are centered around Hamiltonian cycles and cliques. Krarup [26] introduced the 2-peripatetic salesman problem as a generalization of the classic TSP. The maximization version (Max-2-PSP, for short) of this peripatetic problem is defined as follows.

Max-2-PSP

Instance: A set P of n points in Cartesian space $\mathbb{R}^s$; a unit-ball $\mathcal{B}$.
Goal: Find two edge-disjoint round-trips of maximum total length through the points in P, where distances are measured according to the norm with unit-ball $\mathcal{B}$.

In the general (non-geometric) Max-2-PSP, the distances between the points are specified explicitly as part of the input. The general Max-2-PSP is NP-hard, and it is an easy exercise to establish its APX-hardness. The strongest known approximation algorithm for the general version has a worst case guarantee of 7/9 and is due to Glebov and Zambalaeva [18]. It might be possible to reach a worst case guarantee of 4/5 by adapting the ideas of Dudycz, Marcinkowski, Paluch and Rybicki [17] from the Max-TSP to the Max-2-PSP. De Brey and Volgenant [15] discuss some tractable special cases of the general (non-geometric) 2-PSP. The geometric versions of Max-2-PSP are easy to approximate: Baburin and Gimadi [2] design a PTAS for the Euclidean norm in fixed dimension, and Shenmaier [34] extends this to arbitrary norms in fixed dimension. The approaches in [2] and [34] are heavily based on the geometric ideas of Serdyukov [32,33], and do not follow the purely combinatorial approach outlined in Sects. 3 and 4. A central open problem in this area concerns the complexity under polyhedral norms.

Open problem 8. *For every fixed polyhedral unit-ball $\mathcal{B}$, construct a polynomial time algorithm for Max-2-PSP, if distances are measured according to the norm with unit-ball $\mathcal{B}$.*

Of course, a positive solution to this problem would yield the existence of a PTAS for Max-2-PSP under any fixed norm, and hence provide another proof for the results in [2,34]. The complexity of the Euclidean Max-2-PSP should be closely related to the complexity of the Euclidean Max-TSP:

Open problem 9. *For every fixed integer $s \geq 2$, establish NP-hardness of the Euclidean Max-2-PSP in $\mathbb{R}^s$.*

Finally, it would be interesting to find a PTAS for the Euclidean Min-2PSP in $\mathbb{R}^2$ (that is, for the Euclidean <u>minimization</u> version of the 2-PSP).

Let us move on to the next problem. Motivated by problems in medical imaging and in manufacturing, Arkin, Chiang, Mitchell, Skiena and Yang [1] introduced a max-min variant of the TSP that they called the Maximum Scatter TSP (Max-Scatter-TSP, for short).

Max-Scatter-TSP

Instance: A set P of n points in Cartesian space $\mathbb{R}^s$; a unit-ball $\mathcal{B}$.
Goal: Find a round-trip through the points in P that maximizes the length of the shortest edge (where distances are measured according to the norm with unit-ball $\mathcal{B}$).

In the general (non-geometric) Max-Scatter-TSP, the distances between the points are specified explicitly as part of the input. The general Max-Scatter-TSP is NP-hard and does not allow a constant-factor approximation algorithm unless P=NP (Arkin et al. [1]). For the cases where the distances satisfy the triangle inequality, there exists a polynomial time approximation algorithm with worst case guarantee $1/2$, and unless P=NP no better worst case guarantee can be reached (Arkin et al. [1]).

The geometric versions of the Max-Scatter-TSP are easy to approximate: Kozma and Mömke [25] derive a PTAS for every norm in fixed dimension. In fact, the result in [25] even works for all so-called *doubling metrics* and even yields an EPTAS (that is, a particularly nice PTAS, where the exponent of n in the running time is a fixed constant that does not depend on the desired precision of approximation). The Euclidean Max-Scatter-TSP is polynomially solvable in one-dimensional space, is NP-hard for any fixed dimension $s \geq 3$ [5], and is open for dimension $s = 2$:

Open problem 10. *Settle the complexity of the Max-Scatter-TSP in two-dimensional Euclidean space.*

The complexity of the Max-Scatter-TSP in fixed dimension is also open for all non-Euclidean unit-balls. Perhaps the most fundamental open question in this direction is the following.

Open problem 11. *Settle the complexity of the Max-Scatter-TSP under polyhedral norms in fixed dimension.*

Finally, we turn to the remote clique problem (RCP, for short) which originates from the work of Tamir [35]. The RCP asks for a subset of points that are diverse and well-dispersed. It has applications in web-based search, document summarization, facility location, portfolio management and other areas.

> **RCP**
> Instance: A set P of n points in Cartesian space $\mathbb{R}^s$; a positive integer k; a unit-ball $\mathcal{B}$.
> Goal: Select a subset of k points from P that maximizes the sum of all pairwise distances (where distances are measured according to the norm with unit-ball $\mathcal{B}$).

In the general (non-geometric) RCP, the distances between the points are specified explicitly as part of the input. The general RCP is NP-hard and does not allow a constant-factor approximation algorithm under the exponential time hypothesis (Manurangsi [28]). For the cases where the distances satisfy the triangle inequality, Hassin, Rubinstein and Tamir [19] design a polynomial time approximation algorithm with worst case guarantee $1/2$. Birnbaum and Goldman [6] give an elegant analysis of a greedy approach that also yields a worst case guarantee of $1/2$. Borodin, Jain, Lee and Ye [7] prove that the approximation factor of $1/2$ is best possible under the assumption that the planted-clique problem is hard; it would be interesting to get this inapproximability result under some weaker assumption.

The geometric versions of the RCP are easy to approximate: Cevallos, Eisenbrand and Morell [11] give a PTAS for the RCP under any fixed norm. Cevallos, Eisenbrand and Zenklusen [12] establish the NP-hardness of the RCP under the Manhattan norm, if the dimension is part of the input. Unfortunately, the complexity of the RCP in fixed dimension is poorly understood, and there is not a single hardness result in the literature. It is quite possible that the PTAS in [11] could be strengthened to a polynomial time algorithm. Perhaps the most fundamental open questions are the following.

Open problem 12. *For every fixed $s \geq 2$, determine the complexity of the Euclidean RCP in $\mathbb{R}^s$.*

A positive answer to the following problem would yield (as a cheap by-product) the existence of a PTAS for any norm in fixed dimension.

Open problem 13. *For every fixed polyhedral norm in fixed dimension, construct a polynomial time algorithm for the RCP.*

Acknowledgement. This work is supported by the DFG RTG 2236 "UnRAVeL". I thank Stefan Lendl for discussions, and I thank Thomas Erlebach for a number of comments that helped to improve the presentation of the paper.

References

1. Arkin, E.M., Chiang, Y.J., Mitchell, J.S.B., Skiena, S., Yang, T.C.: On the maximum scatter traveling salesperson problem. SIAM J. Comput. **29**, 515–544 (1999)
2. Baburin, A.E., Gimadi, E.K.: On the asymptotic optimality of an algorithm for solving the maximum m-PSP in a multidimensional Euclidean space. Proc. Steklov Inst. Math. **270**, 1–13 (2011)
3. Bandelt, H.J., Crama, Y., Spieksma, F.C.R.: Approximation algorithms for multidimensional assignment problems with decomposable costs. Discrete Appl. Math. **49**, 25–50 (1994)
4. Barvinok, A.: Two algorithmic results for the traveling salesman problem. Math. Oper. Res. **21**, 65–84 (1996)
5. Barvinok, A.I., Fekete, S.P., Johnson, D.S., Tamir, A., Woeginger, G.J., Woodroofe, R.: The geometric maximum travelling salesman problem. J. ACM **50**, 641–664 (2003)
6. Birnbaum, B.E., Goldman, K.J.: An improved analysis for a greedy remote-clique algorithm using factor-revealing LPs. Algorithmica **55**, 42–59 (2009)
7. Borodin, A., Jain, A., Lee, H.C., Ye, Y.: Max-sum diversification, monotone submodular functions, and dynamic updates. ACM Trans. Algorithms **13**, 41:1–41:25 (2017)
8. Burkard, R.E., van Dal, R., Deineko, V.G., van der Veen, J., Woeginger, G.J.: Well-solvable special cases of the traveling salesman problem: a survey. SIAM Rev. **40**, 496–546 (1998)
9. Burkard, R.E., Dell'Amico, M., Martello, S.: Assignment Problems. SIAM, Philadelphia (2009)
10. Burkard, R.E., Rudolf, R., Woeginger, G.J.: Three-dimensional axial assignment problems with decomposable cost coefficients. Discrete Appl. Math. **65**, 123–140 (1996)

11. Cevallos, A., Eisenbrand, F., Morell, S.: Diversity maximization in doubling metrics. In: Proceedings of the 29th International Symposium on Algorithms and Computation (ISAAC-2018) (2018)
12. Cevallos, A., Eisenbrand, F., Zenklusen, R.: Max-sum diversity via convex programming. In: Proceedings of the 32nd Symposium on Computational Geometry (SoCG-2016), pp. 26:1–26:14 (2016)
13. Crama, Y., Spieksma, F.C.R.: Approximation algorithms for three-dimensional assignment problems with triangle inequalities. Eur. J. Oper. Res. **60**, 273–379 (1992)
14. Custic, A., Klinz, B., Woeginger, G.J.: Geometric versions of the 3-dimensional assignment problem under general norms. Discrete Optim. **18**, 38–55 (2015)
15. De Brey, M.J.D., Volgenant, A.: Well-solved cases of the 2-peripatetic salesman problem. Optimization **39**, 275–293 (1997)
16. Deineko, V.G., Woeginger, G.J.: The maximum travelling salesman problem on symmetric Demidenko matrices. Discrete Appl. Math. **99**, 413–425 (2000)
17. Dudycz, S., Marcinkowski, J., Paluch, K., Rybicki, B.: A 4/5 - approximation algorithm for the maximum traveling salesman problem. In: Eisenbrand, F., Koenemann, J. (eds.) IPCO 2017. LNCS, vol. 10328, pp. 173–185. Springer, Cham (2017). https://doi.org/10.1007/978-3-319-59250-3_15
18. Glebov, A.N., Zambalaeva, D.Z.: A polynomial algorithm with approximation ratio 7/9 for the maximum 2-peripatetic salesmen problem. J. Appl. Ind. Math. **6**, 69–89 (2012)
19. Hassin, R., Rubinstein, S., Tamir, A.: Approximation algorithms for maximum dispersion. Oper. Res. Lett. **21**, 133–137 (1997)
20. Itai, A., Papadimitriou, C.H., Swarcfiter, J.L.: Hamiltonian paths in grid graphs. SIAM J. Comput. **11**, 676–686 (1982)
21. Jansen, B.M.P.: Question posed at the ALGO-2018 conference held in Helsinki, Finland, 20–24 August 2018 (2018)
22. Kalmanson, K.: Edgeconvex circuits and the travelling salesman problem. Can. J. Math. **27**, 1000–1010 (1975)
23. Kann, V.: Maximum bounded 3-dimensional matching is MAX SNP-complete. Inf. Process. Lett. **37**, 27–35 (1991)
24. Kowalik, L., Mucha, M.: Deterministic 7/8-approximation for the metric maximum TSP. Theor. Comput. Sci. **410**, 5000–5009 (2009)
25. Kozma, L., Mömke, T: Maximum scatter TSP in doubling metrics. In: Proceedings of the 28th Annual ACM-SIAM Symposium on Discrete Algorithms (SODA-2017), pp. 143–153 (2017)
26. Krarup, J.: The peripatetic salesman and some related unsolved problems. In: Roy, B. (ed.) Combinatorial Programming: Methods and Applications, vol. 19, pp. 173–178. Springer, Dordrecht (1975). https://doi.org/10.1007/978-94-011-7557-9_8
27. Lawler, E.L., Lenstra, J.K., Rinnooy Kan, A.H.G., Shmoys, D.B. (eds.): The Travelling Salesman Problem. Wiley, Chichester (1985)
28. Manurangsi, P.: Almost-polynomial ratio ETH-hardness of approximating densest k-subgraph. In: Proceedings of the 49th Annual ACM SIGACT Symposium on Theory of Computing (STOC-2017), pp. 954–961 (2017)
29. Papadimitriou, C.H., Yannakakis, M.: The traveling salesman problem with distances one and two. Math. Oper. Res. **18**, 1–11 (1993)
30. Pferschy, U., Rudolf, R., Woeginger, G.J.: Some geometric clustering problems. Nordic J. Comput. **1**, 246–263 (1994)
31. Polyakovskiy, S., Spieksma, F.C.R., Woeginger, G.J.: The three-dimensional matching problem in Kalmanson matrices. J. Comb. Optim. **26**, 1–9 (2013)

32. Serdyukov, A.I.: An asymptotically optimal algorithm for the maximum travelling salesman problem in Euclidean space in finite-dimensional normed spaces. Upravlyaemye sistemy (Novosibirsk) **27**, 79–87 (1987). (in Russian)
33. Serdyukov, A.I.: Asymptotic properties of optimal solutions of extremal permutation problems in finite-dimensional normed spaces. Metody Diskretnogo Analiza **51**, 105–111 (1991). (in Russian)
34. Shenmaier, V.V.: Asymptotically optimal algorithms for geometric Max-TSP and Max-m-PSP. Discrete Appl. Math. **163**, 214–219 (2014)
35. Tamir, A.: Obnoxious facility location on graphs. SIAM J. Discrete Math. **4**, 550–567 (1991)

Regular Papers

Deterministic Min-Cost Matching
with Delays

Yossi Azar$^{(\boxtimes)}$ and Amit Jacob Fanani

School of Computer Science, Tel Aviv University, Tel Aviv, Israel
azar@tau.ac.il, amitj@mail.tau.ac.il

Abstract. We consider the online Minimum-Cost Perfect Matching with Delays (MPMD) problem introduced by Emek et al. (STOC 2016), in which a general metric space is given, and requests are submitted in different times in this space by an adversary. The goal is to match requests, while minimizing the sum of distances between matched pairs in addition to the time intervals passed from the moment each request appeared until it is matched.

In the online Minimum-Cost Bipartite Perfect Matching with Delays (MBPMD) problem introduced by Ashlagi et al. (APPROX/RANDOM 2017), each request is also associated with one of two classes, and requests can only be matched with requests of the other class.

Previous algorithms for the problems mentioned above, include randomized $O(\log n)$-competitive algorithms for known and finite metric spaces, n being the size of the metric space, and a deterministic $O(m)$-competitive algorithm, m being the number of requests.

We introduce $O\left(m^{\log\left(\frac{3}{2}+\epsilon\right)}\right)$-competitive deterministic algorithms for both problems and for any fixed $\epsilon > 0$. In particular, for a small enough ϵ the competitive ratio becomes $O\left(m^{0.59}\right)$. These are the first deterministic algorithms for the mentioned online matching problems, achieving a sub-linear competitive ratio. Our algorithms do not need to know the metric space in advance.

Keywords: Matching · Bipartite matching · Delayed service
Online algorithm · Competitive analysis

1 Introduction

In the algorithmic graph theory, a *Perfect Matching* is a subset of graph edges, in which each vertex of the graph is incident on exactly one edge of the subset, and the weight of the matching is the sum of the weights of the edges of the subset. In the well known *Minimum-Cost Perfect Matching* problem a weighted graph is given, and a *Perfect Matching* of minimum weight is to be found. The *Blossom*

Y. Azar—Supported in part by the Israel Science Foundation (grant No. 1506/16) and by the ICRC Blavatnik Fund.

© Springer Nature Switzerland AG 2018
L. Epstein and T. Erlebach (Eds.): WAOA 2018, LNCS 11312, pp. 21–35, 2018.
https://doi.org/10.1007/978-3-030-04693-4_2

Algorithm due to Edmonds [9] is the first algorithm to solve this problem in polynomial time.

Many versions of the *Minimum-Cost Perfect Matching* problem have been studied over the last few decades, some of the noticeable variants are online versions of the problem (e.g. *Minimum-Cost Perfect Matchings with Online Vertex Arrival* due to Kalyanasundaram and Pruhs [14]).

In this paper we suggest a deterministic algorithm for the *Minimum-Cost Perfect Matching with Delays* (MPMD) variant, which was introduced by Emek et al. [10], and a similar deterministic algorithm for another variation of the problem - the *Minimum-Cost Bipartite Perfect Matching with Delays* (MBPMD) problem, which was introduced by Ashlagi et al. [2].

To illustrate the MPMD problem, imagine players logging in through a server to an online game at different times, unknown a priori to the server they have connected through. The server then needs to match between the players while maximizing their satisfaction from playing the game. Players feel satisfied when they play against players at a level similar to their own. Therefore, when pairing players, the server needs to consider the difference in levels between the players, called the *connection cost*.

Once logged in, a player doesn't necessarily start playing instantly, as the server can postpone the decision regarding with whom to match the player, until a good match is found (i.e. another player at a similar level logs in to the game). This is a poor strategy since players are unhappy when forced to wait too long until they start playing. The time a player has to wait until the game starts is called the *delay cost*.

More formally, an adversary presents requests at points in a general metric space, in an online manner. The goal is to produce a minimum-cost perfect matching when the cost of an edge is the sum of its *connection cost* (the distance between the two points in the metric space) and the *delay cost* of the two requests matched by the edge. All requests have to be matched by the server after a finite time from the moment they have arrived.

The MBPMD problem is an extension of the MPMD problem (due to Ashlagi et al. [2]), in which each of the requests may take one of two colors, and each edge of the matching, must be incident on one request from each color. The MBPMD problem has many applications, such as matching drivers to passengers (Uber, Lyft), job finding platforms, etc.

Background. The standard method used to measure an online algorithm's performance is its competitive ratio. We use this method when comparing the performance of matching algorithms for both MPMD and MBMPD. An algorithm is α-competitive if the maximum ratio between the cost of the algorithm to the cost of the optimum solution, over all inputs, is bounded by α.

The first algorithm for MPMD was developed by Emek et al. [10] with an expected competitive ratio $O\left(\log^2 n + \log \Delta\right)$ on a finite metric space of size n, where Δ is the aspect-ratio of the metric space (the ratio of the maximum distance to the minimum distance between any two points in the metric space).

Azar et al. [3] improved the competitive ratio to $O(\log n)$, and showed a lower bound of $\Omega\left(\sqrt{\log n}\right)$ (both deterministic and randomized). Ashlagi et al. [2] improved this lower bound to $\Omega\left(\frac{\log n}{\log\log n}\right)$ (both deterministic and randomized). They also gave an $O(\log n)$-competitive randomized algorithm for MBPMD.

All mentioned above algorithms are randomized (on a general finite metric). In online algorithms where one cannot repeat the algorithm in case the cost is high, a deterministic algorithm is preferable. Bienkowski et al. [7] provided the first deterministic algorithm for MPMD on general metrics, with a competitive-ratio of $O\left(m^{2.46}\right)$, m being the number of requests. While the previous algorithms require the metric space to be known a priori, their algorithm does not, and is also applicable when the metric space is revealed in an online manner. Bienkowski et al. also noted that the algorithm of [3] can be used to provide an $O(n)$-competitive deterministic algorithm for a general known metric space. Recently, Bienkowski et al. [6] provided a new primal-dual deterministic algorithm for MPMD on general metrics, with a competitive-ratio of $O(m)$, m being the number of requests.

Prior to our result there was no deterministic sub-linear competitive algorithm, neither in n nor in m.

Our Contribution. In this paper we introduce deterministic algorithms for both versions of the problem, both with a competitive ratio $O\left(\frac{1}{\epsilon}m^{\log\left(\frac{3}{2}+\epsilon\right)}\right)$. When the constant ϵ is small enough, this becomes $O\left(m^{0.59}\right)$. Our algorithms do not need to know the metric space in advance.

We present a simple algorithm, which is an adaptation of the greedy algorithm for the *Minimum-Cost Perfect Matching* problem by Reingold and Tarjan [21] to an online environment. In our algorithm, requests grow hemispheres around them in a metric that is the Cartesian product of the original metric and the time axis (also called the *time-augmented metric space*). The hemispheres radii grow slowly in the negative direction of the time axis. Once a request is found on the boundary of another request's hemisphere, they are matched by the algorithm. Our analysis is inspired by the analysis of the original greedy algorithm by Reingold and Tarjan.

In the bipartite case, the algorithm is essentially the same, but requests are matched only if they are of different colors.

Related Work. First we consider related work **with delays**. Since Emek et al. [10] introduced the notion of online problems with delayed service, there has been a growing number of works studying such problems (e.g. *Online Service with Delays* [4], *Minimum-Cost Bipartite Perfect Matching with Delays* [2], *Minimum-Cost Perfect Matching with Delays for Two Sources* [11]). Works dealing with the *Minimum-Cost Perfect Matching with Delays* and *Minimum-Cost Bipartite Perfect Matching with Delays* problems, such as the papers by Emek et al. [10], Azar et al. [3], Ashlagi et al. [2] and Bienkowski et al. [7], are the most closely related to this work. As mentioned above, Emek et al. [10] provided a randomized $O\left(\log^2 n + \log \Delta\right)$-competitive algorithm for MPMD on general metrics, in which n is the size of the metric space and Δ is the aspect ratio. They consider the randomized embeddings of the general metric space into a distribution over metrics given by hierarchically separated full binary trees, with distortion $O(\log n)$, and give a randomized algorithm for the hierarchically separated trees metrics.

Subsequently, Azar et al. [3] provided a randomized $O(\log n)$-competitive algorithm for the same problem, thus improving the original upper bound. They used randomized embedding of the general metric space into a distribution over metrics given by hierarchically separated trees of height $O(\log n)$, with distortion $O(\log n)$. Then they give a deterministic $O(1)$-space-competitive (that is the competitive ratio associated with the *connection cost*) and $O(h)$-time-competitive (that is the competitive ratio associated with the *delay cost*) algorithm over tree metrics, where h is the height of the tree. This yields a competitive ratio of $O(\log n)$. Moreover, they provided a randomized $\Omega\left(\sqrt{\log n}\right)$ lower bound, confirming a conjecture made by Emek et al. [10] that the competitive ratio of any online algorithm for the problem must depend on n.

Ashlagi et al. [2] improved the lower bound on the competitive ratio to $\Omega\left(\frac{\log n}{\log \log n}\right)$, almost matching the upper bound of Azar et al. of $O(\log n)$. The rest of the paper focuses on the bipartite version of the problem, providing an $O(\log n)$-competitive ratio by the adaptation of the algorithm of Azar et al. [3] to the bipartite case.

In order to provide a *deterministic* algorithm, Bienkowski et al. [7] used a different approach for the problem - they used a semi-greedy scheme of a ball-growing algorithm. In their analysis, they fix an optimal matching, and charge the cost of each matching-edge generated by their algorithm against the cost of an existing matching-edge of the optimal matching. As mentioned above, their algorithm achieves a competitive ratio of $O\left(m^{2.46}\right)$, where m is the number of requests.

Bienkowski et al. improved this result in [6] by providing a new $O(m)$-competitive LP-based algorithm. Briefly, their algorithm maintains a primal relaxation of the matching problem and its dual (the programs evolve in time as more requests arrive). Dual variables are increased along time, until a dual constraint (corresponding to a pair of requests) becomes tight, which results in the algorithm connecting the pair. They also proved that their analysis is tight (the competitive-ratio of their algorithm is $\Omega(m)$). Recall that our algorithm achieves a sub-linear competitive-ratio (in m).

Next we consider related work **without delays**. The *Online Minimum Weighted Bipartite Matching* (OMM) problem due to [14,16] is another important online version of the *Minimum-Cost Perfect Matching* problem, in which k vertices are given a priori, and k additional vertices are revealed at different times, together with the distances from the first k vertices. The algorithm then needs to match the later k vertices to the first k vertices, while trying to minimize the total weight of the produced matching. In this version, delay of the algorithm's decision is not available. Kalyanasundaram and Pruhs [14] and Khuller et al. [16] showed independently a tight upper and lower bounds of $2k-1$ on the deterministic competitive ratio of the problem.

The first sub-linear competitive randomized algorithm for the problem, was given by Meyerson et al. [18] using randomized embeddings into trees, with a competitive ratio of $O(\log^3 k)$. Consequently, Bansal et al. [5] improved this upper bound by providing a $O(\log^2 k)$-competitive randomized algorithm. In addition, they showed an $\Omega(\log k)$ lower bound on the competitive ratio for randomized algorithms.

The special case of line-metrics is argued to be the most interesting instance of OMM (e.g. [17]). Kalyanasundaram and Pruhs conjectured in 1998 [15] that there exists a 9-competitive deterministic algorithm for OMM on line-metrics, but in 2003 Fuchs et al. [12] disproved the conjecture, proving a lower bound of 9.001 for deterministic algorithms. This is the best known lower bound thus far.

Antoniadis et al. [1] presented the first sub-linear deterministic algorithm for line-metrics, with a competitive ratio of $O\left(\frac{1}{\epsilon}k^{\log\left(\frac{3}{2}+\epsilon\right)}\right)$. Recently, Nayyar and Raghvendra [19] improved this upper bound to $O(\log^2 k)$ by careful analysis of the deterministic algorithm present in [20]. Gupta and Lewi [13] provided a randomized $O(\log k)$-competitive algorithm for doubling metrics, hence for line-metrics as well.

To summarize, the best known deterministic upper bound on the competitive ratio for line-metrics is $O(\log^2 k)$, and the best known lower bound is 9.001. For randomized algorithms the best known upper bound is $O(\log k)$.

Paper Organization. We describe the algorithm for *Minimum-Cost Perfect Matching with Delays* in Sect. 3 and analyze its performance in Sect. 3.1. In Sect. 4 we present the algorithm for *Minimum-Cost Bipartite Perfect Matching with Delays* and analyze its performance.

2   Preliminaries

A *metric space* $\mathcal{M} = (S, d)$ is a set S and a distance function $d : S \times S \longrightarrow \mathbb{R}^+$ that meets the following conditions: non-negativity, symmetry, the triangle-inequality, and that $d(x, y) = 0$ if and only if $x = y$. When S is finite, we refer to $\mathcal{M}$ as a *finite metric space*, and an *infinite metric space* otherwise.

2.1   Model

In the online *Minimum-Cost Perfect Matching with Delays* problem on a metric space $\mathcal{M} = (S, d)$ (known a priori to the algorithm), an input instance $\mathcal{I} = \langle r_i \rangle_{i=1}^m$ is presented to the algorithm in an online fashion, so that each request r_i is revealed to the algorithm at time $t(r_i)$ at the location $x(r_i) \in S$. The number of requests m is even and unknown a priori to the algorithm.

The online algorithm should produce a perfect matching in real time. Formally, two requests p, q can be matched by the algorithm at any time $t \geq \max(t(p), t(q))$, if they have not been matched yet by the algorithm.

Let $\langle p_i, q_i, t_i \rangle_{i=1}^{\frac{m}{2}}$ be the set of pairs of requests matched by the algorithm, and their matching times (p_i and q_i were matched by the algorithm at t_i), then the cost of the matching produced by the algorithm is

$$\sum_{i=1}^{\frac{m}{2}} d(x(p_i), x(q_i)) + |t_i - t(p_i)| + |t_i - t(q_i)|$$

In other words, the cost is the sum of the connection cost of all matched pairs in addition to the sum of the delay cost of all requests. The goal of the algorithm is to minimize this cost.

The *Minimum-Cost Bipartite Perfect Matching with Delays* is virtually the same problem as the *Minimum-Cost Perfect Matching with Delays* problem, except that each request r_i is associated with one of two classes, so that each request r_i can be matched to a request r_j if and only if $class(r_i) \neq class(r_j)$.

2.2 The Time-Augmented Metric Space

Given a metric space $\mathcal{M} = (S, d)$ define the *time-augmented metric space* as $\mathcal{M}_T = (S \times \mathbb{R}, D)$ where D is a distance function defined as

$$D\left((l_1, t_1), (l_2, t_2)\right) = d(l_1, l_2) + |t_1 - t_2|$$

assuming $(l_1, t_1), (l_2, t_2) \in S \times \mathbb{R}$. That is, the time axis was added as another dimension in the metric space. One can easily verify that D indeed defines a metric.

The following lemma shows that for offline algorithms, solving the Minimum-Cost Perfect Matching **with Delays** problem in the metric space $\mathcal{M}$ is equivalent to solving the Minimum-Cost Perfect Matching problem in $\mathcal{M}_T$.

Lemma 1. *Assume* $\mathcal{I} = \langle r_i \rangle_{i=1}^{m}$ *is an instance of MPMD then OPT can be computed as the weight of an optimal solution for the Minimum Metric Perfect Matching problem on the instance* $\mathcal{I}$ *as points in the time-augmented metric space* $\mathcal{M}_T$.

Proof. Let OPT* be an optimal solution for *Minimum Metric Perfect Matching* over the instance $\mathcal{I}$. We show that OPT = OPT*.

Let A be the solution for *Minimum Metric Perfect Matching* over the instance $\mathcal{I}$, which matches the pairs corresponding to those matched by OPT. The cost of A is at most the cost of OPT, since for a given pair (u, v) matched by OPT at time $t_{uv} \geq \max(t(u), t(v))$, OPT would pay $t_{uv} - t(u) + t_{uv} - t(v) + d(x(u), x(v))$, while A would pay $D(u, v) = |t(u) - t(v)| + d(x(u), x(v))$ which cannot be larger. Therefore OPT* $\leq A \leq$ OPT.

For the other direction we define an online algorithm B which matches the pairs corresponding to those matched by OPT^*, as soon as the two end-points arrive. For a given pair of requests (p, q) matched by B, it pays

$$\max(t(p), t(q)) - t(p) + \max(t(p), t(q)) - t(q) + d(x(p), x(q)) =$$
$$|t(p) - t(q)| + d(x(p), x(q))$$

Therefore the cost paid by B is the same as the cost paid by OPT^*. Hence $\text{OPT} \le B = \text{OPT}^*$. □

3 A Deterministic Algorithm for MPMD on General Metrics

Our algorithm ($\text{ALG}(\epsilon)$) is parametrized with a constant $\epsilon \in \mathbb{R}$. Upon the arrival of a request $p \in S \times \mathbb{R}$, the algorithm begins to grow a hemisphere surrounding p in the negative direction of the time axis, such that the radius growth rate is ϵ. Therefore, at time t, a request $q \in S \times \mathbb{R}$ is on the hemisphere's boundary if and only if $\epsilon\,(t - t\,(p)) = D(p, q)$ and $t(q) \le t(p)$, where D is the distance function defined by the time-augmented metric space $\mathcal{M}_T$. The algorithm matches a request q to a request p as soon as q is found on the boundary of p's hemisphere.

Note that the algorithm does not need to know the metric space in advance, but it only requires that together with any arriving request p, it learns the distances from p to all previous requests.

Algorithm 1. A Deterministic Algorithm for MPMD on General Metrics

---

```
 1: procedure ALG(ε)
 2:     At every moment t:
 3:     Add the new requests that arrive at time t
 4:     for each unmatched request p do
 5:         for each unmatched request q ≠ p do
 6:             if t(p) ≥ t(q) and t = t(p) + D(p,q)/ε then
 7:                 match(p, q)
 8:             end if
 9:         end for
10:     end for
11: end procedure
```

The algorithm is described as a continuous process but can be easily discretized using priority queues over anticipated matching events for each pair.

The algorithm breaks ties arbitrarily (i.e. a request that is on multiple hemispheres at the same time, or multiple requests that are on the same hemisphere). Note that for the analysis of the algorithm we may assume that there are no ties, as an adversary might slightly perturb the points so that the algorithm would choose the worse option.

3.1   Analysis

Theorem 1. ALG(ϵ) *is* $O\left(\frac{1}{\epsilon}m^{\log\left(\frac{3+\epsilon}{2}\right)}\right)$-*competitive.*

Given $\epsilon \in \mathbb{R}$ we run ALG(ϵ) over the instance $\mathcal{I} = \langle r_i \rangle_{i=1}^m$, that is with a hemisphere growth rate of ϵ.

For the analysis, we denote ALG_{ON} to be the cost paid by ALG(ϵ), and ALG_{OFF} to be the weight of the matching produced by ALG(ϵ), when viewing $\mathcal{I}$ as points in the time-augmented metric space $\mathcal{M}_T$. OPT is the cost of an optimal solution for MPMD over the instance $\mathcal{I}$.

Consider the last two pairs of requests to be matched by ALG. They consist of four requests, name them a, b, c, d, such that (a, b) is one pair, and (c, d) is the second pair. Assume w.l.o.g that (a, b) were matched at time t_{ab}, and (c, d) at $t_{cd} \geq t_{ab}$. Also, assume w.l.o.g that $t(a) \leq t(b)$.

Lemma 2.

1. $D(a, b) \leq (1 + \epsilon)D(a, c)$ *and* $D(a, b) \leq (1 + \epsilon)D(a, d)$
2. $D(a, b) \leq (1 + \epsilon)D(b, c)$ *and* $D(a, b) \leq (1 + \epsilon)D(b, d)$

Proof. We only prove $D(a, b) \leq (1 + \epsilon)D(a, c)$ and $D(a, b) \leq (1 + \epsilon)D(b, c)$ since there is no difference between c and d.

To prove 1, we look at two cases, that are $t(c) \geq t(a)$, and $t(c) < t(a)$.

Case $t(c) \geq t(a)$: Upon the arrival of c and b, the algorithm begins to grow hemispheres surrounding them, and in particular a might be on their boundaries. Since (a, b) was the first pair to be matched, a was on b's hemisphere before it was on c's hemisphere (otherwise (a, c) should have been matched first). Therefore $t(b) + \frac{D(a,b)}{\epsilon} \leq t(c) + \frac{D(a,c)}{\epsilon}$, and we conclude

$$D(a, b) \leq D(a, c) + \epsilon(t(c) - t(b)) \leq D(a, c) + \epsilon(t(c) - t(a)) \leq (1 + \epsilon)D(a, c)$$

Case $t(c) < t(a)$: Upon the arrival of a and b, the algorithm begins to grow hemispheres surrounding them. In particular, a might be on the boundary of b's hemisphere, and c might be on the boundary of a's hemisphere. Since (a, b) was the first pair to be matched, a was on b's hemisphere before c was on a's hemisphere (otherwise (a, c) should have been matched first). Therefore $t(b) + \frac{D(a,b)}{\epsilon} \leq t(a) + \frac{D(a,c)}{\epsilon}$. Thus, we conclude that

$$D(a, b) \leq D(a, c) + \epsilon(t(a) - t(b)) = D(a, c) - \epsilon(t(b) - t(a)) \leq D(a, c) \leq (1 + \epsilon)D(a, c)$$

To prove 2, we look at the two cases $t(c) \geq t(b)$, and $t(c) < t(b)$.

Case $t(c) \geq t(b)$: Upon the arrival of c and b, the algorithm begins to grow hemispheres surrounding them. In particular, a might be on the boundary of b's hemisphere, and b might be on the boundary of c's hemisphere. Since (a, b) was the first pair to be matched, a was on b's hemisphere before b was on c's hemisphere (otherwise (b, c) should have been matched first). Therefore $t(b) + \frac{D(a,b)}{\epsilon} \leq t(c) + \frac{D(b,c)}{\epsilon}$. Thus, we conclude that

$$D(a, b) \leq D(b, c) + \epsilon(t(c) - t(b)) \leq D(b, c) + \epsilon D(b, c) = (1 + \epsilon)D(b, c)$$

Case $t(c) < t(b)$: Upon b's arrival, the algorithm begins to grow a hemisphere surrounding it, and in particular a and c might be on its boundary. Since (a, b) was the first pair to be matched, a was on b's hemisphere before c was (otherwise (b, c) should have been matched first). Therefore $t(b) + \frac{D(a,b)}{\epsilon} \le t(b) + \frac{D(b,c)}{\epsilon}$. Thus, we conclude that

$$D(a, b) \le D(b, c) \le (1 + \epsilon)D(b, c)$$

$\square$

We use the following well known observation.

Observation 1. *The union of any two matchings is a set of vertex-disjoint cycles. In every such cycle, the edges alternate between the two matchings. Note that two parallel edges are considered a cycle.*

Let $\mathcal{C} = \{C_1, \ldots, C_k\}$ be the set of cycles (vertices and edges) generated from taking the union of the matchings produced by ALG and OPT. Define $l_1, \ldots, l_k \in \mathbb{R}$ such that l_i is the total length of edges of ALG in C_i. Define similarly $l_1^*, \ldots, l_k^* \in \mathbb{R}$ for edges of OPT.

Lemma 3. $\frac{\text{ALG}_{\text{OFF}}}{\text{OPT}} \le \max_i \frac{l_i}{l_i^*}$

Lemma 4. *Denote $\hat{l}_i^*$ the cost paid by an optimal algorithm for Minimum Metric Perfect Matching on the instance constructed from the vertices of C_i, and $\hat{l}_i$ the cost of running ALG over the vertices of C_i. Then $\hat{l}_i^* = l_i^*$ and $\hat{l}_i = l_i$.*

The proofs of Lemmas 3 and 4 are omitted and can be found in the full paper.

Corollary 1. *By virtue of Lemmas 3 and 4 it suffices to consider $\frac{\text{ALG}_{\text{OFF}}}{\text{OPT}}$ when the union of the matchings produced by ALG and OPT forms a single cycle.*

Lemma 5. *Let $\gamma \in \mathbb{R}$ s.t. $\gamma > 2$ and let $f : \mathbb{N} \to \mathbb{R}$ satisfy the recurrence relation*

$$f(2k) = \min_{1 \le i \le k-1} \left\{ f(2i), \frac{1}{\gamma} (f(2i) + f(2k - 2i)) \right\}, \ f(2) = 1$$

Then,

$$f(n) = \Omega \left(\frac{1}{n^{\log\left(\frac{\gamma}{2}\right)}} \right)$$

Proof. We prove by induction on k that $f(2k) \ge \left(\frac{2}{\gamma}\right)^{\log k}$.

Base Case $(k = 1)$: $f(2) = 1$, and $\left(\frac{2}{\gamma}\right)^{\log 1} = \left(\frac{2}{\gamma}\right)^0 = 1$.

Inductive step: Assume the claim holds for all $j < k$. By the induction hypothesis for every $j < k$ it holds that $f(2j) \geq \left(\frac{2}{\gamma}\right)^{\log j} > \left(\frac{2}{\gamma}\right)^{\log k}$. Therefore, from the definition of f

$$f(2k) \geq \min\left(\left(\frac{2}{\gamma}\right)^{\log k}, \frac{1}{\gamma}(f(2) + f(2k-2)), \frac{1}{\gamma}(f(4) + f(2k-4)), \ldots\right)$$

Define $h(j) = \frac{1}{\gamma}(f(2j) + f(2k-2j))$, so

$$f(2k) \geq \min\left(\left(\frac{2}{\gamma}\right)^{\log k}, \min_{1 \leq j \leq k-1}\{h(j)\}\right)$$

By the induction hypothesis,

$$h(j) \geq \frac{1}{\gamma}\left(\left(\frac{2}{\gamma}\right)^{\log j} + \left(\frac{2}{\gamma}\right)^{\log k-j}\right) \geq \min_{x \in \mathbb{R}} \frac{1}{\gamma}\left\{\left(\frac{2}{\gamma}\right)^{\log x} + \left(\frac{2}{\gamma}\right)^{\log k-x}\right\}$$

Note that $\left(\frac{2}{\gamma}\right)^{\log x} + \left(\frac{2}{\gamma}\right)^{\log k-x}$ is symmetric about $x = \frac{k}{2}$. Moreover, it is a concave function as it is the sum of two concave functions, thus the minimum point occurs at $x = \frac{k}{2}$.

We found that $h(j) \geq \frac{1}{\gamma}\left(\left(\frac{2}{\gamma}\right)^{\log \frac{k}{2}} + \left(\frac{2}{\gamma}\right)^{\log \frac{k}{2}}\right) = \left(\frac{2}{\gamma}\right)^{\log \frac{k}{2}+1} = \left(\frac{2}{\gamma}\right)^{\log k}$

Hence, we conclude

$$f(2k) \geq \min\left(\left(\frac{2}{\gamma}\right)^{\log k}, \left(\frac{2}{\gamma}\right)^{\log k}\right) = \left(\frac{2}{\gamma}\right)^{\log k} = \frac{1}{k^{\log \frac{\gamma}{2}}}$$

$\square$

Lemma 6. $\text{ALG}_{\text{OFF}} \leq O\left(m^{\log\left(\frac{3+\epsilon}{2}\right)}\right) \text{OPT}$

Proof. We view the requests as if they were in the time-augmented metric space $\mathcal{M}_T$, and analyze the performance of ALG in an offline manner. By Corollary 1 we analyze the performance of ALG when $G = (\mathcal{I}, E)$, the union of the matchings produced by ALG and OPT, forms a single cycle.

Denote E_O the subset of edges matched by OPT, and E_A the subset of edges matched by ALG. Consider again the last two pairs of requests to be matched by ALG, that is (a, b) and (c, d), and assume that $t_{ab} \leq t_{cd}$ and $t(b) \geq t(a)$ (t_{ab} is the time that ALG matched (a, b), and t_{cd} is the time that ALG matched (c, d)). Denote $T = \sum_{e \in E \setminus \{(c,d)\}} D(e)$, and let $O = \sum_{e \in E_O} D(e)$. From the triangle inequality we have that $D(c, d)$ is smaller than T, therefore

$$\frac{\text{ALG}_{\text{OFF}}}{\text{OPT}} = \frac{D(c, d) + T - O}{O} \leq \frac{2T - O}{O} = 2\frac{T}{O} - 1 \tag{1}$$

We will bound $\frac{Q}{T}$ from below, by developing and solving a recurrence relation similar to the one developed in [21], thus giving an upper bound on $\frac{\text{ALG}_{\text{OFF}}}{\text{OPT}}$.

Scale the distances so that $T = 1$. Of course, $\frac{Q}{T}$ stays the same. Let $f(m)$ be the minimal value of $\frac{Q}{T}$ over all possible inputs of size m ($|\mathcal{I}| = m$), when the union of the matchings produced by ALG and OPT forms a single cycle.

For the sake of this analysis consider Fig. 1.

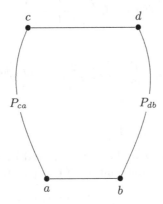

Fig. 1. The cycle formed by the union of the matchings produced by ALG and OPT. The length of P_{ca} is α, and the length of P_{db} is β.

Let P_{ca} be the alternating path from c to a, and P_{db} be the alternating path from d to b. Denote $\alpha = \sum_{e \in P_{ca}} D(e)$, and $\beta = \sum_{e \in P_{db}} D(e)$. Then, by the triangle inequality

$$\alpha \geq D(a, c) \tag{2}$$

From Lemma 2 we have

$$(1 + \epsilon)D(a, c) \geq D(a, b) \tag{3}$$

It follows from Eqs. (2) and (3) that

$$1 - \alpha - \beta = D(a, b) \leq (1 + \epsilon)\alpha \tag{4}$$

Similarly $1 - \alpha - \beta \leq (1 + \epsilon)\beta$.

Let $2i$ be the number of points on P_{ca}, then $f(m)$ satisfies the recurrence relation

$$f(m) = \min_{\substack{1 \leq i < \frac{m}{2} - 1 \\ 0 < 1-\alpha-\beta \leq (1+\epsilon)\alpha \\ 0 < 1-\alpha-\beta \leq (1+\epsilon)\beta}} \{\alpha f(2i) + \beta f(m - 2i)\} \tag{5}$$

Conditioning on t, both $f(t)$ and $f(m - t)$ are constant, therefore $\alpha f(t) + \beta f(m - t)$ becomes a linear function in α and β, so its minimum must occur at a vertex of the polyhedron defined by the minimization constraints (see for example [8]).

The vertices of this polyhedron are $(1,0), (0,1), (\frac{1}{3+\epsilon}, \frac{1}{3+\epsilon})$, so

$$f(m) = \min_{1 \le i \le \frac{m}{2}-1} \left\{ f(2i), \frac{1}{3+\epsilon} (f(2i) + f(m-2i)) \right\} \qquad (6)$$

Also note that $f(2) = 1$, since there is only one way to match two points, so $T = O$. The conditions of Lemma 5 are met with $\gamma = 3 + \epsilon$, thus

$$f(m) = \Omega \left(\frac{1}{m^{\log(\frac{3+\epsilon}{2})}} \right)$$

Finally, from 1 we conclude

$$\frac{\text{ALG}_{\text{OFF}}}{\text{OPT}} \le 2\frac{T}{O} - 1 \le \frac{2}{f(m)} = O\left(m^{\log(\frac{3+\epsilon}{2})} \right)$$

$\square$

Lemma 7. $\text{ALG}_{\text{ON}} = \Theta\left(\frac{1}{\epsilon}\right) \text{ALG}_{\text{OFF}}$

Proof. Assume two requests p and q were matched by ALG at time t. Assume w.l.o.g that $t(p) \ge t(q)$. The contribution of this pair to ALG_{ON}, is

$$t - t(p) + t - t(q) + d(x(p), x(q)) =$$
$$t - t(p) + t - t(p) + t(p) - t(q) + d(x(p), x(q)) = 2(t - t(p)) + D(p, q)$$

On the contrary, the contribution of this pair to ALG_{OFF}, is just $D(p, q)$.

Note that t is the time that q was on p's hemisphere, so $t = t(p) + \frac{D(p,q)}{\epsilon}$, hence the ratio between ALG_{ON} and ALG_{OFF} for this pair is

$$\frac{2\frac{D(p,q)}{\epsilon} + D(p,q)}{D(p,q)} = 1 + \frac{2}{\epsilon}$$

Summing over all matched pairs we get $\frac{\text{ALG}_{\text{ON}}}{\text{ALG}_{\text{OFF}}} = 1 + \frac{2}{\epsilon} = \Theta\left(\frac{1}{\epsilon}\right)$. $\square$

Finally we prove Theorem 1 using the inequalities proven in the previous lemmas.

Proof (Proof of Theorem 1). Combining Lemmas 1, 6 and 7 we have

$$\text{ALG}_{\text{ON}} \le O\left(\frac{1}{\epsilon}\right) \text{ALG}_{\text{OFF}} \le O\left(\frac{1}{\epsilon} m^{\log(\frac{3+\epsilon}{2})}\right) \text{OPT}$$

Hence, $\text{ALG}(\epsilon)$ is $O\left(\frac{1}{\epsilon} m^{\log(\frac{3+\epsilon}{2})}\right)$-competitive. $\square$

4   The Bipartite Case

For the bipartite case, we suggest the same algorithm as in the monochromatic case. The only difference is that we match a request q to a request p as soon as q is found on the boundary of p's hemisphere, **and that q and p do not belong to the same class.**

Algorithm 2. A Deterministic Algorithm for MBPMD on General Metrics

---

1: **procedure** ALG-B(ϵ)
2:     **At every moment** t:
3:         Add the new requests that arrive at time t
4:         **for each** unmatched request p **do**
5:             **for each** unmatched request $q \neq p$ **do**
6:                 **if** $t(p) \geq t(q)$ **and** $t = t(p) + \frac{D(x(p),x(q))}{\epsilon}$ **and** $class(q) \neq class(p)$ **then**
7:                     **match**(p, q)
8:                 **end if**
9:             **end for**
10:         **end for**
11: **end procedure**

4.1 Analysis

We prove the following theorem:

Theorem 2. ALG-B(ϵ) is $O\left(\frac{1}{\epsilon}m^{\log\left(\frac{3+\epsilon}{2}\right)}\right)$-competitive.

Observation 1, Lemmas 3 and 4 hold for the bipartite case as well, therefore using Corollary 1 we may assume that the union of ALG-B and OPT forms a single cycle.

The key difference in the analysis for this case, is that when we consider the last four requests to be matched, not every two of them could have been matched to each other. Therefore Lemma 2 does not hold, but a weaker yet similar result does.

Consider the last two pairs of requests to be matched by ALG-B. Name them (a, b) and (c, d), and assume w.l.o.g that (a, b) were matched at time t_{ab}, and (c, d) at $t_{cd} \geq t_{ab}$. Also, assume w.l.o.g that $t(a) \leq t(b)$.

Lemma 8. If $class(a) = class(d) \neq class(b) = class(c)$ then

1. $D(a, b) \leq (1 + \epsilon)D(a, c)$
2. $D(a, b) \leq (1 + \epsilon)D(b, d)$

We omit the proof of this lemma as it is the same as the proof of Lemma 2 for the relevant cases.

Considering Fig. 1 we have the following lemma.

Lemma 9. $class(a) = class(d) \neq class(b) = class(c)$

Proof. From the alternation property of Observation 1 we have that the number of edges along P_{ca} must be odd (since the number of OPT edges along P_{ca} must be one more than ALG-B edges along P_{ca}). Moreover, the classes of the requests along P_{ca} alternate as well (since every edge must match requests of different classes). Since there are odd number of edges along P_{ca}, there are odd number of class alternations along P_{ca}, so the class of the last request along P_{ca} (that is $class(c)$) must be different from the class of the first request along P_{ca} (that is $class(a)$). Thus $class(c) \neq class(a)$ and of course $class(a) \neq class(b)$, $class(c) \neq class(d)$, so $class(a) = class(d) \neq class(b) = class(c)$. □

Using Lemmas 9 and 8 we repeat the proof of Lemma 6 and achieve the following result:

Lemma 10. ALG-B$_{OFF}$ $\leq O\left(m^{\log\left(\frac{3+\epsilon}{2}\right)}\right)$ OPT

The main theorem for the bipartite case now follows:

Proof (Proof of Theorem 2). Lemmas 7 and 1 hold for ALG-B as well, thus from Lemma 10 we have

$$\text{ALG-B}_{ON} \leq O\left(\frac{1}{\epsilon}\right) \text{ALG-B}_{OFF} \leq O\left(\frac{1}{\epsilon}m^{\log\left(\frac{3+\epsilon}{2}\right)}\right) \text{OPT}$$

Hence, ALG-B(ϵ) is $O\left(\frac{1}{\epsilon}m^{\log\left(\frac{3+\epsilon}{2}\right)}\right)$-competitive. □

5   Concluding Remarks and Open Problems

In this paper we presented the first sub-linear competitive deterministic algorithm for *Minimum-Cost Perfect Matching with Delays* as a function of m, the number of requests. We also provided a similar algorithm for the problem of *Minimum-Cost Bipartite Perfect Matching with Delays* achieving the same competitive ratio.

One open problem is to decide if a deterministic algorithm with a better competitive ratio exists, in particular a polylog(m)-competitive one, by showing a lower bound or providing an algorithm for the problem. In addition, the problem of finding a sub-linear in n competitive deterministic algorithm is still open.

References

1. Antoniadis, A., Barcelo, N., Nugent, M., Pruhs, K., Scquizzato, M.: A $o(n)$-competitive deterministic algorithm for online matching on a line. In: Bampis, E., Svensson, O. (eds.) WAOA 2014. LNCS, vol. 8952, pp. 11–22. Springer, Cham (2015). https://doi.org/10.1007/978-3-319-18263-6_2
2. Ashlagi, I., et al.: Min-cost bipartite perfect matching with delays. In: Approximation, Randomization, and Combinatorial Optimization. Algorithms and Techniques, pp. 1:1–1:20 (2017)
3. Azar, Y., Chiplunkar, A., Kaplan, H.: Polylogarithmic bounds on the competitiveness of min-cost perfect matching with delays. In: Proceedings of the Twenty-Eighth Annual ACM-SIAM Symposium on Discrete Algorithms, pp. 1051–1061 (2017)
4. Azar, Y., Ganesh, A., Ge, R., Panigrahi, D.: Online service with delay. In: Proceedings of the 49th Annual ACM SIGACT Symposium on Theory of Computing, pp. 551–563 (2017)
5. Bansal, N., Buchbinder, N., Gupta, A., Naor, J.: A randomized $O(log^2 k)$-competitive algorithm for metric bipartite matching. Algorithmica **68**(2), 390–403 (2014)

6. Bienkowski, M., Kraska, A., Liu, H.H., Schmidt, P.: A primal-dual online deterministic algorithm for matching with delays. CoRR abs/1804.08097 (2018). http://arxiv.org/abs/1804.08097

7. Bienkowski, M., Kraska, A., Schmidt, P.: A match in time saves nine: deterministic online matching with delays. In: Solis-Oba, R., Fleischer, R. (eds.) WAOA 2017. LNCS, vol. 10787, pp. 132–146. Springer, Cham (2018). https://doi.org/10.1007/978-3-319-89441-6_11

8. Dantzig, G.B.: Linear programming and extensions. Princeton University Press, Princeton (1963)

9. Edmonds, J.: Paths, trees, and flowers. Can. J. Math. **17**, 449–467 (1965)

10. Emek, Y., Kutten, S., Wattenhofer, R.: Online matching: haste makes waste! In: Proceedings of the 48th Annual ACM SIGACT Symposium on Theory of Computing, pp. 333–344 (2016)

11. Emek, Y., Shapiro, Y., Wang, Y.: Minimum cost perfect matching with delays for two sources. In: Fotakis, D., Pagourtzis, A., Paschos, V. (eds.) CIAC 2017. LNCS, vol. 10236, pp. 209–221. Springer, Cham (2017). https://doi.org/10.1007/978-3-319-57586-5_18

12. Fuchs, B., Hochstättler, W., Kern, W.: Online matching on a line. Electron. Notes Discrete Math. **13**, 49–51 (2003)

13. Gupta, A., Lewi, K.: The online metric matching problem for doubling metrics. In: Czumaj, A., Mehlhorn, K., Pitts, A., Wattenhofer, R. (eds.) ICALP 2012. LNCS, vol. 7391, pp. 424–435. Springer, Heidelberg (2012). https://doi.org/10.1007/978-3-642-31594-7_36

14. Kalyanasundaram, B., Pruhs, K.: Online weighted matching. J. Algorithms **14**(3), 478–488 (1993)

15. Kalyanasundaram, B., Pruhs, K.: On-line network optimization problems. In: Fiat, A., Woeginger, G.J. (eds.) Online Algorithms. LNCS, vol. 1442, pp. 268–280. Springer, Heidelberg (1998). https://doi.org/10.1007/BFb0029573. The State of the Art (the book grow out of a Dagstuhl Seminar)

16. Khuller, S., Mitchell, S.G., Vazirani, V.V.: On-line algorithms for weighted bipartite matching and stable marriages. In: Albert, J.L., Monien, B., Artalejo, M.R. (eds.) ICALP 1991. LNCS, vol. 510, pp. 728–738. Springer, Heidelberg (1991). https://doi.org/10.1007/3-540-54233-7_178

17. Koutsoupias, E., Nanavati, A.: The online matching problem on a line. In: Solis-Oba, R., Jansen, K. (eds.) WAOA 2003. LNCS, vol. 2909, pp. 179–191. Springer, Heidelberg (2004). https://doi.org/10.1007/978-3-540-24592-6_14

18. Meyerson, A., Nanavati, A., Poplawski, L.J.: Randomized online algorithms for minimum metric bipartite matching. In: Proceedings of the Seventeenth Annual ACM-SIAM Symposium on Discrete Algorithms, pp. 954–959 (2006)

19. Nayyar, K., Raghvendra, S.: An input sensitive online algorithm for the metric bipartite matching problem. In: 58th IEEE Annual Symposium on Foundations of Computer Science, pp. 505–515 (2017)

20. Raghvendra, S.: A robust and optimal online algorithm for minimum metric bipartite matching. In: Approximation, Randomization, and Combinatorial Optimization. Algorithms and Techniques, pp. 18:1–18:16 (2016)

21. Reingold, E.M., Tarjan, R.E.: On a greedy heuristic for complete matching. SIAM J. Comput. **10**(4), 676–681 (1981)

Sequential Metric Dimension

Julien Bensmail[1], Dorian Mazauric[2], Fionn Mc Inerney[1(✉)], Nicolas Nisse[1], and Stéphane Pérennes[1]

[1] Université Côte d'Azur, Inria, CNRS, I3S, 2004 route des Lucioles, 06902 Sophia Antipolis, France
{julien.bensmail,fionn.mc-inerney,nicolas.nisse, stephane.perennes}@inria.fr
[2] Université Côte d'Azur, Inria, 2004 route des Lucioles, 06902 Sophia Antipolis, France
dorian.mazauric@inria.fr

Abstract. Seager introduced the following game in 2013. An invisible and immobile target is hidden at some vertex of a graph G. Every step, one vertex v of G can be probed which results in the knowledge of the distance between v and the target. The objective of the game is to minimize the number of steps needed to locate the target, wherever it is.

We address the generalization of this game where $k \geq 1$ vertices can be probed at every step. Our game also generalizes the notion of the *metric dimension* of a graph. Precisely, given a graph G and two integers $k, \ell \geq 1$, the LOCALIZATION Problem asks whether there exists a strategy to locate a target hidden in G in at most ℓ steps by probing at most k vertices per step. We show this problem is NP-complete when k (resp., ℓ) is a fixed parameter.

Our main results are for the class of trees where we prove this problem is NP-complete when k and ℓ are part of the input but, despite this, we design a polynomial-time $(+1)$-approximation algorithm in trees which gives a solution using at most one more step than the optimal one. It follows that the LOCALIZATION Problem is polynomial-time solvable in trees if k is fixed.

Keywords: Games in graphs · Metric dimension · Complexity

1 Introduction

Localization (or *Identification*) problems consist of distinguishing the vertices of a connected graph $G = (V, E)$ using a smallest subset $R \subseteq V$ of its vertices. Many variants have been studied depending on how a subset of vertices allows to identify other vertices. For instance, *identifying codes* [16], *adaptive identifying*

This work has been partially supported by ANR program "Investments for the Future" under reference ANR-11-LABX-0031-01, the Inria Associated Team AlDyNet. Due to lack of space, several proofs have been omitted and can be found in [3].

L. Epstein and T. Erlebach (Eds.): WAOA 2018, LNCS 11312, pp. 36–50, 2018.
https://doi.org/10.1007/978-3-030-04693-4_3

codes [2], and *locating dominating sets* [21] ask for the vertices to be distinguished by their neighbourhood in R. Another well studied example is the one of a *resolving set* [13,20] which aims at distinguishing each vertex of a graph by its distance to each vertex of this set. Given a graph G, the main problem is to compute a resolving set with minimum size, called the *metric dimension* of G [13,20]. The corresponding decision problem (first shown to be NP-complete in [12]) is NP-complete in planar graphs [8] and in graphs of diameter 2 [11], and W[2]-hard (parameterized by the solution's size) [14]. On the positive side, the problem is FPT in the class of graphs with bounded treelength [1]. Bounds on the metric dimension have also been determined for various graph classes [10]. In this paper, we address a *sequential* variant of this problem.

Let us consider a graph $G = (V, E)$ where an unknown vertex $t \in V$ hosts a hidden (invisible) and immobile target. *Probing* one vertex $v \in V$ results in the knowledge of the distance between t and v, denoted by $d_G(v, t)$. Probing a set $R \subseteq V$ of vertices results in the distance vector $(d_G(v, t))_{v \in R}$ and a set is a *resolving set* if the distance vectors are pairwise distinct for every $t \in V$. The *metric dimension* of G, denoted by $MD(G)$, is then the minimum number of vertices that must be probed simultaneously (in one step) to determine the location t of the target wherever it is. For instance, in the case of a path, probing one of its ends is sufficient to locate the target, *i.e.*, $MD(P) = 1$ for every path P. Another example (that we use throughout the paper) is the case of a star (tree with a universal vertex) with n leaves, denoted by S_n, for which it is necessary and sufficient to probe every leaf but one, *i.e.*, $MD(S_n) = n - 1$.

If less than $MD(G)$ vertices can be probed at once, it is natural to allow more than one step. Obviously, if at most $1 \leq k < MD(G)$ vertices can be probed at once, it is always feasible to locate an immobile target in $\lceil MD(G)/k \rceil$ steps by sequentially probing k different vertices of a smallest resolving set at each step. However, there are graphs for which the target can be located much faster. In [18], Seager initiated the study of the following sequential locating game: at each step, one vertex of a graph can be probed, and the objective is to minimize the number of steps required to locate the target, wherever it is. Seager gave bounds and exact values on this minimum number of steps in particular subclasses of trees (e.g., subdivisions of caterpillars) [18] but left the problem open in trees in general. In this paper, we study the generalization of this game where $k \geq 1$ vertices can be probed at each step.

Precisely, let $k \geq 1$ be an integer and let $G = (V, E)$ be a graph hosting an invisible and immobile target hidden at $t \in V$. A *k-strategy* is allowed to probe at most k vertices at each step of the game (where the choice of the probed vertices at some step may obviously depend on the results of the probes during previous steps) until the location t of the target is uniquely determined. Let $\lambda_k(G)$ denote the minimum integer h such that there exists a k-strategy that locates the target in G in at most h steps, wherever it is. Given G, k and $\ell \geq 1$, the LOCALIZATION Problem asks whether $\lambda_k(G) \leq \ell$. We also consider the dual parameter $\kappa_\ell(G)$ defined as the minimum integer h such that there exists an h-strategy that locates the target in G in at most ℓ steps. Note that, for every graph G, $\kappa_1(G)$ is exactly the metric dimension $MD(G)$ of G, and $\lambda_k(G) \leq \ell$ if and only if $\kappa_\ell(G) \leq k$. We are interested in the complexity of the LOCALIZATION Problem in general graphs and particularly in trees. Note that by the remarks above, the LOCALIZATION Problem and METRIC DIMENSION Problem (for which $\ell = 1$) behave very differently, so knowing that METRIC DIMENSION Problem is NP-complete does not imply the same for the LOCALIZATION Problem.

1.1   Related Work

The literature related to localizing a target in a graph is vast. Below, we focus on the related work that we find the most relevant.

Moving Target. Sequential games related to resolving sets have first been introduced and mainly studied in the case of a mobile target. That is, at every step, some vertices may be probed and, if the target has not been located yet, it may move to one of its neighbours (sometimes, it is required that the target cannot move to a vertex that has been probed during the previous step which is called the "no-backtrack condition") [17]. In this setting, locating the target may not be feasible. For instance, it is not possible to locate a moving target in a triangle when probing one vertex per step if the target may "backtrack". The question of how many times all the edges of a graph must be subdivided to ensure locating a moving target probing 1 vertex (resp., k vertices) per step has been addressed in [7] (resp. [15]). Let a graph be called *locatable* if there exists a 1-strategy for locating the target in a finite number of steps with the "no-backtrack condition". The case of trees with the "no-backtrack condition" has first been studied in [17] where it was shown that all trees T are locatable, and in [6], the upper bound on the number of steps it takes to locate the target in T was improved. In [19], the case of trees where the target may "backtrack" was considered. Let $\zeta(G)$ be the minimum integer k such that there exists a k-strategy for locating a moving target in G. In [5], it was shown that deciding whether $\zeta(G) \leq k$ is NP-hard and that $\zeta(G)$ is not bounded in the class of graphs with treewidth 2. Moreover, $\zeta(G) \leq 3$ for any outerplanar graph G [4].

Relative Distance and Centroidal Dimension. Foucaud *et al.* defined a variant of resolving sets, called *centroidal basis*, where the vertices of a graph must be distinguished by their *relative* distance to the probed vertices [9]. In this setting, given an integer $k \geq 2$, probing a set $B = \{v_1, \ldots, v_k\}$ of vertices results in the vector $(\delta_{i,j}(t))_{1 \leq i < j \leq k}$ where, for every $1 \leq i < j \leq k$, $\delta_{i,j}(t) = 0$ if $d_G(t, v_i) = d_G(t, v_j)$, $\delta_{i,j}(t) = 1$ if $d_G(t, v_i) > d_G(t, v_j)$ and $\delta_{i,j}(t) = -1$ otherwise. In other words, probing any two vertices returns the information of which one is closer to the target or whether they are equidistant from it. The set B is a *centroidal basis* if the vectors of relative distances for every $t \in V$ are pairwise distinct. The *centroidal dimension* of a graph G, denoted by $CD(G) \geq 2$, is the minimum size of a centroidal basis of G [9] (this is well defined since, clearly, V is a centroidal basis of G). The decision problem associated to the centroidal dimension is NP-complete and almost tight bounds on the centroidal dimension of paths have been computed [9].

A sequential variant of the centroidal basis can naturally be defined. This variant has been studied in the case of a moving target in [4].

Here, we also initiate the study of this variant when the target is immobile. Let $k \geq 2$ be an integer and G be a graph. Let $\lambda_k^{rel}(G)$ denote the minimum integer h such that there exists a k-strategy that locates (using relative distances) a hidden immobile target in G in at most h steps, whatever be the location of the target. Given G, k, and ℓ, the RELATIVE-LOCALIZATION Problem asks whether $\lambda_k^{rel}(G) \leq \ell$. The dual parameter $\kappa_\ell^{rel}(G)$ is defined as the minimum integer h such that there exists an h-strategy (with relative distances) that locates the target in G in at most ℓ steps. Note that, for every graph G, $\kappa_1^{rel}(G)$ is exactly the centroidal dimension $CD(G)$ of G, and $\lambda_k^{rel}(G) \leq \ell$ if and only if $\kappa_\ell^{rel}(G) \leq k$.

1.2   Our Results

In the whole paper, G denotes a connected undirected simple graph. We consider the computational complexity of the LOCALIZATION Problem. In Sect. 2, we show that it is polynomial-time solvable when both k and ℓ are fixed parameters but that it is NP-complete when only one of those two parameters is fixed. Precisely:

– Let $k \geq 1$ and $\ell \geq 1$ be two fixed integers. Given a graph G as an input, the problem of deciding whether $\lambda_k(G) \leq \ell$ is polynomial-time solvable (in time $n^{O(k\ell)}$) (Theorem 1).
– Let $k \geq 1$ be a fixed integer. Given a graph G with a universal vertex and an integer $\ell \geq 1$ as inputs, the problem of deciding whether $\lambda_k(G) \leq \ell$ is NP-complete (Theorem 2).
– Let $\ell \geq 1$ be a fixed integer. Given a graph G with a universal vertex and an integer $k \geq 1$ as inputs, the problem of deciding whether $\kappa_\ell(G) \leq k$ is NP-complete (Theorem 4).

On the way, we also show that the RELATIVE-LOCALIZATION Problem is polynomial-time solvable when both k and ℓ are fixed parameters (Theorem 1) but that it is NP-complete when only one of those two parameters is fixed (Theorems 3 and 5).

In Sect. 3, we then focus on the LOCALIZATION Problem in the class of trees. Surprisingly, in trees, the complexity of the LOCALIZATION Problem only comes from the first step. We show that, after the first step, the problem becomes polynomial-time solvable. This allows us to design a polynomial-time approximation algorithm for the problem. More precisely, we show that

– deciding whether $\lambda_k(T) \leq \ell$ is NP-complete in the class of trees T when both k and ℓ are part of the input (Theorem 6);
– there exists an algorithm that computes, in time $O(n \log n)$ (independent of k), a k-strategy for locating a target in at most $\lambda_k(T) + 1$ steps in any n-node tree (possibly edge-weighted) (Theorem 9);
– deciding whether $\lambda_k(T) \leq \ell$ can be solved in time $O(n^{k+2} \log n)$ (independent of ℓ) in the class of n-node trees (possibly edge-weighted) (Corollary 1).

2   Complexity of the LOCALIZATION Problem

This section is devoted to prove that the (RELATIVE) LOCALIZATION Problem is polynomial-time solvable when both k and ℓ are fixed parameters but that it is NP-complete when only one of those two parameters is fixed. The proof when ℓ is fixed is an almost straightforward reduction from the METRIC DIMENSION Problem. In the case when k is fixed, it is a much more involved reduction from the 3-DIMENSIONAL MATCHING Problem. The proof that the (RELATIVE) LOCALIZATION Problem is in NP is given as a separate claim (Claim 1) as it is used in all of the NP-completeness proofs.

Theorem 1. *Let* $k \geq 1$ *(*$k \geq 2$ *for the* RELATIVE LOCALIZATION *Problem) and* $\ell \geq 1$ *be two fixed integers. The (*RELATIVE*)* LOCALIZATION *Problem is polynomial-time solvable (in time* $n^{O(k\ell)}$*).*

Proof. Let G be any n-node graph. Let us consider the following tree $\mathcal{T}$ that will be used to represent all possible strategies that probe exactly k vertices per step and last at most ℓ steps in G.

The tree $\mathcal{T}$ is rooted in r and all leaves are at distance 2ℓ from the root. The two types of vertices of $\mathcal{T}$ are labelled by subsets of vertices of $V(G)$. For any vertex $v \in V(\mathcal{T})$ at even distance from r, its label $L(v) \subseteq V(G)$ represents the set of possible locations of the target at this moment. For any vertex $v \in V(\mathcal{T})$ at odd distance from r, its label $L(v) \subseteq V(G)$, of size k, represents the set of vertices that are probed at this moment.

Precisely, $\mathcal{T}$ is defined as follows. Its root r is labelled with $L(r) = V(G)$ (initially, the target may be anywhere). Then, given a vertex $v \in V(\mathcal{T})$ at even distance from r and such that $L(v) = S \subseteq V(G)$, the node v has exactly $\binom{n}{k}$ children labelled by each of the subsets of size k of $V(G)$. Then, for every $Q \in V(G)^k$, let w be the child of v such that $L(w) = Q$. The at most n children of w are defined as follows. Let $(S_1, \cdots, S_q)$ be the partition of S such that, for any $x, y \in S$, the vertices x and y belong to the same S_i if and only if probing the vertices of Q knowing that the target is in S gives the same answer (distance vector) for x and y. Then, w has exactly q children $s_1, \ldots, s_q$ such that $L(s_i) = S_i$ for every $1 \leq i \leq q$. Intuitively, each child of w corresponds to the possible locations of the target in response to the probing of the vertices of Q.

First, note that $|V(\mathcal{T})|$ is polynomial in n when k and ℓ are fixed. Precisely, since $\mathcal{T}$ has at most $(\binom{n}{k}n)^\ell$ leaves (due to the degree of the nodes and the height of $\mathcal{T}$) and all leaves are at distance 2ℓ from r, $|V(\mathcal{T})|$ is upper bounded by $O(2\ell(\binom{n}{k}n)^\ell) = n^{O(k\ell)}$.

Secondly, every strategy (of length ℓ and probing k vertices per turn) is "contained" in $\mathcal{T}$. Indeed, any subtree T' of $\mathcal{T}$ built as follows represents a strategy: start with T' reduced to the root r, then while possible, for any leaf v of T', if v is at an even distance from r, choose a single child of v and add it to T' (this is the probing that the strategy performs in this situation), otherwise, if v is at odd distance from r, add all its children to T'. It is easy to see that, in this way, any strategy, winning (locating the target in at most ℓ turns, wherever it is) or not, can be represented.

By the same reasoning, for every node v at even distance $2(\ell - \ell')$ from r, the subtree of $\mathcal{T}$ rooted in v "contains" all strategies of length ℓ' and probing k vertices per turn, assuming that, initially, the target occupies a vertex in $L(v)$. Let us say that v is *valid* if it contains at least one such winning strategy.

To find out if there is a winning strategy in G, let us proceed by dynamic programming, bottom-up from the leaves of this tree to the root. A leaf v of $\mathcal{T}$ is valid if and only if $L(v)$ is a singleton (indeed, the leaves of $\mathcal{T}$ represent strategies without any probe so the location of the target must be uniquely identified). Then, a vertex v at odd distance from the root is valid if and only if all its children are valid (after a probing, there must be a winning strategy, whatever be the answer). Finally, a vertex v at even distance from the root is valid if and only if at least one of its children is valid. Indeed, the subtree rooted at v contains a winning strategy if, knowing that the target is in $L(v)$, there exists at least one possible probing (one set of k vertices to be probed) that leads toward a winning strategy, whatever be the answer to this probing.

Therefore, there is a winning strategy for G if and only if the root is valid which can be decided in time $|V(\mathcal{T})| = n^{O(k\ell)}$. □

Claim 1. The (RELATIVE) LOCALIZATION Problem is in NP.

Proof of Claim. The proof is done for the LOCALIZATION Problem. The certificate is a k-strategy which can be described by a rooted decision tree T as follows. The nodes of T are labelled by sets of k vertices (the vertices to be probed at a given step) and its edges are labelled by sets of vertices representing the possible locations of the target. Precisely, the root node represents the first k vertices to be probed in G according to the k-strategy. For every node $v \in V(T)$ (but the root), the label $L_e \subseteq V(G)$ of the parent-edge e of v represents the current possible locations of the target and the label $L_v \subseteq V(G)$, $|L_v| \leq k$, is the set of vertices to be probed according to the strategy, given that the target occupies a vertex in L_e. Then, every child w of v corresponds to a possible outcome (after probing the vertices in L_v). That is, L_{vw} is the new set of possible locations after having probed L_v (given that the target was in L_e). Note that, clearly, $L_{vw} \subseteq L_e$. Moreover, we may restrict our attention to *progressive* strategies, *i.e.*, strategies for which, for every non-root vertex v with parent-edge e, and for every child-edge f of v, $L_f \subset L_e$. Indeed, otherwise, the vertices probed in L_v are not relevant and a better choice would be any subset containing at least one vertex of L_e (two vertices of L_e in the case of the RELATIVE LOCALIZATION Problem, where by definition $k \geq 2$, and this is the only part of the proof that differs between the two problems).

The previous remark shows that we can restrict ourselves to k-strategies represented by rooted trees where all non-leaf nodes have at least two children. Moreover, any such tree representing a winning strategy (a k-strategy that locates the target) has exactly $|V(G)|$ leaves since there is a one-to-one correspondence between a path from the root to a leaf of T with the location of the target in G. A trivial induction on $|V(T)|$ allows to show that any rooted tree with n leaves and where all non-leaf nodes have at least two children, has at most $2n$ nodes. Thus, any winning k-strategy may be encoded polynomially and the LOCALIZATION Problem is in NP.   ◇

2.1   When the Number k of Probed Vertices per Step Is Fixed

Let $k \geq 1$ be a fixed integer. The k-PROBE LOCALIZATION Problem takes a graph G and an integer $\ell \geq 1$ as inputs and asks whether $\lambda_k(G) \leq \ell$.

Theorem 2. *Let $k \geq 1$ be a fixed integer. The k-PROBE LOCALIZATION Problem is NP-complete in the class of graphs with a universal vertex.*

Sketch of proof. The problem is in NP by Claim 1. Let us prove it is NP-hard by a reduction from the 3-DIMENSIONAL MATCHING Problem (3DMP) which is a well known NP-hard problem. The 3DMP takes a set $\mathcal{X} = I_1 \cup I_2 \cup I_3$ of $3n$ elements ($|I_1| = |I_2| = |I_3| = n$) and a set $\mathcal{S}$ of triples $(x, y, z) \in I_1 \times I_2 \times I_3$ as inputs and asks whether there are n triples of $\mathcal{S}$ that are pairwise disjoint.

Let $k \geq 1$ be a fixed integer and let $\mathcal{I} = (\mathcal{X}, \mathcal{S})$ be an instance of 3DMP. First, we may assume that $|\mathcal{X}| = 3kn$ since, if not, it is sufficient to take k disjoint copies of $(\mathcal{X}, \mathcal{S})$. Moreover, we may assume that $m = |\mathcal{S}|$ is such that $2m - 1 \equiv 0$ mod k (for instance by adding dummy triples if needed). Let $\mathcal{X} = \{x_1, \dots, x_{3kn}\}$ and $\mathcal{S} = \{S_1, \dots, S_m\}$.

Let us build the graph $G = (V, E)$ as follows. Let the vertex-set $V = X \cup X'' \cup S \cup \{s\} \cup \{q\}$ be such that $X = X^1 \cup \dots \cup X^{k+2}$ with $X^i = \{x_1^i, \dots, x_{3kn}^i\}$ for every $1 \leq i \leq k + 2$; $X'' = \{x_1'', \dots, x_{(k+2)m}''\}$; and $S = S^1 \cup \dots \cup S^{k+2}$ with $S^i = \{s_j^i, 1 \leq j \leq m\}$ for every $i \in [\![1, k + 2]\!]$. The vertex s is universal (*i.e.*, adjacent to every other vertex), the vertex q is adjacent to every vertex in $X \cup X''$ and, for every $j \in [\![1, 3kn]\!]$ and every $g \in [\![1, m]\!]$ such that $x_j \in S_g$, there is an edge between x_j^i and s_g^i for every $i \in [\![1, k + 2]\!]$. Intuitively, X^i is a "copy" of $\mathcal{X}$ and S^i is a "copy" of $\mathcal{S}$ for every $1 \leq i \leq k + 2$.

Let $p = \frac{m(k+2)-1}{k} \in \mathbb{N}$. We prove the theorem by showing that $\mathcal{I} = (\mathcal{X}, \mathcal{S})$ admits a 3DM if and only if $\lambda_k(G) \leq (k + 2)n + p + 1$.   □

The same proof also works for the case with relative distances. Hence,

Theorem 3. *Let $k \geq 2$ be a fixed integer. Given a graph G with a universal vertex and $1 \leq \ell \in \mathbb{N}$, the problem of deciding if $\lambda_k^{rel}(G) \leq \ell$ is NP-complete.*

2.2   When the Number ℓ of Steps Is Fixed

Let $\ell \geq 1$ be a fixed integer. The ℓ-STEP LOCALIZATION Problem takes a graph G and an integer $k \geq 1$ as inputs and asks whether $\kappa_\ell(G) \leq k$.

Theorem 4. *Let $\ell \geq 1$ be a fixed parameter. The ℓ-STEP LOCALIZATION Problem is NP-complete in the class of graphs with a universal vertex.*

Sketch of proof. For $\ell = 1$, the result follows from the fact that $\kappa_1(G)$ is exactly the metric dimension and from its NP-completeness [8].

Let $\ell \geq 2$ be fixed. The problem is in NP by Claim 1. To prove the NP-hardness, let us reduce the METRIC DIMENSION Problem restricted to the class of graphs that contain a universal vertex, which is known to be NP-hard [11]. Let $<G, k>$ be an instance of METRIC DIMENSION where G contains a universal vertex. We construct, in polynomial time, an instance $<G', k>$ of the ℓ-STEP LOCALIZATION Problem such that $MD(G) \leq k$ if and only if $\kappa_\ell(G') \leq k$.

The construction of G' is as follows. Start from $k(\ell - 1) + 1$ disjoint copies $G_1, \ldots, G_{k(\ell-1)+1}$ of G. Let v be a universal vertex of G, and for $1 \leq i \leq k(\ell - 1) + 1$, let v_i denote the copy of v in G_i. Finally, add a universal vertex u to the graph. □

A similar proof (but based on a reduction of CENTROIDAL DIMENSION) works for the case with relative distances. Hence,

Theorem 5. *Let $\ell \geq 1$ be a fixed integer. Given a graph G with a universal vertex and $2 \leq k \in \mathbb{N}$, the problem of deciding if $\kappa_\ell^{rel}(G) \leq k$ is NP-complete.*

3 The LOCALIZATION Problem in Trees

This section is devoted to the study of the LOCALIZATION Problem in the class of trees. Note that, if the number of steps is $\ell = 1$, the problem is equivalent to the one of METRIC DIMENSION which can easily be solved in polynomial-time in trees [13, 20]. We first show that, if k and ℓ are part of the input, deciding whether $\lambda_k(T) \leq \ell$ is NP-complete in the class of trees T. Our reduction actually shows that the difficulty of the problem comes from the choice of the vertices to be probed during the first step. Surprisingly, we show that the first step is actually the only source of complexity. More precisely, our main result is that, if the first step is given (intuitively, either given by an oracle or imposed by an adversary), then an optimal strategy (according to this first pre-defined step) can be computed in polynomial-time. This allows us to design a $(+1)$-approximation algorithm for the LOCALIZATION Problem in trees and to prove that, in contrast with general graphs (Theorem 2), the k-PROBE LOCALIZATION Problem is polynomial-time solvable in the class of trees (Corollary 1).

3.1 NP-Hardness

Theorem 6. *The LOCALIZATION Problem is NP-complete in the class of trees.*

Sketch of proof. Again, the problem is in NP by Claim 1. To prove the NP-hardness, let us reduce the HITTING-SET Problem. The inputs are an integer $k \geq 1$, a ground-set $B = (b_1, \ldots, b_n)$, and a set $\mathcal{S} = \{S_1, \ldots, S_m\}$ of subsets of B, i.e., $S_i \subseteq B$ for every $i \leq m$. The HITTING-SET Problem aims at deciding if there exists a set $H \subseteq B$, $|H| \leq k$, and $H \cap S_i \neq \emptyset$ for every $1 \leq i \leq m$.

Adding one new element to the ground-set and adding this element to one single subset clearly does not change the solution. Therefore, by adding some dummy elements (each one belonging to a single subset), we may assume that all the subsets are of the same size σ and that $\sigma - 1 \equiv 0 \mod k$.

Let γ be any integer such that $\gamma - 1 \equiv 0 \mod k$ and $\gamma > n - k - 1$.

The instance T of the LOCALIZATION Problem is built as follows. Let us start with n vertex-disjoint paths $B_1, \cdots, B_n$ (the *branches*) of length $2m$, where $B_i = (b_1^i, \ldots, b_{2m+1}^i)$ for each $1 \leq i \leq n$. Then, let us add one new vertex c adjacent to b_1^i for all $1 \leq i \leq n$. For every $1 \leq j \leq m$ and for every $1 \leq i \leq n$ such that $b_i \in S_j$, let us add γ new vertices adjacent to b_{2j}^i. The subgraph induced by b_{2j}^i and by the γ leaves adjacent to it is referred to as the *star* representing the element i in the set S_j (or representing the set S_j in the branch i). The obtained tree T is depicted in Fig. 1.

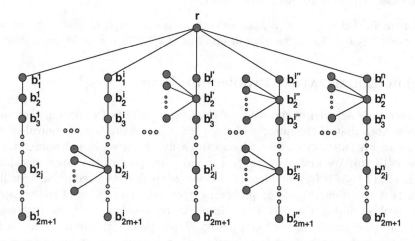

Fig. 1. An example of a tree T built from an instance $(k, B, \mathcal{S})$ of HITTING SET in the proof of Theorem 6. The elements $b_{i'}, b_{i''}$, and b_n belong to the set S_1 (but not the elements b_1 and b_i) as figured by the three "stars" at level 2. The elements b_i and $b_{i''}$ belong to S_j (stars at level $2j$), but not the elements $b_1, b_{i'}$, and b_n.

Intuitively, it will always be better for the target to be located in a leaf of some star because γ is "huge". During the first turn of any strategy, the *level* (roughly, the distance to the root) of the target can be identified. Each even level $2j$ corresponds to a set S_j. If, during the first turn, one star corresponding to each even level can be eliminated from the possible locations (which corresponds to hitting every subset), then the strategy finishes one step earlier than if all subsets cannot be hit (if so, all stars would have to be checked).

Precisely, we show that $\lambda_k(T) \leq 1 + \frac{\sigma-1}{k} + \frac{\gamma-1}{k}$ if and only if there is a hitting set of size at most k. $\qquad\square$

3.2   Algorithm in Trees

The proof above actually shows that, in our reduction, choosing the vertices to be probed during the first step to ensure an optimal strategy is equivalent to finding a minimum hitting set. We show below that the first step is actually the only "part" of the problem that is difficult.

The key argument is the following easy remark. Let us consider a tree T where a target is hidden and assume that a single vertex $r \in V(T)$ is probed. After this single probe, the distance $d \in \mathbb{N}$ between the target and r is known. Therefore, from the second step, the instance becomes equivalent to a tree T' (a subtree of T) rooted in r, with all leaves the same distance d from r, and where the target is known to occupy some leaf of T'. We first present an algorithm that computes in polynomial-time (independent of k and ℓ) an optimal strategy to locate the target in such instances.

Let $\mathcal{T}$ be the set of rooted trees with all leaves the same distance from the root. Given a rooted tree $(T, r) \in \mathcal{T}$ (in what follows, we omit r when it is clear from the context), let $\lambda_k^L(T)$ be the minimum integer h such that there exists a k-strategy that locates a target in at most h steps knowing *a priori* that the target occupies some leaf of T. The next claim is a key argument for why the problem is easier in the class $\mathcal{T}$ when the target is known to occupy a leaf.

Claim 2. Let $(T, r) \in \mathcal{T}$ rooted in r, let v be a child of r and T_v the subtree rooted in v. If the target is known to occupy a leaf of T, then probing any vertex in T_v allows to learn if the target occupies a leaf of T_v or a leaf of $T \setminus T_v$.

Proof of Claim. Let d be the distance between r and the leaves of T. Let w be any vertex of T_v and let d' be the distance between w and r. The target occupies a leaf of T_v if and only if its distance to w is strictly less than $d + d'$.   ◇

Let $T \in \mathcal{T}$ rooted in r, let v be a child of r, and let us assume that the secret location of the target is some leaf of T_v. Note that $(T_v, v) \in \mathcal{T}$. Let us assume that T_v is not a path and let s be the first step of an optimal strategy ϕ in T that probes some vertex of T_v (such a step s must exist since otherwise the target would never be detected in T_v). By Claim 2, it is sufficient to probe a single vertex of T_v to learn whether the target occupies a leaf of T_v. Then, applying an optimal strategy ϕ_v in T_v will locate the target in a total of $s + \lambda_k^L(T_v) - 1$ steps if the first step of ϕ_v only requires probing a single vertex of T_v and $s + \lambda_k^L(T_v)$ steps otherwise. So, it may be possible to do better. Indeed, probing several vertices of T_v during the sth step of ϕ may serve not only to detect the target in T_v but also to "play" the first step of ϕ_v. Doing so, the strategy will take only $s + \lambda_k^L(T_v) - 1$ steps. So, elaborating, an optimal strategy will consist of doing a tradeoff between probing one single vertex in a subtree (and detecting "quickly" in which subtree the target is hidden since several subtrees are considered simultaneously) and probing more vertices in a subtree in order to get a head start for the strategy in the case the target is in this subtree.

For any tree T, let $\pi(T)$ be the minimum integer q such that there exists a k-strategy that locates a target in at most $\lambda_k^L(T)$ steps, knowing *a priori* that

---

Algorithm 1. $\mathcal{A}_1(k, (T, r))$.

Require: An integer k and a tree $T \in \mathcal{T}$ rooted in r with children $v_1, \ldots, v_{d^*}$
Ensure: $(\lambda_k^L(T), \pi(T))$
1: **if** (T, r) is a rooted path **then**
2:   **return** $(0, 0)$
3: **for** $i = 1$ to d^* **do**
4:   Let $(\lambda_i, \pi_i) = \mathcal{A}_1(k, (T[i], v_i))$
5: Sort the $(\lambda_i, \pi_i)_{1 \leq i \leq d^*}$ in non-increasing lexicographical order
6: **return** $\mathcal{A}_2(k, (T, r), (\lambda_i, \pi_i)_{1 \leq i \leq d^*})$

---

the target occupies some leaf of T, and such that at most q vertices are probed during the first step.

To illustrate the need of a tradeoff, let us consider the following simple example utilizing π. Consider two children v_1 and v_2 of r such that $(\lambda_k^L(T_{v_1}), \pi(T_{v_1})) = (6, 4)$ and $(\lambda_k^L(T_{v_2}), \pi(T_{v_2})) = (6, 3)$. Let $k = 6$. Then, at the first step, we cannot probe $\pi(T_{v_1}) + \pi(T_{v_2}) = 7$ vertices. W.l.o.g., let us assume that at most $3 < \pi(T_{v_1})$ vertices of T_{v_1} have been probed during the first step. Thus, by definition of π, a total of $\lambda_k^L(T_{v_1}) + 1 = 7$ steps are necessary if we learn at the first step that the target occupies some leaf of T_{v_1}.

Let $T \in \mathcal{T}$ rooted in r and let $v_1, \ldots, v_{d^*}$ be the children of r. From previous arguments, the computation of an optimal strategy for T consists of determining, for each subtree T_{v_i} ($1 \leq i \leq d^*$), the first step for which a vertex of T_{v_i} will be probed (if the target has not been located in a different subtree at a previous step). If 1 vertex is probed during this step, then $\lambda_k^L(T_{v_i})$ extra steps are needed if the target occupies some leaf of T_{v_i} (unless $\pi(T_{v_i}) = 1$ in which case $\lambda_k^L(T_{v_i}) - 1$ extra steps are needed). If $\pi(T_{v_i})$ vertices of T_{v_i} are probed during this step, then $\lambda_k^L(T_{v_i}) - 1$ extra steps are needed if the target occupies some leaf of T_{v_i}.

Description of Algorithm 1. The main algorithm $\mathcal{A}_1(k, (T, r))$ takes an integer $k \geq 1$ and a rooted tree $(T, r) \in \mathcal{T}$ as inputs and computes $(\lambda_k^L(T), \pi(T))$ and a corresponding k-strategy. It proceeds bottom-up by dynamic programming from the leaves to the root. Precisely, let $v_1, \ldots, v_{d^*}$ be the children of r. For any $1 \leq i \leq j \leq d^*$, let $T[i] = T_{v_i}$ be the subtree rooted at v_i, and let $T[i, j] = \{r\} \cup T_{v_i} \cup \cdots \cup T_{v_j}$ ($T[i, j] = \emptyset$ if $i > j$). To lighten the notations, let us set $\lambda_i = \lambda_k^L(T[i])$ and $\pi_i = \pi(T[i])$ for every $1 \leq i \leq d^*$. Assume that, $(\Lambda, \Pi) = (\lambda_i, \pi_i)_{1 \leq i \leq d^*}$ have been computed recursively and sorted in non-increasing lexicographical order. Then, $\mathcal{A}_2(k, (T, r), (\Lambda, \Pi))$, described in Algorithm 2, takes the integer $k \geq 1$, the rooted tree $(T, r) \in \mathcal{T}$, and the sorted tuple (Λ, Π) as inputs and computes $(\lambda_k^L(T), \pi(T))$ and a corresponding strategy.

Description of Algorithm 2. We now informally describe $\mathcal{A}_2(k, (T, r), (\Lambda, \Pi))$. First, Line 2 to Line 5 deals with the subtrees $T_{v_{d+1}}, \ldots, T_{v_{d^*}}$ that are rooted paths (path rooted at one of its vertices of degree one, the other vertex is the leaf). In other words, it concerns all the subtrees T_{v_i} such that $(\lambda_i, \pi_i) = (0, 0)$. Indeed, this case is somehow pathologic. Claim 3 proves that Line 2 to Line 5 computes $(\lambda_k^L(T[v_{d+1}, d^*]), \pi(T[v_{d+1}, d^*]))$. Let us define $S \subset T$ as the set of subdivided

---

Algorithm 2. $\mathcal{A}_2(k, (T, r), (\Lambda, \Pi))$.

Require: $k \in \mathbb{N}^*$, a rooted tree (T, r) with $v_1, \ldots, v_{d^*}$ the children of r such that
$(\Lambda, \Pi) = (\lambda_i, \pi_i)_{1 \leq i \leq d^*}$ is sorted in non-increasing lexicographical order.
1: $l \leftarrow 1$, $p \leftarrow k$, $d \leftarrow d^*$
2: **if** $T[d^*]$ is a rooted path **then**
3:    $d \leftarrow z$ with $0 \leq z < d^*$ the smallest integer such that $T[z + 1]$ is a rooted path
4:    $l \leftarrow 1 + \lceil \frac{d^* - d - 1}{k} \rceil$
5:    $p \leftarrow k + k(\lceil \frac{d^* - d - 1}{k} \rceil - \lceil \frac{d^* - d - 1}{d^* - d} \rceil) - (d^* - d - 1)$
6: **for** $i = d$ down to 1 **do**
7:    **if** $p = 0$ or $l < \lambda_i + 1$ **then**
8:       $p \leftarrow k$, $l \leftarrow \max(l + 1, \lambda_i + 1)$
9:       $\alpha \leftarrow \pi_i - (\pi_i - 1)\lceil (l - (\lambda_i + 1))/l \rceil$
10:    **if** $\alpha \leq p$ **then**
11:       $p \leftarrow p - \alpha$
12:    **else**
13:       $p \leftarrow k - 1$, $l \leftarrow l + 1$
14: **return** $(l - 1, k - p)$

---

stars S (*i.e.*, trees with at most one vertex of degree at least 3) with all leaves the same distance from the root, where the root of S is the (unique) vertex with degree > 2 or one of the two ends if S is a path.

Claim 3. Let $S \in \mathcal{S}$ and let δ be the degree of the root r. Then, $\lambda_k^L(S) = \lceil \frac{\delta - 1}{k} \rceil$ and $\pi(S) = -k(\lceil \frac{\delta - 1}{k} \rceil - \lceil \frac{\delta - 1}{\delta} \rceil) + (\delta - 1)$.

We are now able to detail the second part of the algorithm (from Line 6). Informally, $\mathcal{A}_2(k, (T, r), (\Lambda, \Pi))$ recursively builds, for $i = d$ down to 1, an optimal k-strategy ϕ for $T[i, d^*]$ from an optimal k-strategy ϕ' of $T[i + 1, d^*]$ and from an optimal k-strategy ϕ'' of $T[i]$ (the latter one being given as input through (λ_i, π_i)). In other words, $(\lambda_k^L(T[i, d^*]), \pi(T[i, d^*]))$ is computed from $(\lambda_k^L(T[i + 1, d^*]), \pi(T[i + 1, d^*]))$ and (λ_i, π_i). For every $1 \leq i \leq d + 1$, let l_i (resp., p_i) denote the value of l (resp. of p) just before the $(d + 2 - i)^{th}$ iteration of the for-loop (so, l_1 and p_1 are the final values of l and p). Intuitively, let us assume that optimal strategy for $T[i + 1, d^*]$ has been computed, takes at most $l_{i+1} - 1$ steps and requires $k - p_{i+1} = \pi(T[i + 1, d^*])$ vertices to be probed during its first step. Roughly, there are five cases to be considered.

- If $\pi_i \leq p_{i+1}$ and $\lambda_i = l_{i+1} - 1$, the strategy ϕ follows ϕ' but, in addition, probes π_i vertices of $T[i]$ during its first step. If the target is in $T[i]$, then ϕ follows ϕ'' (and takes a total of at most λ_i steps), otherwise, it proceeds as ϕ' (and takes a total of at most $l_{i+1} - 1$ steps). We get $l_i = l_{i+1}$ and $p_i = p_{i+1} - \pi_i$.
- Else if $\pi_i > p_{i+1} > 0$ and $\lambda_i = l_{i+1} - 1$, the first step of ϕ probes a unique vertex in $T[i]$. If the target is in $T[i]$, then ϕ follows ϕ'' (and takes a total of at most $\lambda_i + 1$ steps). Otherwise, it proceeds as ϕ' (and takes a total of at most l_{i+1} steps). We get $l_i = l_{i+1} + 1$ and $p_i = k - 1$.

- Else, if $p_{i+1} = 0$ and $\lambda_i \leq l_{i+1} - 1$, the first step of ϕ probes a unique vertex in $T[i]$. If the target is in $T[i]$, then ϕ follows ϕ'' (and takes a total of at most $\lambda_i + 1$ steps). Otherwise, it proceeds as ϕ' (and takes a total of at most l_{i+1} steps). We get $l_i = l_{i+1} + 1$ and $p_i = k - 1$.
- Else, if $\lambda_i < l_{i+1} - 1$ and $p_{i+1} > 0$, the strategy ϕ follows ϕ' but, in addition, probes one vertex of $T[i]$ during its first step. If the target is in $T[i]$, then ϕ follows ϕ'' (and takes a total of at most $\lambda_i + 1$ steps), otherwise, it proceeds as ϕ' (and takes a total of at most $l_{i+1} - 1$ steps). We get $l_i = l_{i+1}$ and $p_i = p_{i+1} - 1$.
- Else ($\lambda_i > l_{i+1} - 1$), the strategy ϕ probes π_i vertices in $T[i]$ during the first step. If the target is in $T[i]$, then ϕ follows ϕ'' (and takes a total of at most λ_i steps), otherwise, it proceeds as ϕ' (and takes a total of at most l_{i+1} steps). We get $l_i = \lambda_i + 1$ and $p_i = k - \pi_i$.

As the subtrees are sorted in non-increasing lexicographical order (of (λ_i, π_i)), we prove in Lemma 1 that the strategy ϕ described before is optimal for $T[i, d^*]$, that is, it computes $(\lambda_k^L(T[i, d^*]), \pi(T[i, d^*]))$.

Lemma 1. *For every $1 \leq i \leq d+1$, $\lambda_k^L(T[i, d^*]) = l_i - 1$ and $\pi(T[i, d^*]) = k - p_i$.*

Correctness and Complexity of Algorithms 1 and 2. We prove in Theorem 8 that $\mathcal{A}_1(k, (T, r))$ computes $(\lambda_k^L(T), \pi(T))$ and a corresponding k-strategy in time $O(n \log n)$, where n is the number of vertices. To do that, Theorem 7 proves the correctness and the linear (in the number of children of r) time complexity of $\mathcal{A}_2(k, (T, r), (\Lambda, \Pi))$.

Theorem 7. *Let $k \geq 1$, let $(T, r) \in \mathcal{T}$ be a rooted tree, and let $v_1, \ldots, v_{d^*}$ be the children of r such that the tuples $(\Lambda, \Pi) = (\lambda_i, \pi_i)_{1 \leq i \leq d^*}$ are sorted in non-increasing lexicographical ordering. Then, $\mathcal{A}_2(k, (T, r), (\Lambda, \Pi))$ returns $(\lambda_k^L(T), \pi(T))$ and a corresponding strategy. Furthermore, the time-complexity of $\mathcal{A}_2$ is $O(d^*)$ (independent of k).*

Proof. The time-complexity is obvious and the correctness follows from Lemma 1 for $i = 1$. The fact that the strategy is also returned is not explicitly described in Algorithm 2 but directly follows from the proof of Lemma 1. □

Theorem 8. *Let $k \geq 1$, and let $(T, r) \in \mathcal{T}$ be an n-node rooted tree. Then, $\mathcal{A}_1(k, (T, r))$ returns $(\lambda_k^L(T), \pi(T))$ and a corresponding strategy. Furthermore, the time-complexity of $\mathcal{A}_1$ is $O(n \log n)$ (independent of k).*

Proof. The correctness is simply proved by induction and by Theorem 7. For the time-complexity, at every recursive call on a subtree T_v rooted at v (with d_v children), the additional number of operations is $O(d_v \log d_v)$ (sorting) plus $O(d_v)$ (Algorithm $\mathcal{A}_2$, by Theorem 7). Since in a tree, $\sum_{v \in V(T)} d_v = 2(n-1)$, this gives a total complexity of $O(\sum_{v \in V(T)} d_v \log d_v) = O(n \log n)$. Again, the strategy is not explicit in our presentation but can be easily computed. □

Main Results. From $\mathcal{A}_1(k, (T, r))$ presented before, it is easy to get an efficient approximation algorithm when k and ℓ are part of the input and a polynomial-time algorithm when k is fixed.

Theorem 9. *There exists an algorithm that, given any integer $k \geq 1$ and any n-node tree T, computes a k-strategy that locates a target in T in at most $\lambda_k(T)+1$ steps. Furthermore, the time-complexity of the algorithm is $O(n \log n)$.*

Proof. The strategy proceeds as follows. The first step probes any arbitrary vertex r of T. Let d be the distance between r and the target, let $L \subseteq V(T)$ be the set of vertices at distance exactly d from r, and let T^d be the subtree induced by r and every vertex on a path between r and the vertices in L. Note that $(T^d, r) \in \mathcal{T}$ and that the target is occupying a leaf of T^d. Hence, it is sufficient to apply $\mathcal{A}_1(k, (T^d, r))$. By Theorem 8, the above strategy will locate the target in at most $1 + \max_d \lambda_k^L(T^d) \leq 1 + \lambda_k(T)$ steps. $\square$

Corollary 1. *There exists an algorithm that, given any integer $k \geq 1$ and any n-node tree T, computes an optimal k-strategy for locating a target in T in at most $\lambda_k(T)$ steps. Furthermore, the time-complexity of the algorithm is $O(n^{k+2} \log n)$.*

4   Further Work

Our results in trees leave the open question of whether $\lambda_k(T)$ is Fixed Parameter Tractable (in k) in the class of n-node trees T. Moreover, it would be interesting to study the LOCALIZATION Problem in other graph classes such as interval graphs and planar graphs. Also, what is the complexity of the ℓ-STEP LOCALIZATION Problem in trees?

The RELATIVE-LOCALIZATION Problem is much more intricate even for simple topologies. A first step towards a better understanding of this problem would be to fully solve it in the case of paths (*i.e.*, to determine $\kappa_1^{rel}(P)$ for every path P), which has been partially solved in [9], before studying it in the class of trees.

References

1. Belmonte, R., Fomin, F.V., Golovach, P.A., Ramanujan, M.S.: Metric dimension of bounded tree-length graphs. SIAM J. Discrete Math. **31**(2), 1217–1243 (2017)
2. Ben-Haim, Y., Gravier, S., Lobstein, A., Moncel, J.: Adaptive identification in graphs. J. Combin. Theor. Ser. A **115**(7), 1114–1126 (2008)
3. Bensmail, J., Mazauric, D., Mc Inerney, F., Nisse, N., Pérennes, S.: Sequential metric dimension. Technical report, INRIA (2018). https://hal.archives-ouvertes. fr/hal-01717629
4. Bosek, B., Gordinowicz, P., Grytczuk, J., Nisse, N., Sokól, J., Sleszynska-Nowak, M.: Centroidal localization game. CoRR, abs/1711.08836 (2017)
5. Bosek, B., Gordinowicz, P., Grytczuk, J., Nisse, N., Sokól, J., Sleszynska-Nowak, M.: Localization game on geometric and planar graphs. CoRR, abs/1709.05904 (2017)

6. Brandt, A., Diemunsch, J., Erbes, C., LeGrand, J., Moffatt, C.: A robber locating strategy for trees. Discrete Appl. Math. **232**, 99–106 (2017)
7. Carraher, J.M., Choi, I., Delcourt, M., Erickson, L.H., West, D.B.: Locating a robber on a graph via distance queries. Theor. Comput. Sci. **463**, 54–61 (2012)
8. Díaz, J., Pottonen, O., Serna, M.J., van Leeuwen, E.J.: Complexity of metric dimension on planar graphs. J. Comput. Syst. Sci. **83**(1), 132–158 (2017)
9. Foucaud, F., Klasing, R., Slater, P.J.: Centroidal bases in graphs. Networks **64**(2), 96–108 (2014)
10. Foucaud, F., Mertzios, G.B., Naserasr, R., Parreau, A., Valicov, P.: Identification, location-domination and metric dimension on interval and permutation graphs. I. Bounds. Theor. Comput. Sci. **668**, 43–58 (2017)
11. Foucaud, F., Mertzios, G.B., Naserasr, R., Parreau, A., Valicov, P.: Identification, location-domination and metric dimension on interval and permutation graphs. II. Algorithms and complexity. Algorithmica **78**(3), 914–944 (2017)
12. Garey, M.R., Johnson, D.S.: Computers and Intractability - A Guide to NP-Completeness. W.H. Freeman and Company, New York (1979)
13. Harary, F., Melter, R.A.: On the metric dimension of a graph. Ars Comb. **2**, 191–195 (1976)
14. Hartung, S., Nichterlein, A.: On the parameterized and approximation hardness of metric dimension. In: Proceedings of the 28th Conference on Computational Complexity, CCC, pp. 266–276. IEEE Computer Society (2013)
15. Haslegrave, J., Johnson, R.A.B., Koch, S.: Locating a robber with multiple probes. Discrete Math. **341**(1), 184–193 (2018)
16. Karpovsky, M.G., Chakrabarty, K., Levitin, L.B.: On a new class of codes for identifying vertices in graphs. IEEE Trans. Inf. Theor. **44**(2), 599–611 (1998)
17. Seager, S.M.: Locating a robber on a graph. Discrete Math. **312**(22), 3265–3269 (2012)
18. Seager, S.M.: A sequential locating game on graphs. Ars Comb. **110**, 45–54 (2013)
19. Seager, S.M.: Locating a backtracking robber on a tree. Theor. Comput. Sci. **539**, 28–37 (2014)
20. Slater, P.J.: Leaves of trees. In: Congressus Numerantium, vol. 14, pp. 549–559 (1975)
21. Slater, P.J.: Domination and location in acyclic graphs. Networks **17**(1), 55–64 (1987)

A Primal-Dual Online Deterministic Algorithm for Matching with Delays

Marcin Bienkowski[ORCID], Artur Kraska[✉], Hsiang-Hsuan Liu[ORCID],
and Paweł Schmidt

Institute of Computer Science, University of Wrocław, Wrocław, Poland
{marcin.bienkowski,artur.kraska,alison.hhliu,
pawel.schmidt}@cs.uni.wroc.pl

Abstract. In the Min-cost Perfect Matching with Delays (MPMD) problem, $2m$ requests arrive over time at points of a metric space. An online algorithm has to connect these requests in pairs, but a decision to match may be postponed till a more suitable matching pair is found. The goal is to minimize the joint cost of connection and the total waiting time of all requests.

We present an $O(m)$-competitive deterministic algorithm for this problem, improving on an existing bound of $O(m^{\log_2 5.5}) = O(m^{2.46})$. Our algorithm also solves (with the same competitive ratio) a bipartite variant of MPMD, where requests are either positive or negative and only requests with different polarities may be matched with each other. Unlike the existing randomized solutions, our approach does not depend on the size of the metric space and does not have to know it in advance.

Keywords: Online algorithms · Delayed service · Metric matching · Primal-dual algorithms · Competitive analysis

1 Introduction

Consider a gaming platform that hosts two-player games, such as chess, go or Scrabble, where participants are joining in real time, each wanting to play against another human player. The system matches players according to their known capabilities aiming at minimizing their dissimilarities: any player wants to compete against an opponent with comparable skills. A better match for a player can be found if the platform delays matching decisions as meanwhile more appropriate opponents may join the system. However, an excessive delay may also degrade the quality of experience. Therefore, a matching mechanism that runs on a gaming platform has to balance two conflicting objectives: to minimize the waiting time of any player and to minimize dissimilarities between matched players.

The problem informally described above, called *Min-cost Perfect Matching with Delays (MPMD)*, has been recently introduced by Emek et al. [20]. The

Partially supported by Polish National Science Centre grant 2016/22/E/ST6/00499.

L. Epstein and T. Erlebach (Eds.): WAOA 2018, LNCS 11312, pp. 51–68, 2018.
https://doi.org/10.1007/978-3-030-04693-4_4

problem is inherently online[1]: a matching algorithm for the gaming platform has to react in real time, without knowledge about future requests (player arrivals) and make its decision irrevocably: once two requests (players) are paired, they remain paired forever.

The MPMD problem was also considered in a *bipartite* variant, called *Min-cost Bipartite Perfect Matching with Delays (MBPMD)* introduced by Ashlagi et al. [2]. There requests have polarities: one half of them is positive, and the other half is negative. An algorithm may match only requests of different signs. This setting corresponds to a variety of real-life scenarios, e.g., assigning drivers to passengers on ride-sharing platforms or matching patients to donors in kidney transplants. Similarly to the MPMD problem, there is a trade-off between minimizing the waiting time and finding a better match (a closer driver or a more compatible donor).

1.1   Problem Definition

Formally, both in the MPMD and MBPMD problems, there is a metric space $\mathcal{X}$ equipped with a distance function $\mathsf{dist} : \mathcal{X} \times \mathcal{X} \to \mathbb{R}_{\geq 0}$, both known in advance to an online algorithm. An online part of the input is a sequence of $2m$ requests $u_1, u_2, \ldots, u_{2m}$. A request (e.g., a player arrival) u is a triple $u = (\mathsf{pos}(u), \mathsf{atime}(u), \mathsf{sgn}(u))$, where $\mathsf{atime}(u)$ is the arrival time of request u, $\mathsf{pos}(u) \in \mathcal{X}$ is the request location, and $\mathsf{sgn}(u)$ is the polarity of the request.

In the bipartite case, half of the requests are positive and $\mathsf{sgn}(u) = +1$ for any such request u; the remaining half are negative and there $\mathsf{sgn}(u) = -1$. In the non-bipartite case, requests do not have polarities, but for technical convenience we set $\mathsf{sgn}(u) = 0$ for any request u.

In applications described above, the function dist measures the dissimilarity of a given pair of requests (e.g., discrepancy between player capabilities in the gaming platform scenario or the physical distance between a driver and a passenger in the ride-sharing platform scenario). For instance, for chess, a player is commonly characterized by her Elo rating (an integer) [19]. In such case, $\mathcal{X}$ may be simply a set of all integers with the distance between two points defined as the difference of their values.

Requests arrive over time, i.e., $\mathsf{atime}(u_1) \leq \mathsf{atime}(u_2) \leq \cdots \leq \mathsf{atime}(u_{2m})$. We note that the integer m is not known beforehand to an online algorithm. At any time τ, an online algorithm may match a pair of requests (players) u and v that

- have already arrived ($\tau \geq \mathsf{atime}(u)$ and $\tau \geq \mathsf{atime}(v)$),
- have not been matched yet,
- satisfy $\mathsf{sgn}(u) = -\mathsf{sgn}(v)$ (i.e., have opposite polarities in the bipartite case; in the non-bipartite case, this condition trivially holds for any pair).

---

[1] The offline variant of the problem, where all player arrivals are known a priori, can be easily solved in polynomial time.

The cost incurred by such *matching edge* is then $\mathsf{dist}(\mathsf{pos}(u), \mathsf{pos}(v)) + (\tau - \mathsf{atime}(u)) + (\tau - \mathsf{atime}(v))$. That is, it is the sum of the *connection cost* defined as $\mathsf{dist}(\mathsf{pos}(u), \mathsf{pos}(v))$ and the *waiting costs* of u and v, defined as $\tau - \mathsf{atime}(u)$ and $\tau - \mathsf{atime}(v)$, respectively.

The goal is to eventually match all requests and minimize the total cost of all matching edges. We perform worst-case analysis, assuming that the requests are given by an adversary. To measure the performance of an online algorithm ALG for an input instance $\mathcal{I}$, we compare its cost $\mathrm{ALG}(\mathcal{I})$ to the cost $\mathrm{OPT}(\mathcal{I})$ of an optimal offline solution OPT that knows the entire input sequence in advance. The objective is to minimize the competitive ratio [14] defined as $\sup_{\mathcal{I}} \{\mathrm{ALG}(\mathcal{I})/\mathrm{OPT}(\mathcal{I})\}$.

1.2 Previous Work

The MPMD problem was introduced by Emek et al. [20], who presented a *randomized* $O(\log^2 n + \log \Delta)$-competitive algorithm. There, n is the number of points in the metric space $\mathcal{X}$ and Δ is its aspect ratio (the ratio between the largest and the smallest distance in $\mathcal{X}$). The competitive ratio was subsequently improved by Azar et al. [4] to $O(\log n)$. They also showed that the competitive ratio of any randomized algorithm is at least $\Omega(\sqrt{\log n})$. The currently best lower bound of $\Omega(\log n / \log \log n)$ for randomized solutions was given by Ashlagi et al. [2].

Ashlagi et al. [2] adapted the algorithm of Azar et al. [4] to the bipartite setting and obtained a randomized $O(\log n)$-competitive algorithm for this variant. The currently best lower bound of $\Omega(\sqrt{\log n / \log \log n})$ for this variant was also given in [2].

Both lower bounds use $O(n)$ requests. Therefore, they imply that no randomized algorithm can achieve a competitive ratio lower than $\Omega(\log m / \log \log m)$ in the non-bipartite case and lower than $\Omega(\sqrt{\log m / \log \log m})$ in the bipartite one. (Recall that $2m$ is the number of requests in the input.)

The status of the achievable performance of *deterministic* solutions is far from being resolved. No better lower bounds than the ones used for randomized settings are known for deterministic algorithms. The first solution that worked for general metric spaces was given by Bienkowski et al. and achieved an embarrassingly high competitive ratio of $O(m^{\log_2 5.5}) = O(m^{2.46})$ [12]. Roughly speaking, their algorithm is based on growing spheres around not-yet-paired requests. Each sphere is created upon a request arrival, grows with time, and when two spheres touch, the corresponding requests become matched.

Concurrently and independently of our current paper, Azar and Jacob-Fanani [6] improved the deterministic ratio to $O((1/\varepsilon) \cdot m^{\log(3/2+\varepsilon)})$, where $\varepsilon > 0$ is a parameter of their algorithm. When ε is small enough, this ratio becomes $O(m^{0.59})$. Their approach is similar to that of [12], but they grow spheres in a smarter way: slower than time progresses and only in the negative direction of time axis. (Their work also appears in these proceedings.)

Better deterministic algorithms are known only for simple spaces: Azar et al. [4] gave an $O(\text{height})$-competitive algorithm for trees and Emek et al. [21] constructed a 3-competitive deterministic solution for two-point metrics (the latter competitive ratio is best possible).

1.3   Our Contribution

In this paper, we focus on deterministic solutions for both the MPMD and MBPMD problems, i.e., for both the non-bipartite and the bipartite variants of the problem. We present a simple $O(m)$-competitive LP-based algorithm that works in both settings.

In contrast to the previous randomized solutions to these problems [2,4,20], and similarly to other deterministic solutions [6,12], we do not need the metric space $\mathcal{X}$ to be finite and known in advance by an online algorithm. (All previous randomized solutions started by approximating $\mathcal{X}$ by a random HST (hierarchically separated tree) [22] or a random HST tree with reduced height [8].) This approach, which can be performed only in the randomized setting, greatly simplifies the task as the underlying tree metric reveals a lot of structural information about the cuts between points of $\mathcal{X}$ and hence about the structure of an optimal solution. In the deterministic setting, such information has to be gradually learned as time passes. For our algorithm, we require only that, together with any request u, it learns the distances from u to all previous requests.

In contrast to the previous deterministic algorithms [6,12], we base our algorithm on the moat-growing framework, developed originally for (offline) constrained connectivity problems (e.g., for Steiner problems) by Goemans and Williamson [24]. Glossing over a lot of details, in this framework, one writes a primal linear relaxation of the problem and its dual. The primal program has a constraint (connectivity requirement) for any subset of requests and the dual program has a variable for any such subset. The algorithm maintains a family of active sets, which are initially singletons. In the runtime, dual variables are increased simultaneously, till some dual constraint (corresponding to a pair of requests) becomes tight: in such case an algorithm connects such pair and merges the corresponding sets. At the end, an algorithm usually performs pruning by removing redundant edges.

When one tries to adapt the moat-growing framework to online setting, the main difficulty stems from the irrevocability of the pairing decision: the pruning operation performed at the end is no longer an option. Another difficulty is that an algorithm has to combine the concept of *actual* time that passes in an online instance with the *virtual* time that dictates the growth of dual variables. In particular, dual variables may only start to grow once an online algorithm learns about the request and not from the very beginning as they would do in the offline setting. Finally, requests appear online, and hence both primal and dual programs evolve in time. For instance, this means that for badly defined algorithms, appearing dual constraints may be violated already once they are introduced.

We note that $2m$ (the number of requests) is incomparable with n (the number of different points in the metric space $\mathcal{X}$) and their relation depends on the application. Our algorithm is better suited for applications, where $\mathcal{X}$ is infinite or virtually infinite (e.g., it corresponds to an Euclidean plane or a city map for ride-sharing platforms [32]) or very large (e.g., for some real-time online games, where player capabilities are represented as multi-dimensional vectors describing their rank, reflex, offensive and defensive skills, etc. [3]).

1.4 Alternative Deterministic Approaches (That Fail)

A few standard deterministic approaches fail when applied to the MPMD and MBPMD problems. One such attempt is the *doubling technique* (see, e.g., [17]): an online algorithm may trace the cost of an optimal solution OPT and perform a global operation (e.g., match many pending requests) once the cost of OPT increases significantly (e.g., by a factor of two) since the last time when such global operation was performed. This approach does not seem to be feasible here as the total cost of OPT may *decrease* when new requests appear.

Another attempt is to observe that the randomized algorithm by Azar et al. [4] is a deterministic algorithm run on a random tree that approximates the original metric space. One may try to replace a random tree by a deterministically generated tree that spans requested points of the metric space. Such spanning tree can be computed by the standard greedy routine for the online Steiner tree problem [26]. However, it turns out that the competitive ratio of the resulting algorithm is $2^{\Omega(m)}$. (The main reason is that the adversary may give an initial subsequence that forces the algorithm to create a spanning tree with the worst-case stretch of $2^{\Omega(m)}$ and such initial subsequence can be served by OPT with a negligible cost. The details are given in Appendix B.)

1.5 Related Work

Originally, online metric matching problems have been studied in variants where delaying decisions was not permitted. In this variant, m requests with positive polarities are given at the beginning to an algorithm. Afterwards, m requests with negative polarities are presented one by one to an algorithm and they have to be matched *immediately* to existing positive requests. The goal is to minimize the weight of a perfect matching created by the algorithm. For general metric spaces, the best deterministic algorithms achieve the optimal competitive ratio of $2m - 1$ [27,30,36] and the best randomized solution is $O(\log^2 m)$-competitive [7,34]. Better bounds are known for line metrics [1,23,25,31]: here the best deterministic algorithm is $O(\log^2 m)$-competitive [35] and the best randomized one achieves the ratio of $O(\log m)$ [25].

Another strand of research concerning online matching problems arose around a non-metric setting where points with different polarities are connected by graph edges and the goal is to maximize the cardinality or the weight of the produced matching. For a comprehensive overview of these type of problems we refer the reader to a recent survey by Mehta [33].

The M(B)PMD problem is an instance in a broader category of problems, where an online algorithm may delay its decisions, but such delays come with a certain cost. Similar trade-offs were employed in other areas of online analysis: in aggregating orders in supply-chain management [9–11,15,16], aggregating messages in computer networks [18,28,29], or recently for server problems [5,13].

2   Primal-Dual Formulation

We start with introducing a linear program that allows us to lower-bound the cost of an optimal solution. To this end, fix an instance $\mathcal{I}$ of M(B)PMD. Let V be the set of all requests. We call any unordered pair of different requests in $\mathcal{I}$ an edge; let E be the set of all edges that correspond to potential matching pairs, i.e., the set of all edges in the non-bipartite case, and the edges that connect requests of opposite polarities in the bipartite variant. For each set $S \subseteq V$, by $\delta(S)$ we denote the set of all edges from E crossing the boundary of S, i.e., having exactly one endpoint in S.

For any set $S \subseteq V$, we define $\mathsf{sur}(S)$ (*surplus* of set S) as the number of unmatched requests in a maximum cardinality matching of requests within set S.

- In the non-bipartite variant (MPMD), we are allowed to match any two requests. Hence, if S is of even size, then $\mathsf{sur}(S) = 0$. Otherwise, $\mathsf{sur}(S) = 1$ as in any maximum cardinality matching of requests within S exactly one request remains unmatched.
- In the bipartite variant (MBPMD), we can always match two requests of different polarities. Hence, the surplus of a set S is the discrepancy between the number of positive and negative requests inside S, i.e., $\mathsf{sur}(S) = |\sum_{u \in S} \mathsf{sgn}(u)|$.

To describe a matching, we use the following notation. For each edge e, we introduce a binary variable x_e, such that $x_e = 1$ if and only if e is a matching edge. For any set $S \subseteq V$ and any feasible matching (in particular the optimal one), it holds that $\sum_{e \in \delta(S)} x_e \geq \mathsf{sur}(S)$.

Fix an optimal solution OPT for $\mathcal{I}$. If a pair of requests $e = (u, v)$ is matched by OPT, it is matched as soon as both u and v arrive, and hence the cost of matching u with v in the solution of OPT is equal to $\mathsf{opt\text{-}cost}(e) := \mathsf{dist}(\mathsf{pos}(u), \mathsf{pos}(v)) + |\mathsf{atime}(u) - \mathsf{atime}(v)|$. This, together with the preceding observations, motivates the following linear program $\mathcal{P}$:

$$\text{minimize} \quad \sum_{e \in E} \mathsf{opt\text{-}cost}(e) \cdot x_e$$

$$\text{subject to} \quad \sum_{e \in \delta(S)} x_e \geq \mathsf{sur}(S) \qquad \forall S \subseteq V$$

$$x_e \geq 0 \qquad \forall e \in E.$$

As any matching is a feasible solution to $\mathcal{P}$, the cost of the optimal solution of $\mathcal{P}$ lower-bounds the cost of the optimal solution for instance $\mathcal{I}$ of M(B)PMD. Note that there might exist a feasible integral solution of $\mathcal{P}$ that does not correspond to any matching. To exclude all such solutions, we could add constraints

$\sum_{e \in \delta(S)} x_e = 1$ for all singleton sets S. The resulting linear program would then exactly describe the matching problem (cf. Chap. 25 of [37]). However, our main concern is not $\mathcal{P}$, but its dual and its current shape is sufficient for our purposes. The program $\mathcal{D}$, dual to $\mathcal{P}$, is then

$$\text{maximize} \quad \sum_{S \subseteq V} \text{sur}(S) \cdot y_S$$

$$\text{subject to} \quad \sum_{S: e \in \delta(S)} y_S \leq \text{opt-cost}(e) \qquad \forall e \in E$$

$$y_S \geq 0 \qquad \forall S \subseteq V.$$

Note that in any solution, the dual variables y_S corresponding to sets S for which $\text{sur}(S) = 0$, can be set to 0 without changing feasibility or objective value.

The following lemma is an immediate consequence of weak duality.

Lemma 1. *Fix any instance $\mathcal{I}$ of the M(B)PMD problem. Let $\text{OPT}(\mathcal{I})$ be the value of any optimal solution of $\mathcal{I}$ and D be the value of any feasible solution of $\mathcal{D}$. Then $\text{OPT}(\mathcal{I}) \geq D$.*

Proof. Let P^* and D^* be the values of optimal solutions for $\mathcal{P}$ and $\mathcal{D}$, respectively. Since any matching is a feasible solution for $\mathcal{P}$, $\text{OPT}(\mathcal{I}) \geq P^*$. Hence, $\text{OPT}(\mathcal{I}) \geq P^* \geq D^* \geq D$. ◻

Lemma 1 motivates the following approach: We construct an online algorithm GREEDY DUAL (GD), which, along with its own solution, maintains a feasible solution D for $\mathcal{D}$ corresponding to the already seen part of the input instance. This feasible dual solution not only yields a lower bound on the cost of the optimal matching, but also plays a crucial role in deciding which pair of requests should be matched.

Note that since the requests arrive in an online manner, $\mathcal{D}$ evolves in time. When a request arrives, the number of subsets of V increases (more precisely, it doubles), and hence more dual variables y_S are introduced. Moreover, the newly arrived request creates an edge with every existing request and the corresponding dual constraints are introduced. Therefore, showing the feasibility of the created dual solution is not immediate; we deal with this issue in Sect. 4.

3   Algorithm Greedy Dual

The high-level idea of our algorithm is as follows: GREEDY DUAL (GD) resembles moat-growing algorithms for solving constrained forest problems [24]. During its runtime, GD partitions all the requests that have already arrived into *active sets*.[2] If an active set contains any free requests, we call this set *growing*, and *non-growing* otherwise. At any time, for each active growing set S, the algorithm

---

[2] A reader familiar with the moat-growing algorithm may think that active sets are moats. However, not all of them are growing in time.

increases continuously its dual variable y_S until a constraint in $\mathcal{D}$ corresponding to some edge (u, v) becomes tight. When it happens, GD makes both active sets (containing u and v, respectively) inactive, and the set being their union active. In addition, if this happened due to two growing sets, GD matches as many pairs of free requests in these sets as possible: in the non-bipartite variant GD matches exactly one pair of free requests, while in the bipartite variant, GD matches free requests of different polarities until all remaining free requests have the same sign.

3.1   Algorithm Description

More precisely, at any time, GD partitions all requests that arrived until that time into *active* sets. It maintains mapping $\mathcal{A}$, which assigns an active set to each such request. An active set S, whose all requests are matched is called *non-growing*. Conversely, an active set S is called *growing* if it contains at least one free request. GD ensures that the number of free requests in an active set S is always equal to $\mathsf{sur}(S)$. We denote the set of free requests in an active set S by $\mathsf{free}(S)$; if S is non-growing, then $\mathsf{free}(S) = \emptyset$.

When a request u arrives, the singleton $\{u\}$ becomes a new active and growing set, i.e., $\mathcal{A}(u) = \{u\}$. The dual variables of all active growing sets are increased continuously with the same rate in which time passes. This increase takes place until a dual constraint between two active sets becomes tight, i.e., until there exists at least one edge $e = (u, v)$, such that

$$\mathcal{A}(u) \neq \mathcal{A}(v) \quad \text{and} \quad \sum_{S: e \in \delta(S)} y_S = \mathsf{opt\text{-}cost}(e). \tag{1}$$

In such case, while there exists an edge $e = (u, v)$ satisfying (1), GD processes such edge in the following way. First, it *merges* active sets $\mathcal{A}(u)$ and $\mathcal{A}(v)$. By merging we mean that the mapping $\mathcal{A}$ is adjusted to the new active set $S = \mathcal{A}(u) \uplus \mathcal{A}(v)$ for each request of S. Old active sets $\mathcal{A}(u)$ and $\mathcal{A}(v)$ become *inactive*.[3] Second, as long as there is a pair of free requests $u', v' \in S$ that can be matched with each other, GD matches them.

In the non-bipartite variant, GD matches at most one pair as each active set contains at most one free request. In the bipartite variant, GD matches pairs of free requests until all unmatched requests in S (possibly zero) have the same polarity. Observe that in either case, the number of free requests after merge is equal to $\mathsf{sur}(S)$. Finally, GD *marks* edge e. Marked edges are used in the analysis, to find a proper charging of the connection cost to the cost of the produced solution for $\mathcal{D}$. The pseudocode of GD is given in Algorithm 1 and an example execution that shows a partition of requests into active sets is given in Fig. 1.

---

[3] Note that *inactive* is not the opposite of being *active*, but means that the set was active previously: some sets are never active or inactive.

Algorithm 1. Algorithm GREEDY DUAL

1: **Request arrival event:**
2:     **if** a request u arrives **then**
3:         $\mathcal{A}(u) \leftarrow \{u\}$
4:         **for all** sets S such that $u \in S$ **do**
5:             $y_S \leftarrow 0$ ▷ *initialize dual variables for sets containing u*
6:
7: **Tight constraint event:**
8:     **while** exists a tight dual constraint for edge $e = (u, v)$ where $\mathcal{A}(u) \neq \mathcal{A}(v)$ **do**
9:         $S \leftarrow \mathcal{A}(u) \uplus \mathcal{A}(v)$ ▷ *merge two active sets*
10:        **for all** $w \in S$ **do**   ▷ *adjust assignment $\mathcal{A}$ for the new active set S*
11:            $\mathcal{A}(w) \leftarrow S$
12:        **mark** edge e
13:        **while** there are $u', v' \in \mathsf{free}(S)$ such that $\mathsf{sgn}(u') = -\mathsf{sgn}(v')$ **do**
14:            **match** u' with v' ▷ *match as many pairs as possible*
15:
16: **None of the above events occurs:**
17:     **for all** growing active sets S **do**
18:         increase continuously y_S with the same rate in which time passes

---

3.2   Greedy Dual Properties

It is instructive to trace how the set $\mathcal{A}(u)$ changes in time for a request u. At the beginning, when u arrives, $\mathcal{A}(u)$ is just the singleton set $\{u\}$. Then, the set $\mathcal{A}(u)$ is merged at least once with another active set. If $\mathcal{A}(u)$ is merged with a non-growing set, the number of requests in $\mathcal{A}(u)$ increases, but its surplus remains intact. After $\mathcal{A}(u)$ is merged with a growing set, some requests inside the new $\mathcal{A}(u)$ may become matched. It is possible that, in effect, the surplus of the new set $\mathcal{A}(u)$ is zero, in which case the new set $\mathcal{A}(u)$ is non-growing. (In the non-bipartite variant, this is always the case when two growing sets merge.) After $\mathcal{A}(u)$ becomes non-growing, another growing set may be merged with $\mathcal{A}(u)$, and so on. Thus, the set $\mathcal{A}(u)$ can change its state from growing to non-growing (and back) multiple times.

The next observation summarizes the process described above, listing properties of GD that we use later in our proofs.

Observation 1. *The following properties hold during the runtime of* GD.

1. *For a request u, when time passes, $\mathcal{A}(u)$ refers to different active sets that contain u.*
2. *At any time, every request is contained in exactly one active set. If this request is free, then the active set is growing.*
3. *At any time, an active set S contains exactly $\mathsf{sur}(S)$ free requests.*
4. *Active and inactive sets together constitute a laminar family of sets.*
5. *For any two requests u and v, once $\mathcal{A}(u)$ becomes equal to $\mathcal{A}(v)$, they will be equal forever.*

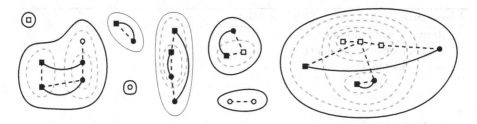

Fig. 1. A partition of requests into active sets created by GD. Different polarities of requests are represented by discs and squares. Free requests are depicted as empty discs and squares, matched requests by filled ones. Active growing sets have bold boundaries and each of them contains at least one free request. Active non-growing sets contain only matched requests. Dashed lines represent marked edges and solid curvy lines represent matching edges. Dashed gray sets are already inactive; the inactive singleton sets have been omitted.

4   Correctness

We now prove that GREEDY DUAL is defined properly. In other words, we show that the dual values maintained by GD always form a feasible solution of $\mathcal{D}$ (Lemma 3) and GD returns a feasible matching of all requests at the end (Lemma 4). From now on, we denote the values of a dual variable y_S at time τ by $y_S(\tau)$.

By the definition, the waiting cost of a request is the time difference between the time it arrives and the time it is matched. In the following lemma, we relate the waiting cost of a request to the dual variables for the active sets it belongs to.

Lemma 2. *Fix any request u. For any time $\tau \geq \text{atime}(u)$, it holds that*

$$\sum_{S:u\in S} y_S(\tau) \leq \tau - \text{atime}(u).$$

The relation holds with equality if u is free at time τ.

Proof. We show that the inequality is preserved as time passes. At time $\tau = \text{atime}(u)$, request u is introduced and sets S containing u appear. Their y_S values are initialized to 0. Therefore, at that time, $\sum_{S:u\in S} y_S(\tau) = 0$ as desired.

Whenever a merging event or an arrival of any other requests occur, new variables y_S may appear in the sum $\sum_{S:u\in S} y_S(\tau)$, but, at these times, the values of these variables are equal to zero, and therefore do not change the sum value.

It remains to analyze the case when time passes infinitesimally by ε and no event occurs within this period. It is sufficient to argue that the sum $\sum_{S:u\in S} y_S(\tau)$ increases exactly by ε if u is free at τ and at most by ε otherwise. Recall that y_S may grow only if S is an active growing set. By Property 2 of Observation 1, the only active set containing u is $\mathcal{A}(u)$. This set is growing if u is free (and then $y_{\mathcal{A}(u)}$ increases exactly by ε) and may be growing or non-growing if u is matched (and then $y_{\mathcal{A}(u)}$ increases at most by ε). $\qquad\square$

The following lemma shows that throughout its runtime, GD maintains a feasible dual solution.

Lemma 3. *At any time, the values y_S maintained by the algorithm constitute a feasible solution to $\mathcal{D}$.*

Proof. We show that no dual constraint is ever violated during the execution of GD.

When a new request u arrives at time $\tau = \mathsf{atime}(u)$, new sets containing u appear and the dual variables y_S corresponding to these sets are initialized to 0.

Each already existing constraint, corresponding to an edge e not incident to u, is modified: new y_S variables for sets S containing both u and exactly one of endpoints of e appear in the sum. However, all these variables are zero, and hence the feasibility of such constraints is preserved.

Moreover, for any edge $e = (u, v)$ where v is an existing request, a new dual constraint for this edge appears in $\mathcal{D}$. We show that it is not violated, i.e., $\sum_{S:e\in\delta(S)} y_S(\tau) \leq \mathsf{opt\text{-}cost}(e)$. As discussed before, $y_S(\tau) = 0$ for the sets S containing u. Therefore,

$$\sum_{S:e\in\delta(S)} y_S(\tau) = \sum_{S:v\in S\wedge u\notin S} y_S(\tau) + \sum_{S:u\in S\wedge v\notin S} y_S(\tau)$$

$$= \sum_{S:v\in S\wedge u\notin S} y_S(\tau) \leq \sum_{S:v\in S} y_S(\tau)$$

$$\leq \mathsf{atime}(u) - \mathsf{atime}(v) \qquad \text{(by Lemma 2)}$$

$$\leq \mathsf{opt\text{-}cost}(e).$$

Now, we prove that once a dual constraint for an edge $e = (u, v)$ becomes tight, the involved y_S values are no longer increased. According to the algorithm definition, $\mathcal{A}(u)$ and $\mathcal{A}(v)$ become merged together. By Property 5 of Observation 1, from this moment on, any active set S contains either both u and v or neither of them. Hence, there is no active set S, such that $(u, v) \in \delta(S)$, and in particular there is no such active growing set. Therefore, the value of $\sum_{S:e\in\delta(S)} y_S$ remains unchanged, and hence the dual constraint corresponding to edge e remains tight and not violated. $\square$

Finally, we prove that GD returns a proper matching. We need to show that if a pair of requests remains unmatched, then appropriate dual variables increase and they will eventually trigger the matching event.

Lemma 4. *For any input for the M(B)PMD problem, GD returns a feasible matching.*

Proof. Suppose for a contradiction that GD does not match some request u. Then, by Property 2 of Observation 1, $\mathcal{A}(u)$ is always an active growing set and by Property 3, $\mathsf{sur}(\mathcal{A}(u)) > 0$. Therefore, the corresponding dual variable $y_{\mathcal{A}(u)}$ always increases during the execution of GD and appears in the objective function of $\mathcal{D}$ with a positive coefficient. By Lemma 3, the solution of $\mathcal{D}$

maintained by GD is always feasible, and hence the optimal value of $\mathcal{D}$ would be unbounded. This would be a contradiction, as there exists a finite solution to the primal program $\mathcal{P}$ (as all distances in the metric space are finite). □

5   Cost Analysis

In this section, we show how to relate the cost of the matching returned by GREEDY DUAL to the value of the produced dual solution. First, we show that the total waiting cost of the algorithm is equal to the value of the dual solution. Afterwards, we bound the connection cost of GD by $2m$ times the dual solution, where $2m$ is the number of requests in the input. This, along with Lemma 1, yields the competitive ratio of $2m + 1$.

5.1   Waiting Cost

In the proof below, we link the generated waiting cost with the growth of appropriate dual variables. To this end, suppose that a set S is an active set for time period of length Δt. By Property 3 of Observation 1, S contains exactly $\mathsf{sur}(S)$ free points, and thus the waiting cost incurred within this time by requests in S is $\Delta t \cdot \mathsf{sur}(S)$. Moreover, in the same time interval, the dual variable y_S increases by Δt, which contributes the same amount, $\mathsf{sur}(S) \cdot \Delta t$, to the growth of the objective function of $\mathcal{D}$. The following lemma formalizes this observation and applies it to all active sets considered by GD in its runtime.

Lemma 5. *The total waiting cost of* GD *is equal to* $\sum_{S \subseteq V} \mathsf{sur}(S) \cdot y_S(T)$, *where* T *is the time when* GD *matches the last request.*

Proof. We define $G(\tau)$ as the family of sets that are active and growing at time τ. By Propertys 2 and 3 of Observation 1, the number of free requests at time τ, henceforth denoted $\mathsf{wait}(\tau)$, is then equal to $\sum_S \mathsf{sur}(S) \cdot \mathbb{1}[S \in G(\tau)]$. The total waiting cost at time T can be then expressed as

$$\int_0^T \mathsf{wait}(\tau)\, d\tau = \int_0^T \sum_S \mathsf{sur}(S) \cdot \mathbb{1}[S \in G(\tau)]\, d\tau$$

$$= \sum_S \mathsf{sur}(S) \int_0^T \mathbb{1}[S \in G(\tau)]\, d\tau = \sum_S \mathsf{sur}(S) \cdot y_S(T),$$

where the last equality holds as at any time, GD increases y_S value if and only if S is active and growing. □

5.2   Connection Cost

Below, we relate the connection cost of GD to the value of the final solution of $\mathcal{D}$, created by GD. We focus on the set of marked edges, which are created by GD in Line 12 of Algorithm 1. We show that for any time, the set of marked edges restricted to an active or an inactive set S forms a "spanning tree" of requests of S. That is, there is a unique path of marked edges between any two requests from S. (Note that this path projected to the metric space may contain cycles as two requests may be given at the same point of $\mathcal{X}$.) We start with a helper observation.

Observation 2. *Fix any set S. If S is active at time τ, then its boundary $\delta(S)$ does not contain any marked edge at time τ.*

Proof. After an edge (u, v) becomes marked, both u and v belong to newly created active set. From now on, by Property 5 of Observation 1, they remain in the same active set till the end of the execution. Therefore, this edge will never be contained in a boundary of an active set.   □

Lemma 6. *At any time, for any active or inactive set S, the subset of all marked edges with both endpoints in S forms a spanning tree of all requests from S.*

Proof. We show that the property holds at time passes. When a new request arrives, a new active growing set containing only one request is created. This set is trivially spanned by an empty set of marked edges.

By the definition of GD, a new active set appears when a dual constraint for some edge $e = (u, v)$ becomes tight. Right before it happens, the active sets containing u and v are $\mathcal{A}(u)$ and $\mathcal{A}(v)$, respectively. At that time, marked edges form spanning trees of sets $\mathcal{A}(u)$ and $\mathcal{A}(v)$ and, by Observation 2, there are no marked edges between these two sets. Hence, these spanning trees together with the newly marked edge e constitute a spanning tree of the requests of $S = \mathcal{A}(u) \uplus \mathcal{A}(v)$. Finally, a set may become inactive only if it was active before, and GD never adds any marked edge inside an already existing active or inactive set.   □

Using the lemma above, we are ready to bound the connection cost of one matching edge by the cost of the solution of $\mathcal{D}$.

Lemma 7. *The connection cost of any matching edge is at most $2 \cdot \sum_{S \subseteq V} \mathsf{sur}(S) \cdot y_S(T)$, where T is the time when GD matches the last request.*

Proof. Fix a matching edge (u, v) created by GD at time τ. Its connection cost is the distance $\mathsf{dist}(\mathsf{pos}(u), \mathsf{pos}(v))$ between the points corresponding to requests u and v in the underlying metric space.

We consider the state of GD right after it matches u with v. By Lemma 6, the active set $S = \mathcal{A}(u) = \mathcal{A}(v)$ containing u and v is spanned by a tree of marked edges. Let P be the (unique) path in this tree connecting u with v. Using the triangle inequality, we can bound $\mathsf{dist}(\mathsf{pos}(u), \mathsf{pos}(v))$ by the length of P projected onto the underlying metric space.

Recall that for any edge $e = (w, w')$, it holds that $\mathsf{dist}(\mathsf{pos}(w), \mathsf{pos}(w')) \leq \mathsf{opt\text{-}cost}(e)$. Moreover, if e is marked, the dual constraint for edge e holds with equality, that is, $\text{opt-cost}(e) = \sum_{S:e\in\delta(S)} y_S(\tau)$. Therefore,

$$\text{dist}(\text{pos}(u),\text{pos}(v)) \le \sum_{(w,w')\in P} \text{dist}(\text{pos}(w),\text{pos}(w')) \le \sum_{e\in P} \text{opt-cost}(e)$$

$$= \sum_{e\in P} \sum_{S:e\in\delta(S)} y_S(\tau) = \sum_{S} |\delta(S)\cap P| \cdot y_S(\tau)$$

$$\le \sum_{S} |\delta(S)\cap P| \cdot \text{sur}(S) \cdot y_S(\tau)$$

$$\le \sum_{S} |\delta(S)\cap P| \cdot \text{sur}(S) \cdot y_S(T).$$

The penultimate inequality holds because a dual variable y_S can be positive only if $\text{sur}(S) \ge 1$. It is now sufficient to prove that for each (active or inactive) set S, it holds that $|\delta(S)\cap P| \le 2$, i.e., the path P crosses each such set S at most twice.

For a contradiction, suppose that there exists an (active or inactive) set S, whose boundary is crossed by path P more than twice. We direct all edges on P towards v (we follow P starting from request u and move towards v). Note that u may be inside or outside of S. Let $e_1 = (w_1, w_2)$ be the first edge on P such that $w_1 \in S$ and $w_2 \notin S$, i.e., the first time when path P leaves S. Let $e_2 = (w_3, w_4) \in P$ be the first edge after e_1, such that $w_3 \notin S$ and $w_4 \in S$, that is, the first time when path P returns to S after leaving it with edge e_1. Edge e_2 must exist as we assumed that P crosses the boundary of S at least three times.

By Lemma 6, a subset of the marked edges constitutes a spanning tree of S. Hence, there exists a path of marked edges contained entirely in S that connects requests w_1 and w_4. Furthermore, a sub-path of P connects w_2 and w_3 outside of S. These two paths together with edges e_1 and e_2 form a cycle of marked edges. However, by Lemma 6 and Observation 2, at any time, the set of marked edges forms a forest, which is a contradiction.   □

5.3   Bounding the Competitive Ratio

Using above results we are able to bound the cost of Greedy Dual.

Theorem 1. Greedy Dual *is* $(2m + 1)$-*competitive for the M(B)PMD problem.*

Proof. Fix any input instance $\mathcal{I}$ and let $\mathcal{D}$ be the corresponding dual program. Let D be the cost of the solution to $\mathcal{D}$ output by GD. By Lemma 5, the total waiting cost of the algorithm is bounded by D and by Lemma 7, the connection cost of a single edge in the matching is bounded by $2 \cdot D$. Therefore,

$$\text{GD}(\mathcal{I}) \le D + m \cdot 2D = (2m + 1) \cdot D \le (2m + 1) \cdot \text{Opt}(\mathcal{I}),$$

where the first inequality holds as there are exactly m matched edges and the last equality follows by Lemma 1.   □

A   Tightness of the Analysis

We can show that our analysis of GREEDY DUAL is asymptotically tight, i.e., the competitive ratio of GREEDY DUAL is $\Omega(m)$.

Theorem 2. *Both for MPMD and MBPMD problems, there exists an instance $\mathcal{I}$, such that $GD(\mathcal{I}) = \Omega(m) \cdot OPT(\mathcal{I})$.*

Proof. Let $m > 0$ be an even integer and $\varepsilon = 1/m$. Let $\mathcal{X}$ be the metric containing two points p and q at distance 2.

In the instance $\mathcal{I}$, requests are released at both points p and q at times $0, 1+\varepsilon, 1+3\varepsilon, 1+5\varepsilon, \ldots, 1+(2m-3)\cdot\varepsilon$. For the MBPMD problem, we additionally specify request polarities: at p, all odd-numbered requests are positive and all even-numbered are negative, while requests issued at q have exactly opposite polarities from those at p.

Regardless of the variant (bipartite or non-bipartite) we solve, GD matches the first pair of requests at time 1, when their active growing sets are merged, forming a new active non-growing set. Every subsequent pair of requests appears exactly ε after the previous pair becomes matched. Therefore, they are matched together ε after their arrival, when their growing sets are merged with the large non-growing set containing all the previous pairs of requests. Hence, the total connection cost of GD is equal to $2m$. On the other hand, observe that the total cost of a solution that matches consecutive requests at each point of the metric space separately is equal to $2 \cdot ((1 + \varepsilon) + 2\varepsilon \cdot (m - 2)/2) = 2 \cdot (1 + (m - 1) \cdot \varepsilon) < 4$. □

B   Derandomization Using a Spanning Tree

In this part, we analyze an algorithm that approximates the metric space by a greedily and deterministically chosen spanning tree of requested points and employs the deterministic algorithm for trees of Azar et al. [4]. We show that such algorithm has the competitive ratio of $2^{\Omega(m)}$. For simplicity, we focus on the non-bipartite variant, but the lower bound can be easily extended to the bipartite case.

More precisely, we define a natural algorithm TREE BASED (TB). TB internally maintains a spanning tree T of metric space points corresponding to already seen requests. That is, whenever TB receives a request u at point $pos(u)$, it executes the following two steps.

1. If there was no previous request at $pos(u)$, TB adds $pos(u)$ to T, connecting it to the closest point from T. The addition is performed immediately, at the request arrival. This part essentially mimics the behavior of the greedy algorithm for the online Steiner tree problem [26].
2. To serve the request u, TB runs the deterministic algorithm of [4] on the tree T.[4]

---

[4] The algorithm must be able to operate on a tree that may be extended (new leaves may appear) in the runtime. The algorithm given by Azar et al. [4] has this property.

Theorem 3. *The competitive ratio of* TREE BASED *is* $2^{\Omega(m)}$.

Proof. The idea of the lower bound is as follows. The adversary first gives $m/2$ requests that force TB to create a tree T with the stretch of $2^{\Omega(m)}$ and then gives another $m/2$ requests, so that the initial m requests can be served with a negligible cost by OPT. Afterwards, the adversary consecutively requests a pair of points that are close in the metric space, but far away in the tree T.

Our metric space $\mathcal{X}$ is a continuous ring and we assume that m is an even integer. Let h be the length of this ring and let $\varepsilon = h/(m \cdot 2^{m-1})$.

In the first part of the input, the adversary gives $m/2$ requests in the following way. The first two requests are given at time 0 at antipodal points (their distance is $h/2$). TB connects them using one of two halves of the ring. From now on, the tree T of TB will always cover a contiguous part of the ring. Each of the next $m/2 - 2$ requests is given exactly in the middle of the ring part not covered by T. For $j \in \{3, 4, \ldots, m/2\}$, the j-th request is given at time $(2 \cdot (j-1)/m) \cdot \varepsilon$.

This way, the ring part not covered by T shrinks exponentially, and after $m/2$ initial requests its length is equal to $h/2^{m/2-1}$. Let p and q be the endpoints (the only leaves) of T. Then, $\mathrm{dist}(p,q) = h/2^{m/2-1}$, but the path between p and q in T is of length $h - \mathrm{dist}(p,q)$ and uses an edge of length $h/2$. As T is built as soon as requests appear, its construction is finished right after the appearance of the $(m/2)$-th request, i.e., before time ε.

In the second part of the input, at time ε, the adversary gives $m/2$ requests at the same points as the requests from the first phase. This way, OPT may serve the first m requests paying nothing for the connection cost and paying at most $(m/2) \cdot \varepsilon = h/2^m$ for their waiting cost.

In the third part of the input, the adversary gives $m/2$ pairs of requests, each pair at points p and q. Each pair is given after the previous one is served by TB. OPT may serve each pair immediately after its arrival, paying $\mathrm{dist}(p,q) = h/2^{m/2-1}$ for the connection cost. On the other hand, TB serves each such pair using a path that connects p and q in the tree T. Before matching p with q, TB waits for a time which is at least the length of the longest edge on this path, $h/2$ (see the analysis in [4]). In total, the cost of TB for the last m requests alone is at least $(m/2) \cdot (h/2)$, while the total cost of OPT for the whole input is at most $h/2^m + (m/2) \cdot h/2^{m/2-1}$. This proves that the competitive ratio of TB is $2^{\Omega(m)}$.   □

References

1. Antoniadis, A., Barcelo, N., Nugent, M., Pruhs, K., Scquizzato, M.: A o(n)-competitive deterministic algorithm for online matching on a line. In: Bampis, E., Svensson, O. (eds.) WAOA 2014. LNCS, vol. 8952, pp. 11–22. Springer, Heidelberg (2014). https://doi.org/10.1007/978-3-319-18263-6_2
2. Ashlagi, I., et al.: Min-cost bipartite perfect matching with delays. In: Proceedings of 20th International Workshop on Approximation Algorithms for Combinatorial Optimization (APPROX), pp. 1:1–1:20 (2017)

3. Avontuur, T., Spronck, P., van Zaanen, M.: Player skill modeling in Starcraft II. In: Proceedings of 9th AAAI Conference on Artificial Intelligence and Interactive Digital Entertainment, AIIDE 2013 (2013)
4. Azar, Y., Chiplunkar, A., Kaplan, H.: Polylogarithmic bounds on the competitiveness of min-cost perfect matching with delays. In: Proceedings of 28th ACM-SIAM Symposium on Discrete Algorithms (SODA), pp. 1051–1061 (2017)
5. Azar, Y., Ganesh, A., Ge, R., Panigrahi, D.: Online service with delay. In: Proceedings of 49th ACM Symposium on Theory of Computing (STOC), pp. 551–563 (2017)
6. Azar, Y., Jacob-Fanani, A.: Deterministic min-cost matching with delays. In: Solis-Oba, R., Fleischer, R. (eds.) WAOA 2018. LNCS, vol. 10787, pp. 132–146. Springer, Heidelberg (2018). https://doi.org/10.1007/978-3-319-89441-6_11
7. Bansal, N., Buchbinder, N., Gupta, A., Naor, J.: A randomized $O(\log^2 k)$-competitive algorithm for metric bipartite matching. Algorithmica $68(2)$, 390–403 (2014)
8. Bansal, N., Buchbinder, N., Madry, A., Naor, J.: A polylogarithmic-competitive algorithm for the k-server problem. J. ACM $62(5)$, 40:1–40:49 (2015)
9. Bienkowski, M., et al.: Online algorithms for multi-level aggregation. In: Proceedings of 24th European Symposium on Algorithms, ESA 2016, pp. 12:1–12:17 (2016)
10. Bienkowski, M., Byrka, J., Chrobak, M., Jeż, L., Nogneng, D., Sgall, J.: Better approximation bounds for the joint replenishment problem. In: Proceedings of 25th ACM-SIAM Symposium on Discrete Algorithms (SODA), pp. 42–54 (2014)
11. Bienkowski, M., Byrka, J., Chrobak, M., Jeż, L., Sgall, J., Stachowiak, G.: Online control message aggregation in chain networks. In: Dehne, F., Solis-Oba, R., Sack, J.-R. (eds.) WADS 2013. LNCS, vol. 8037, pp. 133–145. Springer, Heidelberg (2013). https://doi.org/10.1007/978-3-642-40104-6_12
12. Bienkowski, M., Kraska, A., Schmidt, P.: A match in time saves nine: deterministic online matching with delays. In: Solis-Oba, R., Fleischer, R. (eds.) WAOA 2017. LNCS, vol. 10787, pp. 132–146. Springer, Cham (2018). https://doi.org/10.1007/978-3-319-89441-6_11
13. Bienkowski, M., Kraska, A., Schmidt, P.: Online service with delay on a line. In: Lotker, Z., Patt-Shamir, B. (eds.) SIROCCO 2018. LNCS, vol. 11085, pp. 237–248. Springer, Heidelberg (2018)
14. Borodin, A., El-Yaniv, R.: Online Computation and Competitive Analysis. Cambridge University Press, Cambridge (1998)
15. Buchbinder, N., Feldman, M., Naor, J.S., Talmon, O.: O(depth)-competitive algorithm for online multi-level aggregation. In: Proceedings of 28th ACM-SIAM Symposium on Discrete Algorithms (SODA), pp. 1235–1244 (2017)
16. Buchbinder, N., Kimbrel, T., Levi, R., Makarychev, K., Sviridenko, M.: Online make-to-order joint replenishment model: primal dual competitive algorithms. In: Proceedings of 19th ACM-SIAM Symposium on Discrete Algorithms (SODA), pp. 952–961 (2008)
17. Chrobak, M., Kenyon-Mathieu, C.: Competitiveness via doubling. SIGACT News $37(4)$, 115–126 (2006)
18. Dooly, D.R., Goldman, S.A., Scott, S.D.: On-line analysis of the TCP acknowledgment delay problem. J. ACM $48(2)$, 243–273 (2001)
19. Elo, A.E.: The Rating of Chessplayers, Past and Present. Arco Publishing, London (1978)
20. Emek, Y., Kutten, S., Wattenhofer, R.: Online matching: haste makes waste! In: Proceedings of 48th ACM Symposium on Theory of Computing (STOC), pp. 333–344 (2016)

21. Emek, Y., Shapiro, Y., Wang, Y.: Minimum cost perfect matching with delays for two sources. In: Fotakis, D., Pagourtzis, A., Paschos, V.T. (eds.) CIAC 2017. LNCS, vol. 10236, pp. 209–221. Springer, Cham (2017). https://doi.org/10.1007/978-3-319-57586-5_18

22. Fakcharoenphol, J., Rao, S., Talwar, K.: A tight bound on approximating arbitrary metrics by tree metrics. J. Comput. Syst. Sci. **69**(3), 485–497 (2004)

23. Fuchs, B., Hochstättler, W., Kern, W.: Online matching on a line. Theoret. Comput. Sci. **332**(1–3), 251–264 (2005)

24. Goemans, M.X., Williamson, D.P.: A general approximation technique for constrained forest problems. SIAM J. Comput. **24**(2), 296–317 (1995)

25. Gupta, A., Lewi, K.: The online metric matching problem for doubling metrics. In: Czumaj, A., Mehlhorn, K., Pitts, A., Wattenhofer, R. (eds.) ICALP 2012. LNCS, vol. 7391, pp. 424–435. Springer, Heidelberg (2012). https://doi.org/10.1007/978-3-642-31594-7_36

26. Imase, M., Waxman, B.M.: Dynamic Steiner tree problem. SIAM J. Discrete Math. **4**(3), 369–384 (1991)

27. Kalyanasundaram, B., Pruhs, K.: Online weighted matching. J. Algorithms **14**(3), 478–488 (1993)

28. Karlin, A.R., Kenyon, C., Randall, D.: Dynamic TCP acknowledgement and other stories about e/(e - 1). Algorithmica **36**(3), 209–224 (2003)

29. Khanna, S., Naor, J.S., Raz, D.: Control message aggregation in group communication protocols. In: Widmayer, P., Eidenbenz, S., Triguero, F., Morales, R., Conejo, R., Hennessy, M. (eds.) ICALP 2002. LNCS, vol. 2380, pp. 135–146. Springer, Heidelberg (2002). https://doi.org/10.1007/3-540-45465-9_13

30. Khuller, S., Mitchell, S.G., Vazirani, V.V.: On-line algorithms for weighted bipartite matching and stable marriages. Theoret. Comput. Sci. **127**(2), 255–267 (1994)

31. Koutsoupias, E., Nanavati, A.: The online matching problem on a line. In: Solis-Oba, R., Jansen, K. (eds.) WAOA 2003. LNCS, vol. 2909, pp. 179–191. Springer, Heidelberg (2004). https://doi.org/10.1007/978-3-540-24592-6_14

32. Lowalekar, M., Varakantham, P., Jaillet, P.: Online spatio-temporal matching in stochastic and dynamic domains. In: Proceedings of 30th AAAI Conference on Artificial Intelligence, pp. 3271–3277 (2016)

33. Mehta, A.: Online matching and ad allocation. Found. Trends Theoret. Comput. Sci. **8**(4), 265–368 (2013)

34. Meyerson, A., Nanavati, A., Poplawski, L.J.: Randomized online algorithms for minimum metric bipartite matching. In: Proceedings of 7th ACM-SIAM Symposium on Discrete Algorithms (SODA), pp. 954–959 (2006)

35. Nayyar, K., Raghvendra, S.: An input sensitive online algorithm for the metric bipartite matching problem. In: Proceedings of 58th IEEE Symposium on Foundations of Computer Science (FOCS), pp. 505–515 (2017)

36. Raghvendra, S.: A robust and optimal online algorithm for minimum metric bipartite matching. In: Proceedings of 19th International Workshop on Approximation Algorithms for Combinatorial Optimization (APPROX), pp. 18:1–18:16 (2016)

37. Schrijver, A.: Combinatorial Optimization: Polyhedra and Efficiency. Algorithms and Combinatorics, vol. 24. Springer, Heidelberg (2003)

Advice Complexity of Priority Algorithms

Allan Borodin[1], Joan Boyar[2], Kim S. Larsen[2], and Denis Pankratov[3(✉)]

[1] University of Toronto, Toronto, Canada
bor@cs.toronto.edu
[2] University of Southern Denmark, Odense, Denmark
{joan,kslarsen}@imada.sdu.dk
[3] Concordia University, Montreal, Canada
denis.pankratov@concordia.ca

Abstract. The priority model of "greedy-like" algorithms was introduced by Borodin, Nielsen, and Rackoff in 2002. We augment this model by allowing priority algorithms to have access to advice, i.e., side information precomputed by an all-powerful oracle. Obtaining lower bounds in the priority model without advice can be challenging and may involve intricate adversary arguments. Since the priority model with advice is even more powerful, obtaining lower bounds presents additional difficulties. We sidestep these difficulties by developing a general framework of reductions which makes lower-bound proofs relatively straightforward and routine. We start by introducing the Pair Matching problem, for which we are able to prove strong lower bounds in the priority model with advice. We develop a template for constructing a reduction from Pair Matching to other problems in the priority model with advice – this part is technically challenging since the reduction needs to define a valid priority function for Pair Matching while respecting the priority function for the other problem. Finally, we apply the template to obtain lower bounds for a number of standard discrete optimization problems.

1 Introduction

Greedy algorithms are among the first class of algorithms studied in an undergraduate computer science curriculum. They are among the simplest and fastest algorithms for a given optimization problem, often achieving a reasonably good approximation ratio, even when the problem is NP-hard. In spite of their importance, the notion of a greedy algorithm is not well defined. This might be satisfactory for studying upper bounds; when an algorithm is suggested, it does not matter much whether everyone agrees that it is greedy or not. However, lower bounds (inapproximation results) require a precise definition. Perhaps giving a precise definition for all greedy algorithms is not possible, since one can provide examples that seem to be outside the scope of the given model.

The full version of the paper is available on arXiv [6]. For the first author, research is supported by NSERC. The second and third authors were supported in part by the Independent Research Fund Denmark, Natural Sciences, grant DFF-7014-00041.

L. Epstein and T. Erlebach (Eds.): WAOA 2018, LNCS 11312, pp. 69–86, 2018.
https://doi.org/10.1007/978-3-030-04693-4_5

Setting this philosophical question aside, we follow the model of greedy-like algorithms due to Borodin, Nielsen, and Rackoff [9]. The *fixed priority model* captures the observation that many greedy algorithms work by first sorting the input items according to some priority function, and then, during a single pass over the sorted input, making online irrevocable decisions for each input item. This model is similar to the online algorithm model with an additional preprocessing step of sorting inputs. Of course, if any sorting function is allowed, this would trivialize the model for most applications. Instead, a total ordering on the universe of all possible input items is specified before any input is seen, and the sorting is done according to this ordering, after which the algorithm proceeds as an online algorithm. This model has been adopted with respect to a broad array of topics [2,3,5,8,13,16,18,19]. In spite of its appeal, there are relatively few lower bounds in this model. There does not seem to be a general method for proving lower bounds; that is, the adversary arguments tend to be ad-hoc. In addition, the basic priority model does not capture the notion of side information. The assumption that an algorithm does not know anything about the input is quite pessimistic in practice. This issue has been addressed recently in the area of online algorithms by considering models with advice (see [10] for an overview). In these models, side information, such as the number of input items or a maximum weight of an item, is computed by an all powerful oracle and is available to an algorithm before seeing any of the input. This information is then used to make better online decisions. The goal is to study trade-offs between advice length and the competitive ratio.

We introduce a general technique for establishing lower bounds on priority algorithms with advice. These algorithms are a simultaneous generalization of priority algorithms and online algorithms with advice. Our technique is inspired by the recent success of the binary string guessing problem and reductions in the area of online algorithms with advice. We identify a difficult problem (Pair Matching) that can be thought of as a sorting-resistant version of the binary string guessing problem. Then, we describe the template of gadget reductions from Pair Matching to other problems in the world of priority algorithms with advice. This part turns out to be challenging, mostly because one has to ensure that priorities are respected by the reduction. We then apply the template to a number of classic optimization problems. We restrict our attention to the fixed priority model. We also note that we consider deterministic algorithms unless otherwise specified.

Related Model

Fixed priority algorithms with advice can be viewed in terms of the fixed priority backtracking model of Alekhnovich et al. [1]. That model starts by ordering the inputs using a fixed priority function and then executes a computation tree where different decisions can be tried for the same input item by branching in the tree, and then choosing the best result. The lower bound results generally consider how much width (maximum number of nodes for any fixed depth in the tree) is necessary to obtain optimality where the width proven is often of the form $2^{\Omega(n)}$. In contrast, our results give a parameterized trade-off between the number of advice bits and the approximation (competitive) ratio. However, given an algorithm in the fixed priority backtracking model, the logarithm of the width gives an upper bound on the number of bits of advice needed for the same approximation ratio. Similarly, a lower bound on the advice complexity gives a lower bound on width.

Organization

We give a formal description of the models in Sect. 2. We motivate the study of the priority model with advice in Sect. 3. We introduce and analyze the Pair Matching problem in Sect. 4. We describe the reduction framework for obtaining lower bounds in Sect. 5 and apply it to classic problems in Sect. 6. We conclude in Sect. 7. Omitted proofs can all be found in the full version of the paper [6].

2   Preliminaries

We consider optimization problems for which we are given an objective function to minimize or maximize, and measure our success relative to an optimal offline algorithm.

Online Algorithms with Advice

In an online setting, the input is revealed one item at a time by an adversary. An algorithm makes an irrevocable decision about the current item before the next item is revealed. For more background on online algorithms, we refer the reader to the texts by Borodin and El-Yaniv [7] and Komm [15].

The assumption that an online algorithm does not know anything about the input is quite often too pessimistic in practice. Depending on the application domain, the algorithm designer may have access to knowledge about the number of input items, the largest weight of an input item, some partial solution based on historical data, etc. The advice tape model for online algorithms captures the notion of side information in a purely information-theoretic way as follows. An all-powerful oracle that sees the entire input prepares the infinite advice tape with bits, which are available to the algorithm during the entire process. The oracle and the algorithm work in a cooperative mode – the oracle knows how the algorithm will use the bits and is trying to maximize the usefulness of the advice with regards to optimizing the given objective function. The advice complexity of an algorithm is a function of the input length and is the number of bits read by the algorithm in the worst case for inputs of a given size. For more background on online algorithms with advice, see the survey by Boyar et al. [10].

Fixed Priority Model with Advice

Fixed priority algorithms can be formulated as follows. Let $\mathcal{U}$ be a universe of all possible input items. An input to the problem consists of a finite set of items $\mathcal{I} \subset \mathcal{U}$ satisfying some consistency conditions. The algorithm specifies a total order on $\mathcal{U}$ before seeing the input. Then, a subset of the possible input items is revealed (by an adversary) according to the total order specified by the algorithm. The algorithm makes irrevocable decisions about the items as they arrive.[1] The overall set of decisions is then evaluated according to some objective function. The performance of the algorithm is measured by the asymptotic approximation ratio with respect to the value provided by an optimal offline algorithm. The notion of advice is added to the model as follows. After the algorithm has chosen a total order on $\mathcal{U}$, an all-powerful oracle that has access to the entire input $\mathcal{I}$ creates a tape of infinitely many bits. The algorithm knows how the advice bits are created and has access to them during the online decision phase. Our interest is in how many bits of advice the algorithm uses compared with the result it obtains.

We consider only countable universes $\mathcal{U}$. In this case, having a total order on elements in $\mathcal{U}$ is equivalent (via a simple inductive argument) to having a priority function $P : \mathcal{U} \to \mathbb{R}$. The assumption of the universe being countable is natural, but also necessary for the above equivalence: there are uncountably many totally ordered sets that do not embed into the reals with the standard order.

Definition 1. *Let $\mathcal{U}$ be the universe of input items and let $P : \mathcal{U} \to \mathbb{R}$ be a priority function. For $u_1, u_2 \in \mathcal{U}$, we write $u_1 <_P u_2$ to mean $P(u_1) < P(u_2)$. We will say that larger priority means that the item appears earlier in the input, i.e., $u_1 <_P u_2$ means that u_2 appears before u_1 when the input is given according to P.*

Example. Kruskal's optimal algorithm for the minimum spanning tree problem is a fixed priority algorithm without advice. The universe of items is $\mathcal{U} = \mathbb{N} \times \mathbb{N} \times \mathbb{Q}$. An item $(i, j, w) \in \mathcal{U}$ represents an edge between a vertex i and a vertex j of weight w. The consistency condition on the input is that the edge $\{i, j\}$ can be present at most once in the input. The total order on the universe is specified by all items of smaller weight having higher priority than all items of larger weight, breaking ties, say, by lexicographic order on the names of vertices. Kruskal's algorithm processes input items in the given order and greedily accepts those items that do not result in cycles.

In this paper, we only consider the following input model for graph problems in the priority setting:

Vertex arrival, vertex adjacency: an input item consists of a name of a vertex together with a set of names of adjacent vertices. There is a consistency condition on the entire input: if u appears as a neighbor of v, then v must appear as a neighbor of u.

Binary String Guessing Problem

Later we introduce the Pair Matching problem that can be viewed as a priority model analogue of the following online binary string guessing problem.

---

[1] In the adaptive priority model, the algorithm is allowed to specify a new ordering depending on previous items and decisions before a new input item is presented.

Definition 2. *The Binary String Guessing Problem [4] with known history (2-SGKH) is the following online problem. The input consists of $(n, \sigma = (x_1, \dots, x_n))$, where $x_i \in \{0, 1\}$. Upon seeing $x_1, \dots, x_{i-1}$ an algorithm guesses the value of x_i. The actual value of x_i is revealed after the guess. The goal is to maximize the number of correct guesses.*

Böckenhauer et al. [4] provide a trade-off between the number of advice bits and the approximation ratio for the binary string guessing problem.

Theorem 1 (Böckenhauer et al. [4]). *For the 2-SGKH problem and any $\varepsilon \in (0, \frac{1}{2}]$, no online algorithm reading fewer than $(1 - H(\varepsilon))n$ advice bits can make fewer than εn mistakes for large enough n, where $H(p) = H(1-p) = -p\log(p) - (1-p)\log(1-p)$ is the binary entropy function.*

Competitive and Approximation Ratios

The performance of online algorithms is measured by their competitive ratios. For a minimization problem, an online algorithm ALG is said to be *c-competitive* if there exists a constant α such that for all input sequences I we have $\text{ALG}(I) \leq c\,\text{OPT}(I) + \alpha$, where $\text{ALG}(I)$ denotes the cost of the algorithm on I and $\text{OPT}(I)$ is the value achieved by an offline optimal algorithm. The infimum of all c such that ALG is *c*-competitive is ALG's *competitive ratio*. For a maximization problem, $\text{ALG}(I)$ is referred to as profit, and we require that $\text{OPT}(I) \leq c\,\text{ALG}(I) + \alpha$. In this way, we always have $c \geq 1$ and the closer c is to 1, the better. Priority algorithms are thought of as approximation algorithms and the term (asymptotic) approximation ratio is used (but the definition is the same).

3 Motivation

In this section we present a motivating example for studying the priority model with advice. We present a problem that is difficult in the pure priority setting or in the online setting with advice, but easy in the priority model with advice. Furthermore, the advice is easily computed by an offline algorithm.

The problem of interest is called Greater Than Mean (GTM). In the GTM problem, the input is a sequence $x_1, \dots, x_n$ of rational numbers. Let $m = \sum_i x_i / n$ denote the sample mean of the sequence. The goal of an algorithm is to decide for each x_i whether x_i is greater than the mean or not, answering 1 or 0, respectively. We can also assume that the length of the sequence n is known to the algorithm in advance. We start by noting that there is a trivial optimal priority algorithm with little advice for this problem.

Theorem 2. *For Greater Than Mean, there exists a fixed priority algorithm reading at most $\lceil \log n \rceil$ advice bits, solving the problem optimally.*

Proof. The priority order is such that $x_1 \geq x_2 \dots \geq x_n$. Thus, the numbers arrive in the order from largest to smallest. The advice specifies the earliest index $i \in [n]$ such that $x_i \leq m$. □

In the full version, we show that a priority algorithm without advice has to make many errors.[2]

Theorem 3. *For Greater Than Mean and any $\varepsilon \in (0, \frac{1}{2}]$, no fixed priority algorithm without advice can make fewer than $(1/2 - \varepsilon)n$ mistakes for large enough n.*

Finally, we show that an online algorithm requires a lot of advice to achieve good performance for the GTM problem. The proof is a minor modification of a reduction from 2-SGKH to the Binary Separation Problem (see [11] for details). We present the proof in its entirety for completeness.

Algorithm 1. Reduction from 2-SGKH to GTM

---

  procedure REDUCTION-2-SGKH-TO-GTM
    $\ell_1 \leftarrow 0, u_1 \leftarrow 1$
    for $i = 1$ to n **do**
      $y_i \leftarrow (\ell_i + u_i)/2$
      if A predicts y_i is greater than mean **then**
        predict $x_i = 1$
      else
        predict $x_i = 0$
      receive actual x_i
      if actual $x_i = 1$ **then**
        $u_{i+1} \leftarrow y_i, \ell_{i+1} \leftarrow \ell_i$
      else
        $u_{i+1} \leftarrow u_i, \ell_{i+1} \leftarrow y_i$
    $y_{n+1} \leftarrow \frac{n+1}{2}(\ell_{n+1} + u_{n+1}) - \sum_{i=1}^{n} y_i$

Theorem 4. *For the Greater Than Mean problem and any $\varepsilon \in (0, \frac{1}{2}]$, no online algorithm reading fewer than $(1 - H(\varepsilon))(n - 1)$ advice bits can make fewer than εn mistakes for large enough n.*

Proof. We present a reduction from the 2-SGKH problem to the GTM problem. Let A be an online algorithm with advice for the GTM problem. Our reduction is presented in Algorithm 1. In the course of the reduction, an online input $x_1, \ldots, x_n$ of length n for the 2-SGKH problem is converted into an online input $y_1, \ldots, y_{n+1}$ of length $n + 1$ for the GTM problem with the following properties: The reduction is advice-preserving (the number of advice bits is the same) and for each $i \in [n]$, our algorithm A for 2-SGKH makes a mistake on x_i if and only if A makes a mistake on y_i. This would finish the proof of the theorem.

---

[2] In Theorem 3 and in all of our lower bound advice results, we state the result so as to include $\varepsilon = \frac{1}{2}$, in which case the conditions "fewer than $(1/2 - \varepsilon)$" and "fewer than $(1 - H(\varepsilon))$" make the statements vacuously true.

Let $S = \{i \in [n] \mid x_i = 1\}$ and $T = [n] \setminus S$. The reduction uses a technique similar to binary search to make sure that $\forall i \in S$ and $\forall j \in T$ we have $y_i > y_j$, i.e., all the y_i corresponding to $x_i = 1$ are larger than all the y_j corresponding to $x_j = 0$. Then y_{n+1} is chosen to make sure that the mean of the entire stream $y_1, \ldots, y_{n+1}$ lies between the smallest y_i with $i \in S$ and the largest y_j with $j \in T$. This implies that y_i is greater than the mean if and only if the corresponding $x_i = 1$.

The following invariants are easy to see and are left to the reader: (1) $u_i > \ell_i$; (2) if $x_i = 1$, then $u_i > y_i \geq u_{i+1}$; (3) if $x_i = 0$, then $\ell_i < y_i \leq \ell_{i+1}$.

The required properties of the reduction follow immediately from the invariants. Let $i \in S$ and $j \in T$. Then, $y_i \geq u_{n+1} > \ell_{n+1} \geq y_j$. Finally, observe that y_{n+1} is chosen so that the mean is $\sum_{i=1}^{n+1} y_i/(n+1) = \sum_{i=1}^{n} y_i/(n+1) + y_{n+1}/(n+1) = (1/2)(\ell_{n+1} + u_{n+1})$. This mean correctly separates S from T. $\square$

4   Pair Matching Problem

We introduce an online problem called Pair Matching.[3] The input consists of a sequence of n distinct rational numbers between 0 and 1, i.e., $x_1, \ldots, x_n \in \mathbb{Q} \cap [0,1]$. After the arrival of x_i, an algorithm has to answer if there is a $j \in [n] \setminus \{i\}$ such that $x_i + x_j = 1$, in which case we refer to x_i and x_j as forming a pair and say that x_i has a matching value, x_j. The answer "accept" is correct if x_j exists, and "reject" is correct if it does not. Note that since the x_i are all distinct, if $x_i = \frac{1}{2}$, the correct answer is "reject", since $\frac{1}{2}$ cannot have a matching value. We let $pairs(x_1, \ldots, x_n)$ denote the number of pairs in the input $x_1, \ldots, x_n$.

4.1   Online Setting

Analyzing Pair Matching in the online setting is relatively straightforward for both deterministic and randomized algorithms.

Theorem 5. *For Pair Matching, there exists a 2-competitive algorithm, answering correctly on $n - pairs(x_1, \ldots, x_n)$ input items.*

Theorem 6. *For Pair Matching, no deterministic online algorithm can achieve a competitive ratio less than 2.*

Theorem 7. *For Pair Matching, there exists a randomized online algorithm that in expectation answers correctly on $2n/3$ input items.*

Theorem 8. *For Pair Matching, no randomized online algorithm can achieve a competitive ratio less than $3/2$.*

Lastly, we prove that online algorithms need a lot of advice in order to start approaching a competitive ratio of 1 for Pair Matching.

---

[3] There are similarities to the NP-Complete problems, Numerical Matching with Target Sums and Numerical 3-Dimensional Matching, though these problems ask if permutations of sets of inputs will lead to a complete matching.

Theorem 9. *For Pair Matching and any $\varepsilon \in (0, \frac{1}{2}]$, no deterministic online algorithm reading fewer than $(1 - H(\varepsilon))n/2$ advice bits can make fewer than εn mistakes for large enough n.*

4.2 Priority Setting

In this section, we show that Theorem 9 also holds in the priority setting. The proof becomes a bit more subtle, so we give it in full detail.

Theorem 10. *For Pair Matching and any $\varepsilon \in (0, \frac{1}{2}]$, no fixed priority algorithm reading fewer than $(1 - H(\varepsilon))n/2$ advice bits can make fewer than εn mistakes for large enough n.*

Proof. We prove the statement by a reduction from the *online problem* 2-SGKH. Let A be a priority algorithm solving Pair Matching, and let P be the corresponding priority function. (Note that we assume that the algorithm knows P; this is the case in all of our priority algorithm reductions.) The reduction follows the proof of Theorem 9 closely. The idea is to transform the online input to 2-SGKH into an input to Pair Matching. The difficulty arises from having to present the transformed input in the online fashion while respecting the priority function P.

Let $x_1, \ldots, x_n$ be the input to 2-SGKH. The online reduction works as follows. The online algorithm picks n distinct numbers $y_1, \ldots, y_n$ from $[0, 1]$ and creates a list $z_1, \ldots, z_{2n}$ consisting of y_i and $1 - y_i$ sorted according to P. The algorithm keeps a (max-heap ordered) priority queue Q of elements from z_i as well as a subsequence Z of $z_1, \ldots, z_{2n}$. The reduction always picks the first element z from Z. We maintain the invariant that $1 - z$ appears later in Z according to P. If needed, the reduction algorithm will insert $1 - z$ into Q to be simulated as an input to A at the right time later on.

Initialization. Initially, Q is empty and Z is the entire sequence $z_1, \ldots, z_{2n}$. Before the element x_1 arrives, the algorithm feeds z_1 to A. If A answers that z_1 is a part of a pair, then the online algorithm predicts $x_1 = 1$; otherwise the algorithm predicts $x_1 = 0$. Then the online algorithm finds j such that $z_j = 1 - z_1$ and updates Z by deleting z_1 and z_j. Then x_1 is revealed. If the actual value of x_1 is 1, the algorithm inserts z_j into Q; otherwise the algorithm does not modify Q.

Middle Step. Suppose that the algorithm has processed $x_1, \ldots, x_{i-1}$ and has to guess the value of x_i. The algorithm picks the first element z from the subsequence Z. While the top element of Q has higher priority than z according to P, the algorithm deletes that element from the priority queue and feeds it to A. Then, the algorithm feeds z to A. The next steps are similar to the initialization case. If A answers that z is a part of a pair, then the online algorithm predicts $x_i = 1$; otherwise the algorithm predicts $x_i = 0$. The online algorithm finds z' in Z such that $z = 1 - z'$, and updates Z by deleting z and z'. Then x_i is revealed. If the actual value of x_i is 1, the algorithm inserts z' into Q; otherwise the algorithm does not modify Q.

Post-processing. After the algorithm finishes processing x_n, it feeds the remaining elements (in priority order) from Q to A.

It is easy to see that the online algorithm feeds a subsequence of $z_1, \ldots, z_{2n}$ to A in the correct order according to P. In addition, the online algorithm makes exactly the same number of mistakes as A (assuming that A always answers correctly on the second element of a pair). The statement of the theorem follows since the size of the input to A is at most $2n$. $\square$

5   Reduction Template

Our template is restricted to binary decision problems since the goal is to derive inapproximations based on the Pair Matching problem. In reducing from Pair Matching to a problem B, we assume that we have a priority algorithm ALG with advice for problem B with priorities defined by P. Based on ALG and P, we define a priority algorithm ALG$'$ with advice and a priority function, P', for the Pair Matching problem. Input items $x_1, x_2, \ldots, x_n$ in $\mathbb{Q} \cap [0, 1]$ to Pair Matching arrive in an order specified by the priority function we define, based on P. We assume that we are informed when the input ends and can take steps at that point to complete our computation. Knowing the size n of the input, which one naturally would in many situations after the initial sorting according to P', would of course be sufficient.

Based on the input to the Pair Matching problem, we create input items to problem B, and they have to be presented to ALG, respecting the priority function P. Responses from ALG are then used by ALG$'$ to help it answer "accept" or "reject" for its current x_i. Actually, ALG will always answer correctly for a request $x_j = 1 - x_i$ when $i < j$, so the responses from ALG are only used when this is not the case. The main challenge is to ensure that the input items to ALG are presented in the order determined by P, because the decision as to whether or not they are presented needs to be made in time, without knowing whether or not the matching value will arrive.

Here, we give a high level description of a specific kind of gadget reduction. A gadget G for problem B is simply some constant-sized instance for B, i.e., a collection of input items that satisfy the consistency condition for problem B. For example, if B is a graph problem in the vertex arrival, vertex adjacency model, G could be a constant-sized graph, and the universe then contains all possible pairs of the form: a vertex name coupled with a list of possible neighboring vertex names. Note that each possible vertex name exists many times as a part of an input, because it can be coupled with many different possible lists of vertex names. The consistency condition must apply to the actual input chosen, so for each vertex name u which is listed as a neighbor of v, it must be the case that v is listed as a neighbor of u.

The gadgets used in a reduction will be created in pairs (gadgets in a pair may be isomorphic to each other, so that they are the same up to renaming), one pair for each input item less than or equal to $1/2$ (for $x = 1/2$, the gadget will only be used to assign a priority to $x = 1/2$). One gadget from the pair is presented to ALG when $1 - x$ appears later in the input; and the other gadget when it does not. Using fresh names in the input items for problem B, we ensure that each input item less than $\frac{1}{2}$ to the Pair Matching problem has its own collection of input items for its gadgets for problem B. The pair of gadgets associated with an input item $x \leq 1/2$ can be written (G_x^1, G_x^2). The same universe of input items is used for both of these gadgets.

We write $\max_P G$ to denote the first item according to P from the universe of input items for G, i.e., the highest priority item. For now, assume that ALG responds "accept" or "reject" to any possible input item. This captures problems such as vertex cover, independent set, clique, etc.

For each $x \leq 1/2$, the gadget pair satisfies two conditions: the first item condition, and the distinguishing decision condition. The *first item condition* says that the first input item $m_1(x)$ according to P gives no information about which gadget it is in. To accomplish this, we define the priority function for ALG' as $P'(x) = P(\max_P G_x^1)$ for all $x \leq 1/2$ and set $m_1(x) = \max_P G_x^1 = \max_P G_x^2$ (the second equality holds since we assume the two gadgets have the same input universe). The *distinguishing decision condition* says that the decision with regards to item $m_1(x)$ that results in the optimal value of the objective function in G_x^1 is different from the decision that results in the optimal value of the objective function in G_x^2. This explains why the one gadget is presented to ALG when $1 - x$ appears later in the input sequence and the other when it does not.

Now that the first item of the gadget associated with x is defined, the remaining actual input items in the gadget pair for x must be completely defined according to the distinguishing decision condition. This gives two sets (overlapping, at least in $m_1(x)$) of input items. The item with highest priority among all of the items in the actual gadget pair, ignoring $m_1(x)$, is called $m_2(x)$, and we define $P'(1 - x) = P(m_2(x))$ for $x < 1/2$. Thus, we guarantee the following list of properties: $x < 1/2$ will arrive before $1 - x$ in the input sequence for Pair Matching for ALG', $m_1(x)$ will arrive for algorithm ALG at the same time, ALG's response for $m_1(x)$ can define the response of ALG' to x, and the decision as to which gadget in the pair is presented for x can be made at the time $1 - x$ arrives or ALG' can determine that it will not arrive (because either the input sequence ended or an x' with lower priority than $1 - x$ arrived).

To warm up, we start with an example reduction from Pair Matching to Triangle Finding; a somewhat artificial problem in this context, but well-studied in streaming algorithms [17], for instance. This reduction then serves as a model for the general reduction template.

5.1   Example: Triangle Finding

Consider the following priority problem in the vertex arrival, vertex adjacency model: for each vertex v, decide whether or not v belongs to some triangle (a cycle of length 3) in the entire input graph. The answer "accept" is correct if v belongs to some triangle, and otherwise the answer should be "reject". We refer to this problem as Triangle Finding. This problem might look artificial and it is optimally solvable offline in time $O(n^2)$, but as mentioned above, advice-preserving reductions between priority problems require subtle manipulations of a priority function. The Triangle Finding problem allows us to highlight this issue in a relatively simple setting.

Theorem 11. *For Triangle Finding and any $\varepsilon \in (0, \frac{1}{2}]$, no fixed priority algorithm reading at most $(1 - H(\varepsilon))n/8$ advice bits can make fewer than $\varepsilon n/4$ mistakes.*

Proof. We prove this theorem by a reduction from the Pair Matching problem. Let ALG be an algorithm for the Triangle Finding problem, and let P be the corresponding priority function. Let $x_1, \ldots, x_n$ be the input to Pair Matching. We define a priority function P' and a valid input sequence $v_1, \ldots, v_m$ to Triangle Finding. When $x_1, \ldots, x_n$ is presented according to P' to our priority algorithm for Pair Matching, it is able to construct $v_1, \ldots, v_m$ for ALG, respecting the priority function P. Moreover, our algorithm for Pair Matching will be able to use answers of ALG to answer the queries about $x_1, \ldots, x_n$.

Now, we discuss how to define P'. With each number $x \in \mathbb{Q} \cap [0, 1/2]$, we associate four unique vertices $v_x^1, v_x^2, v_x^3, v_x^4$. The universe consists of all input items of the form $(v_x^i, \{v_x^j, v_x^k\})$ with $i, j, k \in [4]$, $i \notin \{j, k\}$ and $j < k$; there are 12 input items for each x: 4 possibilities for the vertex, and for each of the $\binom{3}{2} = 3$ possibilities for the ordered pair of neighbors. Let $m_1(x)$ be the first item according to P among the 12 items. Using only the input items from the 12 items we are currently considering, we extend this item in two ways, to a 3-cycle C_x^3 and to a 4-cycle C_x^4. When we write C_x^3 or C_x^4, we mean the set of items forming the 3-cycle or 4-cycle, respectively. Now, P' is defined as follows:

$$P'(x) = \begin{cases} P(m_1(x)), & \text{if } x \leq 1/2 \\ \max_{g \in (C_{1-x}^3 \cup C_{1-x}^4) \setminus \{m_1(1-x)\}} P(g), & \text{otherwise} \end{cases}$$

In other words, if $x > 1/2$, we set $P'(x)$ to be the first element other than $m_1(1 - x)$ in $C_{1-x}^3 \cup C_{1-x}^4$. In terms of our high level description given at the beginning of this section, (C_x^3, C_x^4) form the pair of gadgets – a triangle and a square (4-cycle). By construction, this pair of gadgets satisfies the first item condition. By the definition of the problem, the optimal decision for all vertices in C_x^3 is "accept" (belongs to a triangle) and the optimal decision for all vertices in C_x^4 is "reject" (does not belong to a triangle). Thus, these gadgets also satisfy the distinguishing decision condition.

Let $x_1, \ldots, x_n$ denote the order input items are presented to our algorithm as specified by P'. Our algorithm constructs an input to ALG which is consistent with P along the following lines: for each $x \leq 1/2$ that appears in the input, the algorithm constructs either a three-cycle or a four-cycle (disjoint from the rest of the graph). Thus, each $x \leq 1/2$ is associated with one connected component. During the course of the algorithm, each connected component will be in one of the following three states: undecided, committed, or finished. When $x \leq 1/2$ arrives, the algorithm initializes the construction with the item $m_1(x)$ and sets the component status to undecided. It answers "accept" (there will be a matching pair) for x if ALG responds "accept" (triangle) for $m_1(x)$, and it answers "reject" if ALG responds "reject" (square).

Note that for any $x \leq 1/2$, $P'(x) > P'(1-x)$, so if $x' > 1/2$ arrives and $1-x'$ has not appeared earlier, ALG' can simply reject x' and does not need to present anything to ALG. If x has arrived and at some point, $1-x$ arrives, the algorithm commits to constructing the 3-cycle C_x^3. If ALG' had guessed correctly that $1-x$ would arrive, it is because ALG responded "accept" for $m_1(x)$) and also guessed correctly. If ALG' had guessed that $1-x$ would not arrive, it is because ALG guessed that a square would arrive, and both guessed incorrectly. If some x' arrives with $P'(x') < P'(1-x)$ for some $x \neq x'$ and x has arrived earlier, then ALG' can be certain that $1-x$ will not arrive. It commits to constructing the 4-cycle C_x^4. Thus, if ALG' answered "reject" for x, it answered correctly, and a square makes ALG's decision for $m_1(x)$ correct. Similarly, if ALG' answered "accept" for x, it answered incorrectly, so a square makes ALG's decision incorrect.

At the end of the input, ALG' finishes off by checking which values of x have arrived without $1-x$ arriving or some x' with higher priority than $1-x$ arriving, and ALG again commits to the 4-cycle, as in the other case where $1-x$ does not arrive.

Throughout the algorithm, there are several connected components, each of which can be undecided, committed, or finished. Note that an undecided component corresponding to input x consists of a single item $m_1(x)$. Upon receiving an item y, the algorithm first checks whether some undecided components have turned into committed ones: namely if an undecided component consisting of $m_1(x)$ satisfies $P'(1-x) > P'(y)$, it switches the status to a committed component according to the rules described above. Then, the algorithm feeds input items corresponding to committed yet unfinished connected components to ALG and does so in the order of P up until the priority of such items falls below $P'(y)$ (this can be done by maintaining a priority queue). Finally, the algorithm processes the item y by either creating a new component or by turning an undecided component into a decided one. Then, the algorithm moves to the next item. Due to our definition of P' and this entire process, the input constructed for ALG is valid and consistent with P. Observe that the input to ALG' is of size at most $4n$, so the number of advice bits must be divided by four relative to Theorem 10, and the theorem follows.   □

5.2 General Template

In this subsection, we establish two theorems that give general templates for gadget reductions from Pair Matching – one for maximization problems and one for minimization problems. A high level overview is given at the beginning of this section.

We let ALG(I) denote the objective function for ALG on input I. The *size* of a gadget G, denoted by $|G|$, is the number of input items specifying the gadget. We write OPT(G) to denote the best value of the objective function on G. Recall that we focus on problems where a solution is specified by making an accept/reject decision for each input item. We write $\mathrm{BAD}(G)$ to denote the best value of the objective function attainable on G after making the wrong decision for the first item (the item with highest priority, $\max(G)$), i.e., if there is an optimal solution that accepts (rejects) the first item of G, then $\mathrm{BAD}(G)$ denotes the best value of the objective function *given* that the first item was rejected (accepted). We say that the objective function for a problem B is *additive*, if for any two instances I_1 and I_2 to B such that $I_1 \cap I_2 = \emptyset$, we have $\mathrm{OPT}(I_1 \cup I_2) = \mathrm{OPT}(I_1) + \mathrm{OPT}(I_2)$.

Theorem 12. *Let B be a minimization problem with an additive objective function. Let* ALG *be a fixed priority algorithm with advice for B with a priority function P. Suppose that for each $x \in \mathbb{Q} \cap [0, 1/2]$ one can construct a pair of gadgets (G_x^1, G_x^2) satisfying the following conditions:*

The first item condition: $m_1(x) = \max_P G_x^1 = \max_P G_x^2$.

The distinguishing decision condition: *the optimal decision for $m_1(x)$ in G_x^1 is different from the optimal decision for $m_1(x)$ in G_x^2 (in particular, the optimal decision is unique for each gadget). Without loss of generality, we assume $m_1(x)$ is accepted in an optimal solution in G_x^1.*

The size condition: *the gadgets have finite sizes; let $s = \max_x(|G_x^1|, |G_x^2|)$, where the cardinality of a gadget is the number of input items it consists of.*

The disjoint copies condition: *for $x \neq y$ and $i, j \in \{1, 2\}$, input items making up G_x^i and G_y^j are disjoint.*

The optimal/bad condition: *the values $\mathrm{OPT}(G_x^1)$, $\mathrm{BAD}(G_x^1)$ and $\mathrm{OPT}(G_x^2)$, $\mathrm{BAD}(G_x^2)$ are independent of x, and we denote them by $\mathrm{OPT}(G^1)$, $\mathrm{BAD}(G^1)$, $\mathrm{OPT}(G^2)$, and $\mathrm{BAD}(G^2)$; we assume that $\mathrm{OPT}(G^2) \geq \mathrm{OPT}(G^1)$.*

Define $r = \min\left\{ \frac{\mathrm{BAD}(G^1)}{\mathrm{OPT}(G^1)}, \frac{\mathrm{BAD}(G^2)}{\mathrm{OPT}(G^2)} \right\}$. Then for any $\varepsilon \in (0, \frac{1}{2}]$, no fixed priority algorithm reading fewer than $(1 - H(\varepsilon))n/(2s)$ advice bits can achieve an approximation ratio smaller than

$$1 + \frac{\varepsilon(r-1)\,\mathrm{OPT}(G^1)}{\varepsilon\,\mathrm{OPT}(G^1) + (1-\varepsilon)\,\mathrm{OPT}(G^2)}.$$

The following theorem is for maximization problems.

Theorem 13. *Let B be a maximization problem with an additive objective function. Let* ALG *be a fixed priority algorithm with advice for B with a priority function P. Suppose that for each $x \in \mathbb{Q} \cap [0, 1/2]$ one can construct a pair of gadgets (G_x^1, G_x^2) satisfying the conditions in Theorem 12. Then for any $\varepsilon \in (0, \frac{1}{2}]$, no fixed priority algorithm reading fewer than $(1 - H(\varepsilon))n/(2s)$ advice bits can achieve an approximation ratio smaller than*

$$1 + \frac{\varepsilon(r-1)\,\mathrm{OPT}(G^1)}{\varepsilon\,\mathrm{OPT}(G^1) + (1-\varepsilon)r\,\mathrm{OPT}(G^2)},$$

where $r = \min\left\{ \frac{\mathrm{OPT}(G^1)}{\mathrm{BAD}(G^1)}, \frac{\mathrm{OPT}(G^2)}{\mathrm{BAD}(G^2)} \right\}$.

We mostly use Theorems 12 and 13 in the following specialized form.

Corollary 1. *With the set-up from Theorems 12 and 13, we have the following:*

For a minimization problem, if $\mathrm{OPT}(G^1) = \mathrm{OPT}(G^2) = \mathrm{BAD}(G^1) - 1 = \mathrm{BAD}(G^2) - 1$, *then no fixed priority algorithm reading fewer than* $(1 - H(\varepsilon))n/(2s)$ *advice bits can achieve an approximation ratio smaller than* $1 + \frac{\varepsilon}{\mathrm{OPT}(G^1)}$.

For a maximization problem, if $\mathrm{OPT}(G^1) = \mathrm{OPT}(G^2) = \mathrm{BAD}(G^1) + 1 = \mathrm{BAD}(G^2) + 1$, *then no fixed priority algorithm reading fewer than* $(1 - H(\varepsilon))n/(2s)$ *advice bits can achieve an approximation ratio smaller than* $1 + \frac{\varepsilon}{\mathrm{OPT}(G^1) - \varepsilon}$.

Next, we describe a general procedure for constructing gadgets with the above properties. For simplicity, we do it for graph problems in the vertex arrival, vertex adjacency input model. Later we discuss what is required to carry out such general constructions for other combinatorial problems. In the case of graphs, an input item consists of a vertex name with the names of neighbors of that vertex. First, consider defining a single gadget instead of a pair. We define a gadget in several steps. As the first step, we define a graph $G = \left([n], E \subset \binom{[n]}{2}\right)$ over n vertices. Then, when defining a gadget based on input x to Pair Matching, we pick n vertex names V_x and give a bijection $f : V_x \to [n]$. Finally, we read off the resulting input items in the order given by the priority function. Thus, we think of G as giving a topological structure of the instance, and it is converted into an actual instance by assigning new names to the vertices. The reason that the names from the topological structure are not used directly is that we want to define a separate gadget instance for each $x \in \mathbb{Q} \cap [0, 1/2]$. Thus, all gadgets instances are going to have the same topological structure,[4] but will differ in names of vertices.

For graphs in the vertex arrival, vertex adjacency model, we say that two input items are isomorphic if they have the same number of neighbors, i.e., they differ in just the names of the vertices and the names of their neighbors. A topological structure G consisting only of isomorphic items is a regular graph. For any priority function P and any vertex $v \in [n]$, we can force the corresponding item to appear first according to P by naming vertices appropriately. Fix x and consider all possible input items that can be formed from V_x consistently with G. One of those items appears first according to P. Define a bijection f by first mapping that first item to u and its neighbors in G, and extending this one-to-one correspondence to other vertices in G in an arbitrary, consistent manner. In this case, the input item corresponding to u would appear first according to P in the input to the graph problem. Because all items are isomorphic, it is always possible to extend the bijection to all of G.

Now, suppose that two topological structures $G^1 = ([n], E^1)$ and $G^2 = ([m], E^2)$ consist only of isomorphic items. Using a similar idea, for each priority function P, each $x \in \mathbb{Q} \cap [0, 1/2)$, each $u \in [n]$, and each $v \in [m]$, one can

---

[4] However, both gadgets within a pair do not necessarily have the same topological structure. In Triangle Finding, they did not.

assign names to vertices of G^1 and G^2 such that the first input item according to P is associated with u in G^1 and the same item is associated with v in G^2. In particular, this means that as long as the two topological structures are regular, we can always convert them into gadgets satisfying the first item condition.

Suppose that there is a vertex u in G^1 that appears in every optimal solution in G^1, i.e., a "reject" decision leads to non-optimality. Furthermore, suppose that there is a vertex v in G^2 that is excluded from every optimal solution in G^2, i.e., an "accept" decision leads to non-optimality. Then for each x, using the above construction, we can make the first item according to P be associated with u in G^1 and with v in G^2. This means that we can always convert the topological structures into gadgets satisfying the distinguishing decision condition. Finally, observe that the size condition is satisfied with $s = \max(|G^1|, |G^2|)$.

This gadget construction can be carried out in other input models. We need to have a notion of isomorphism between input items, and a notion of the topological structure of a gadget. Once we have the two notions, if we find topological structures consisting only of isomorphic items with uniquely identifiable "reject"/"accept" items in all optimal solutions, then we immediately conclude that the problem requires the trade-off between advice and approximation ratio as outlined in Theorems 12 and 13 and Corollary 1 with parameter s equal to the size of the topological template.

We finish this section by remarking that one can perform similar reductions with gadgets where not all input items are isomorphic. Theorem 17, which is based on a lower bound construction from [5], is proven via a reduction for Vertex Cover using *two* gadget pairs with some vertices of degree 2 and others of degree 3. One simply needs that there is one gadget pair for the case where a vertex of degree 2 has the highest priority and another gadget pair for the case where a vertex of degree 3 has highest priority. For both gadget pairs, $s = 7$, the optimal value is 3, and the minimum possible objective value for the gadget in the pair is 4. Thus, the results of Theorem 12 (or Theorem 13 if it was a maximization problem) and Corollary 1 can be applied. This idea can be extended to other input models where the gadgets have input items which are not isomorphic. For simplicity, we do not restate the two theorems or the corollary for the extension where there are t different classes of isomorphic input items and thus t pairs of gadgets.

6   Reductions to Classic Optimization Problems

In this section, we provide one detailed example of an application of the general reduction template, plus statement of results for other problems. With the exception of bipartite matching, all of these problems are NP-hard, as a consequence of the NP-completeness of their underlying decision problems, as established in the seminal papers by Cook [12] and Karp [14]. Furthermore, these problems are known to have various hardness of approximation bounds.

6.1   Detailed Example: Independent Set

We consider the maximum independent set problem in the vertex arrival, vertex adjacency input model. Consider the topological structure of a gadget in Fig. 1. There are 5 vertices on the top and 3 vertices on the bottom. All top vertices are connected to all bottom vertices. Additionally, the 5 vertices on the top form a cycle. In this way, each vertex has degree 5 and hence all the input items are isomorphic. If we pick any vertex from the top to be in the independent set, then we forgo all the bottom vertices, and we are essentially restricted to picking an independent set from C_5, which has size at most 2. On the other hand, we could pick all 3 vertices from the bottom to form an independent set.

Suppose without loss of generality that the highest priority input item is $(1, \{4, 5, 6, 7, 8\})$. The optimal decision for the first vertex is unique: For G^1, one should accept, and for G^2, reject.

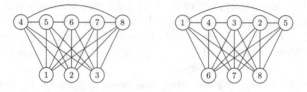

Fig. 1. Topological structure of the gadgets (G^1, G^2) for independent set.

In this case, the maximum number s of input items for a gadget is 8, $\mathrm{OPT}(G^1) = \mathrm{OPT}(G^2) = 3$, and $\mathrm{BAD}(G^1) = \mathrm{BAD}(G^2) = 2$. By Corollary 1, we can conclude the following:

Theorem 14. *For Maximum Independent Set and any $\varepsilon \in (0, \frac{1}{2}]$, no fixed priority algorithm reading fewer than $(1 - H(\varepsilon))n/16$ advice bits can achieve an approximation ratio smaller than $1 + \frac{\varepsilon}{3-\varepsilon}$.*

Theorem 14 is related to but incomparable with the inapproximation bound results on priority algorithms (without advice) of Borodin et al. [5] for weaker models.

6.2   Other Results

Detailed definitions of the problems below, input universes, general details on how to obtain the following results from the general template, and the relationship to the known literature can be found in the full version.

Theorem 15. *For Maximum Bipartite Matching and any $\varepsilon \in (0, \frac{1}{2}]$, no fixed priority algorithm reading fewer than $(1 - H(\varepsilon))n/6$ advice bits can achieve an approximation ratio smaller than $1 + \frac{\varepsilon}{3-\varepsilon}$.*

Theorem 16. *For Maximum Cut and any* $\varepsilon \in (0, \frac{1}{2}]$*, no fixed priority algorithm reading fewer than* $(1 - H(\varepsilon))n/16$ *advice bits can achieve an approximation ratio smaller than* $1 + \frac{\varepsilon}{15-\varepsilon}$.

Theorem 17. *For Minimum Vertex Cover and any* $\varepsilon \in (0, \frac{1}{2}]$*, no fixed priority algorithm reading fewer than* $(1 - H(\varepsilon))n/14$ *advice bits can achieve an approximation ratio smaller than* $1 + \frac{\varepsilon}{3}$.

Theorem 18. *For Maximum 3-Satisfiability and any* $\varepsilon \in (0, \frac{1}{2}]$*, no fixed priority algorithm reading fewer than* $(1 - H(\varepsilon))n/6$ *advice bits can achieve an approximation ratio smaller than* $1 + \frac{\varepsilon}{8-\varepsilon}$.

Theorem 19. *For Job Scheduling of Unit Time Jobs with Precedence Constraints and any* $\varepsilon \in (0, \frac{1}{2}]$*, no fixed priority algorithm reading fewer than* $(1 - H(\varepsilon))n/18$ *advice bits can achieve an approximation ratio smaller than* $1 + \frac{\varepsilon}{6-\varepsilon}$.

7   Concluding Remarks

We have developed a general framework for showing linear lower bounds on the number of advice bits required to get a constant approximation ratio for fixed priority algorithms with advice. The framework relies on reductions from the Pair Matching problem—analogue of the Binary String Guessing problem from the online world, resistant to universe orderings. Many problems remain open:

- Can our framework (or a modification of it) show non-constant inapproximation results with large advice, for example, for independent set?
- In vertex coloring, any decision for the first item can be completed to an optimal solution. Can our framework be modified to handle such problems? For example, see an argument for the makespan problem in [19].
- An interesting goal is to study the "structural complexity" of online and priority algorithms. Can one define analogues of classes such as NP, NP-Complete, ♯P, etc. for online/priority problems? If so, are complete problems for these classes natural?

Acknowledgments. Part of the work was done when the first author was visiting Toyota Technological Institute at Chicago. The work was initiated while the second and third authors were visiting the University of Toronto. Most of the work was done when the fourth author was a postdoc at the University of Toronto.

References

1. Alekhnovich, M., Borodin, A., Buresh-Oppenheim, J., Impagliazzo, R., Magen, A., Pitassi, T.: Toward a model for backtracking and dynamic programming. Comput. Complex. **20**(4), 679–740 (2011)
2. Angelopoulos, S., Borodin, A.: On the power of priority algorithms for facility location and set cover. Algorithmica **40**(4), 271–291 (2004)

3. Besser, B., Poloczek, M.: Greedy matching: guarantees and limitations. Algorithmica **77**(1), 201–234 (2017)
4. Böckenhauer, H.-J., Hromkovič, J., Komm, D., Krug, S., Smula, J., Sprock, A.: The string guessing problem as a method to prove lower bounds on the advice complexity. Theor. Comput. Sci. **554**, 95–108 (2014)
5. Borodin, A., Boyar, J., Larsen, K.S., Mirmohammadi, N.: Priority algorithms for graph optimization problems. Theor. Comput. Sci. **411**(1), 239–258 (2010)
6. Borodin, A., Boyar, J., Larsen, K.S., Pankratov, D.: Advice complexity of priority algorithms. ArXiv arXiv:1806.06223 [cs.DS] (2018)
7. Borodin, A., El-Yaniv, R.: Online Computation and Competitive Analysis. Cambridge University Press, Cambridge (1998)
8. Borodin, A., Lucier, B.: On the limitations of greedy mechanism design for truthful combinatorial auctions. ACM Trans. Econ. Comput. **5**(1), 2:1–2:23 (2016)
9. Borodin, A., Nielsen, M.N., Rackoff, C.: (Incremental) priority algorithms. Algorithmica **37**(4), 295–326 (2003)
10. Boyar, J., Favrholdt, L.M., Kudahl, C., Larsen, K.S., Mikkelsen, J.W.: Online algorithms with advice: a survey. ACM Comput. Surv. **50**(2), 19:1–19:34 (2017)
11. Boyar, J., Kamali, S., Larsen, K.S., López-Ortiz, A.: Online bin packing with advice. Algorithmica **74**(1), 507–527 (2016)
12. Cook, S.A.: The complexity of theorem-proving procedures. In: 3rd Annual ACM Symposium on Theory of Computing (STOC), pp. 151–158. ACM (1971)
13. Davis, S., Impagliazzo, R.: Models of greedy algorithms for graph problems. Algorithmica **54**(3), 269–317 (2009)
14. Karp, R.M.: Reducibility among combinatorial problems. In: Miller, R.E., Thatcher, J.W., Bohlinger, J.D. (eds.) Complexity of Computer Computations. The IBM Research Symposia Series, pp. 85–103. Springer, Boston (1972). https://doi.org/10.1007/978-1-4684-2001-2_9
15. Komm, D.: An Introduction to Online Computation - Determinism, Randomization, Advice. Texts in Theoretical Computer Science. An EATCS Series. Springer, Cham (2016). https://doi.org/10.1007/978-3-319-42749-2
16. Lesh, N., Mitzenmacher, M.: Bubblesearch: a simple heuristic for improving priority-based greedy algorithms. Inf. Process. Lett. **97**(4), 161–169 (2006)
17. McGregor, A.: Graph stream algorithms: a survey. ACM SIGMOD Rec. **43**(1), 9–20 (2014)
18. Poloczek, M.: Bounds on greedy algorithms for MAX SAT. In: Demetrescu, C., Halldórsson, M.M. (eds.) ESA 2011. LNCS, vol. 6942, pp. 37–48. Springer, Heidelberg (2011). https://doi.org/10.1007/978-3-642-23719-5_4
19. Regev, O.: Priority algorithms for makespan minimization in the subset model. Inf. Process. Lett. **84**(3), 153–157 (2002)

Approximating Node-Weighted k-MST on Planar Graphs

Jarosław Byrka[1], Mateusz Lewandowski[1(✉)], and Joachim Spoerhase[2]

[1] Institute of Computer Science, University of Wrocław, Wrocław, Poland
{jby,mlewandowski}@cs.uni.wroc.pl
[2] Lehrstuhl für Informatik I, Universität Würzburg, Würzburg, Germany
joachim.spoerhase@uni-wuerzburg.de

Abstract. We study the problem of finding a minimum weight connected subgraph spanning at least k vertices on planar, node-weighted graphs. We give a $(4 + \varepsilon)$-approximation algorithm for this problem. We achieve this by utilizing the recent Lagrangian-multiplier preserving (LMP) primal-dual 3-approximation for the node-weighted prize-collecting Steiner tree problem by Byrka et al. (SWAT'16) and adopting an approach by Chudak et al. (Math. Prog. '04) regarding Lagrangian relaxation for the edge-weighted variant. In particular, we improve the procedure of picking additional vertices (tree merging procedure) given by Sadeghian (2013) by taking a constant number of recursive steps and utilizing the limited guessing procedure of Arora and Karakostas (Math. Prog. '06).

More generally, our approach readily gives a $(4/3 \cdot r + \varepsilon)$-approximation on any graph class where the algorithm of Byrka et al. for the prize-collecting version gives an r-approximation. We argue that this can be interpreted as a generalization of an analogous result by Könemann et al. (Algorithmica '11) for partial cover problems. Together with a lower bound construction by Mestre (STACS'08) for partial cover this implies that our bound is essentially best possible among algorithms that utilize an LMP algorithm for the Lagrangian relaxation as a black box. In addition to that, we argue by a more involved lower bound construction that even using the LMP algorithm by Byrka et al. in a *non-black-box* fashion could not beat the factor $4/3 \cdot r$ when the tree merging step relies only on the solutions output by the LMP algorithm.

1 Introduction

We consider the node-weighted variant of the well-studied k-MST problem. Given a graph $G = (V, E)$ with non-negative node weights $c \colon V \to \mathbb{R}_+$ and a positive integer k, we consider the problem of finding a minimum cost connected subgraph of G spanning at least k vertices. In analogy to the edge-weighted case, we call this problem node-weighted k-MST (NW-k-MST) because the solution can be assumed to be a tree. In fact, we focus on the rooted variant in which a given vertex r has to be included in the final solution. To obtain the unrooted version, simply use the resulting algorithm for each choice of root vertex.

© Springer Nature Switzerland AG 2018
L. Epstein and T. Erlebach (Eds.): WAOA 2018, LNCS 11312, pp. 87–101, 2018.
https://doi.org/10.1007/978-3-030-04693-4_6

It was already observed that this problem is $\Omega(\log n)$-hard to approximate [17]. However, the problem becomes easier when we restrict G to be a planar graph. It is still NP-hard, as the edge-weighted variant is NP-hard even on planar graphs [18]. To this end, consider the following reduction from edge-weighted variant to the node-weighted variant. Each original vertex gets weight 0. Now, each edge e is replaced with a new vertex v_e of weight equal to the cost of e. Moreover, v_e is connected by two edges with original endpoints of e. Finally, each original vertex is connected to l new leaves of weight 0 where l is a parameter. It is easy to see, that for $l > |E|$, solutions for k-MST instances correspond to solutions to node-weighted $(k \cdot l + k - 1)$-MST instances after reduction and vice-versa.

The above reduction preserves planarity. Therefore, the focus of this work is to provide an approximation algorithm with small factor for planar NW-k-MST.

1.1   Related Work

Edge-Weighted k-MST. The standard, edge-weighted k-MST problem has been thoroughly studied. In a sequence of papers [1,9,10] the 2-approximation algorithm for prize-collecting Steiner tree problem [11] was used to finally obtain a 2-approximation algorithm for k-MST. These results can be, to some extent, explained as in the work of Chudak et al. [7] in terms of Lagrangian Relaxation.

In particular, a 5-approximation algorithm follows the framework known mostly from Jain and Vazirani's work on the k-median problem [12]. In these algorithms, the Lagrangian multiplier preserving (LMP) property plays a crucial role. The LMP property is also satisfied by the Goemans-Williamson algorithm for the prize-collecting Steiner tree problem (PC-ST). Intuitively, the LMP property of an α-approximation algorithm for some prize-collecting problem, means that the solutions it produces would also be not more expensive than α times optimum value even if we would have to pay α times more for penalties.

Node-Weighted k-MST. The NW-k-MST problem was already studied in the more general quota setting, where each node has an associated profit, and the goal is to find the minimum cost connected set of vertices having total profit at least Π. In particular, an $O(\log n)$-approximation was given in [17]. However, this result was based on their invalid $O(\log n)$-approximation for NW-PC-ST. Recently, Chekuri et al. [6] and also independently Bateni et al. [2] proposed correct algorithms for generalizations of NW-PC-ST, but without LMP guarantee. The result on the quota problem was finally restored by Könemann et al. [14] who developed an LMP algorithm. In the related master thesis [19], Sadeghian gives also an alternative way of picking vertices[1] in the reduction for the quota problem. In these results, the constant lost in the process was not optimized.

---

[1] by picking vertices we mean augmenting the smaller solution with some vertices of larger solution. This is an important ingredient for the Lagrangian Relaxation technique.

Node-Weighted Planar Steiner Problems. Recently, the planar variants of Steiner problems received increased attention. In particular, Demaine et al. [8] obtained a 6-approximation for the node-weighted Steiner forest problem. The factor was further improved to 3 by Moldenhauer [16]. Both results rely on the moat-growing algorithm similar to that of Goemans and Williamson [11]. Currently the best result for this problem is the 2.4 approximation by Berman and Yaroslavtsev [3] who use a different oracle for determining violated sets.

More general network design problems on planar graphs where also studied by Chekuri et al. [5]. Finally, the result of Moldenhauer was generalized to the prize-collecting variant by Byrka et al. [4], resulting in an LMP 3-approximation for NW-PC-ST on planar graphs. We note that our result highly relies on this last algorithm.

Partial Cover. Below, we argue that our problem on arbitrary graphs generalizes the *partial cover* problem. In this problem we are given a set cover instance along with a positive integer k. The objective is to cover at least k ground elements by a family of sets of minimum cost. In the *prize-collecting* version of the problem every element has a penalty and the objective is to minimize the sum of costs of the chosen sets and the penalties of the elements that are not covered. Könemann et al. [13] describe a unified framework for partial cover. They show how to obtain an approximation algorithm for a class $\mathcal{I}$ of partial cover instances if there is an r-approximate LMP algorithm for the corresponding prize-collecting version. In particular, their result implies a $(\frac{4}{3} + \varepsilon)r$-approximation algorithm for the class $\mathcal{I}$. Mestre [15] shows that no algorithm that uses an LMP algorithm as a black box can obtain a ratio better than $\frac{4}{3}r$ so these results are essentially optimal.

1.2 Our Result and Techniques

We give a polynomial-time $(4 + \varepsilon)$-approximation algorithm for the NW-k-MST problem on planar graphs. Our result extends to an algorithm for the quota node-weighted Steiner tree problem on planar graphs with the same factor.

The main technique we use is the Lagrangian relaxation framework (as mentioned in the section above) where two solutions—one with fewer and the other with more than k nodes—are combined to obtain a feasible tree. The overview of our algorithm is as follows:

1. guess a skeleton and prune the instance
2. using the LMP algorithm [4], find trees T_1, T_2 with $\leq k$ and $\geq k$ nodes, respectively
3. combine T_1 and T_2 into a single tree with exactly k vertices.

This is the standard design (although guessing step is not always necessary) of algorithms based on Lagrangian relaxation framework. However, in order to optimize the constant we employ additional ideas and techniques.

The first guessing step bears some similarities to that of Arora and Karakostas [1] where they improve Garg's 3-approximation for edge-weighted k-MST to $2 + \varepsilon$. This additional guessing allows them to pay $\varepsilon \cdot$ OPT instead of OPT for connecting a single set of vertices to the rest of the solution. Here, we provide a node-weighted variant of this idea and also use it more extensively, because we have to buy multiple (but still a constant number of) such connections. In our approach, we guess a set of vertices from optimum solution and call it a skeleton. Then, we can safely prune the instance ensuring that each remaining node will be not too far away from the skeleton. The guessing step is described in Sect. 2.

For the second step, we have to slightly modify the primal-dual LMP 3-approximation algorithm [4], so it returns solutions containing the guessed skeleton. This modification is technical and is described—together with the method used to find suitable T_1 and T_2—in Sect. 4.

In the third step, we combine T_1 and T_2 by extending the procedure of picking vertices of Sadeghian [19]. He finds some cost-effective subset of vertices, which is two times larger than needed. We show that by picking vertices in certain order and applying recursion a *constant* number of times, we are able to pick almost exactly the number of nodes that is needed. Although, the number of components of this set of nodes might be arbitrary, we need to buy only a constant number of connections to restore connectivity. This is our main contribution and is described in the Sect. 3.

The resulting approximation factor of our algorithm is $(4 + \varepsilon)$. Additionally, we show some evidence that our combining step is in some sense optimal. More precisely, we show that no other algorithm, using LMP 3-approximation as a black-box and which does not use planarity can give better constant than 4. This is obtained by interpreting our algorithm in terms of the results for the partial cover problem. The optimality of our algorithm within this framework is discussed in Sect. 5.

2   Pruning the Instance

First, we assume that we know OPT up to a factor $1 + \varepsilon$ by using standard guessing techniques [9]. A node v is called ε-*distant* to a node set $U \subseteq V$ if there exists a path P in G from v to a node $u \in U$ of node weight $c(V(P) \setminus \{u\}) \leq \varepsilon \cdot$ OPT.

Lemma 1. *Consider an optimum solution T and an $\varepsilon > 0$. Then there exists a set $W \subseteq V(T)$ of size at most $1/\varepsilon$ such that each node in T is ε-distant to $W \cup \{r\}$.*

Proof. Consider T as a tree rooted at r. For any node u in this tree let T_u denote the subtree hanging from u. A subtree T_u is called *good* if for any node in T_u the total weight of the unique path from this node to u within T_u (including the weight of the end nodes) is at most $\varepsilon \cdot$ OPT.

We traverse T in a bottom-up fashion starting with the leaves. We maintain the invariant (by removing subtrees) that for all nodes u visited so far and still being in T, the subtree T_u is good. To this end, when we encounter a node u such that T_u is good we just continue with the traversal. If T_u is bad, however, then there must be a path P within T_u ending in u of node weight $c(P) \geq \varepsilon \cdot \text{OPT}$. We include u into W and assign P as a *witness* to u. Because of our invariant for all (if any) children v of u, we have that T_v is good. This means in particular that for all nodes z in T_u the node weight (*excluding* the weight of u) of the path from z to u is at most $\varepsilon \cdot \text{OPT}$. Finally, remove T_u from T and continue with the traversal. We stop when we reach the root r at which point we remove the remaining tree (for the sake of analysis).

First, note that the set W has cardinality at most $1/\varepsilon$ because we assigned to each node in W a witness path of weight at least $\varepsilon \cdot \text{OPT}$ and because the witness paths are pairwise node-disjoint. Second, observe that whenever we removed a node z from T as part of a subtree T_u, the node weight (excluding the weight of u) of the path from z to u was at most $\varepsilon \cdot \text{OPT}$. Hence, for every node in T there exists such a path to a node in $W \cup \{r\}$ at the end of the tree traversal since every node was removed. □

In the sequel, we will call such a set W whose existence is provided by the above lemma an ε-*skeleton*.

In a pre-processing, we iterate over all $n^{\mathcal{O}(1/\varepsilon)}$ many sets $W' \subseteq V$ with $|W'| \leq 1/\varepsilon$ thereby guessing the ε-skeleton W whose existence is guaranteed by the above lemma. Moreover, we prune all nodes u from the instance that are not ε-distant to $W \cup \{r\}$.

3 The $(4 + \varepsilon)$-Approximation Algorithm

Sadeghian [19, Chap. 3] describes a $O(\log n)$ approximation for node-weighted quota Steiner tree problem. His result is established using a framework of [7], repeated also in [17] where a primal-dual LMP approximation algorithm for the prize-collecting Steiner tree problem can be used along with the Lagrangian relaxation method to obtain an approximation algorithm for the quota version of the problem. Sadeghian loses some large constant factor in the process. Direct application of his result would yield two digit approximation factor for our problem.

We now show that carefully injecting the LMP 3-approximation algorithm for NW-PC-ST on planar graphs given in [4] into his analysis yields a $(4 + \varepsilon)$-approximation. However, in the process, we need a more efficient way to pick additional vertices. We show that it is possible to pick a cheap set of these vertices. Although it will not be connected, only a *constant* number of additional ε-distant vertices will suffice to connect the picked vertices.

For ease of the presentation, we will focus on the NW-k-MST problem. The algorithm for quota version can be then easily deduced by arguments of Bateni et al. [2]

The analysis relies on the following lemma.

Lemma 2. *We can produce trees T_1 and T_2 containing all the vertices W from the ε-skeleton and the root r of sizes $|T_1| \leq k \leq |T_2|$, such that for $\alpha_1, \alpha_2 \geq 0$ with $\alpha_1 + \alpha_2 = 1$ and $\alpha_1|T_1| + \alpha_2|T_2| = k$ we have that*

$$\alpha_1 c(T_1) + \alpha_2 c(T_2) \leq (3+\varepsilon)OPT$$

The construction of these trees T_1 and T_2 and the proof of above lemma is described in Sect. 4.

Let now $q = k - |T_1|$ be the number of vertices that are missing from the tree T_1. We will now show, that these vertices can be picked from $T_2 \setminus T_1$ without paying too much.

Lemma 3. *It is possible to find a (not necessarily connected) set S of at least q vertices in $T_2 \setminus T_1$ of cost at most $(1+\varepsilon_2)\alpha_2 c(T_2)$, which can be connected to T_1 by connecting additionally $\mathcal{O}(\log(1/\varepsilon_2))$ many ε-distant vertices to the ε-skeleton, where ε_2 is any constant.*

Proof. Here, we substantially extend the analysis in [19]. Consider a graph T_2' constructed from T_2 by contracting all vertices from $T_1 \cap T_2$ to a single vertex r'. Define the cost of this vertex r' to 0 (we will buy T_1 anyway). From now on, whenever we count the cardinality of some subset S of vertices in T_2', we do not count vertex r'.

Definition 1. *A subset of vertices S is cost-effective if $\frac{c(S)}{|S|} \leq \frac{c(T_2')}{|T_2'|}$.*

Lemma 4. *If cost-effective set S has size $(1 + \varepsilon_2)q$ then its cost is at most $(1+\varepsilon_2)\alpha_2 c(T_2)$.*

Proof.

$$c(S) \leq |S|\frac{c(T_2')}{|T_2'|} \leq (1+\varepsilon_2)q\frac{c(T_2)}{|T_2| - |T_1|} \leq (1+\varepsilon_2)\alpha_2 c(T_2),$$

where we used the fact that $\alpha_2 = \frac{k-|T_1|}{|T_2|-|T_1|}$.   □

So now, our goal is to find a cost-effective set S in T_2' of size only slightly larger that q. First, we start with a procedure for picking at most $2q$ vertices as in [19]. Initialize graph H with any spanning tree of T_2'. Observe that H is cost-effective. Consider any edge e of H. Let X and Y be the two components that would be created after removing the edge e from H. At least one of these two components must be cost-effective. For any cost-effective component from this two, say X, do the following. If X has enough vertices, *i.e.* $|X| \geq q$, remove Y from H and continue. Otherwise, contract vertices of X to a single super vertex and set its cost to the sum of all vertices in X. We consider that the new super-vertex has *super-cardinality* equal to $|X|$.

It can be seen that after repeating this procedure as many times as possible, the graph H will be a star graph with super-cardinality of each leaf at most q.

Let p be the number of leaves of H. In the case when $p \leq 1$ it is easy to see, that taking the whole graph H would result in a cost-effective set of vertices of size at most $2q$. Therefore, assume now that $p \geq 2$. Then, there exists a central vertex of the star graph H, call it c, which is not a super vertex. Moreover, every leaf v must be cost-effective (otherwise either we would remove v, or H would consist of two nodes). Observe also, that the super-cardinality of each leaf is at most q. Hence adding leaves to S one by one, would eventually lead to the set S with super-cardinality at most $2q$ (and at least q). Finally, S could be connected to T_1 by a single path from vertex c.

We now modify this procedure of adding leaves. First, consider them in the order of decreasing super-cardinalities. To this end, let $v_1, v_2, \ldots v_p$ be leaves of H and $s_1 \geq s_2 \geq \cdots \geq s_p$ be the corresponding super-cardinalities. Find the smallest i such that $\sum_{j=1}^{i} s_j + s_{i+1} \geq q$. If $s_{i+1} = 1$, then the desired set S consist of all vertices in $v_1, v_2, \ldots v_{i+1}$ and it has exactly q vertices. Otherwise, add the first i leaves to the set S. Let $t = \sum_{j=1}^{i} s_j$ be the number of vertices added to S. Now, instead of adding to S all vertices in the super vertex s_{i+1}, we expand this super vertex back to the original graph and repeat the above process with the new number of vertices to pick equal to $q' = q - t$. Observe that, because of sorting we have that $t \geq \frac{1}{2}q$, which also implies that $q' \leq \frac{1}{2}q$. This process is repeated recursively up to l times—where l is a parameter— but in the last call we take the last leaf completely.

Let now $q_1, q_2, \ldots, q_l$ be the numbers of vertices to pick in respective recursive calls (note that $q_1 = q$ and $q_j \leq \frac{1}{2}q_{j-1}$). The total number of picked vertices is then at most $q + 2q_l \leq (1 + 2^{-l+2})q$. Therefore, to find the desired set S of at most $(1 + \varepsilon_2)q$ vertices, we need only a constant number of recursive calls—parameter l is only $\mathcal{O}(\log(1/\varepsilon_2))$. Moreover all the vertices of S can be connected to T_1 by buying paths from the central nodes of all the l star graphs that appeared in the process. This finishes the proof. $\square$

To construct a feasible solution, take the set S guaranteed by the above lemma and connect it to T_1 by the $\mathcal{O}(\log(1/\varepsilon_2))$ shortest paths to the ε-skeleton. Denote this solution by SOL_1. Let also SOL_2 be the entire tree T_2. Our algorithm outputs cheaper of the two solutions SOL_1 and SOL_2.

This enables us to prove the following.

Lemma 5. *Assuming $\varepsilon \leq 1$, the cost of the cheaper of the two solutions SOL_1 and SOL_2 is $(4 + O(\sqrt{\varepsilon})) \cdot \mathrm{OPT}$.*

Proof. To bound the cost of the cheaper of two solutions SOL_1 and SOL_2 we employ the following Lemma by Könemann et al. [13].

Lemma 6 ([13]). *For any $r > 1$ and $\delta > 0$, we have*

$$\max_{\substack{\alpha \in (0,1) \\ \beta \in [0,r]}} \min \left\{ \frac{r(1+\delta) - (1-\alpha)\beta}{\alpha}, r(1+\delta) + \alpha\beta \right\} = \left(\frac{4}{3} + O(\sqrt{\delta}) \right) r .$$

Now, let $\alpha = \alpha_2$ and $\beta = \frac{c(T_1)}{OPT}$. With this notation we obtain in a similar way as Könemann et al. [13]

$$c(\text{SOL}_1) \leq c(T_1) + (1 + \varepsilon_2)\alpha \cdot c(T_2) + \varepsilon \cdot \mathcal{O}(\log(1/\varepsilon_2)) \cdot \text{OPT}$$
$$\leq \alpha \cdot c(T_1) + (1 - \alpha) \cdot c(T_1) + (1 + \varepsilon_2)\alpha \cdot c(T_2) + \varepsilon \cdot \mathcal{O}(\log(1/\varepsilon_2)) \cdot \text{OPT}$$
$$\leq (3(1 + \varepsilon_2) + \alpha\beta) \cdot \text{OPT} + \varepsilon \cdot \mathcal{O}(\log(1/\varepsilon_2)) \cdot \text{OPT},$$

and

$$c(\text{SOL}_2) = c(T_2)$$
$$= \frac{\alpha \cdot c(T_2)}{\alpha}$$
$$\leq \frac{(3 + \varepsilon)\text{OPT} - (1 - \alpha)c(T_1)}{\alpha}$$
$$\leq \frac{3(1 + \varepsilon) - (1 - \alpha)\beta}{\alpha} \cdot \text{OPT}.$$

By setting $r = 3$ and $\delta = \varepsilon = \varepsilon_2$ we obtain via Lemma 6 that the better of the two solutions has cost no more than $(4 + O(\sqrt{\varepsilon} + \varepsilon \log 1/\varepsilon)) \cdot \text{OPT} = (4 + O(\sqrt{\varepsilon})) \cdot \text{OPT}$ completing the proof. □

4   Lagrangian Relaxation and Moat Growing on Planar Graphs

In this section we prove Lemma 2. The proof utilizes Lagrangian Relaxation and follows a framework similar to the one in [7].

We start with the following LP relaxation for the NW-k-MST problem, where solutions are additionally constrained to contain all guessed vertices W of the ε-skeleton. For each vertex v we have the x_v variable indicating whether we will include this vertex in the solution. The z variables are indexed by sets of vertices not containing the root and the guessed vertices. There exists optimum integral solution, such that only the one z_X variable is set to 1. This would be for the set X of vertices not included in the final solution.

$$\min \sum_{v \in V \setminus \{r\}} x_v c_v \qquad\qquad (LP)$$

$s.t.$

$$\sum_{v \in \Gamma(S)} x_v + \sum_{\substack{X : S \subseteq X \\ X \cap W = \emptyset}} z_X \geq 1 \qquad\qquad \forall S \subseteq V \setminus \{r\}$$

$$x_v + \sum_{\substack{X : v \in X \\ X \cap W = \emptyset}} z_X \geq 1 \qquad\qquad \forall v \in V \setminus \{r\}$$

$$\sum_{X \subseteq V \setminus \{r\}} |X| z_X \leq n - k \qquad\qquad (1)$$

$$x_v \geq 0 \qquad\qquad \forall v \in V \setminus \{r\}$$

$$z_X \geq 0 \qquad\qquad \forall X \subseteq V \setminus \{r\}$$

The first two types of constraints guarantee connectivity of the solution to the root vertex and skeleton W. The $\Gamma(S)$ denotes the neighborhood of the set S, *i.e.* the set of vertices that are not in S, but have a neighboring vertex in S.

The constraint (1) ensures that the final solution will have at least k vertices and introduces difficulties. Therefore, we move it to the objective function obtaining the following Lagrangian Relaxation:

$$\min \quad \sum_{v \in V \setminus \{r\}} x_v c_v + \lambda \left(\sum_{X \subseteq V \setminus \{r\}} |X| z_X - (n - k) \right) \qquad (LR(\lambda))$$

$$s.t.$$

$$\sum_{v \in \Gamma(S)} x_v + \sum_{\substack{X : S \subseteq X \\ X \cap W = \emptyset}} z_X \geq 1 \qquad \forall S \subseteq V \setminus \{r\}$$

$$x_v + \sum_{\substack{X : v \in X \\ X \cap W = \emptyset}} z_X \geq 1 \qquad \forall v \in V \setminus \{r\}$$

$$x_v \geq 0 \qquad \forall v \in V \setminus \{r\}$$

$$z_X \geq 0 \qquad \forall X \subseteq V \setminus \{r\}$$

The above LP (ignoring the constant $-\lambda(n - k)$ term in the objective function) is exactly the LP for the node-weighted prize-collecting Steiner tree (NW-PC-ST in which the penalty of each vertex in $V' = V \setminus W$ is equal to the parameter λ) with a slight modification that the subset of vertices W is required to be in the solution.

Consider now, the dual of the $LR(\lambda)$:

$$\max \quad \sum_{S \subseteq V \setminus \{r\}} y_S + \sum_{v \in V \setminus \{r\}} p_v - \lambda(n - k) \qquad (DLR(\lambda))$$

$$s.t.$$

$$\sum_{S : v \in \Gamma(S)} y_S + p_v \leq c_v \qquad \forall v \in V \setminus \{r\}$$

$$\sum_{X \subseteq S} y_X + \sum_{v \in S} p_v \leq \lambda |S| \qquad \forall S \subseteq V' \setminus \{r\}$$

$$y_S \geq 0 \qquad \forall S \subseteq V \setminus \{r\}$$

Now, the slightly modified primal-dual LMP 3-approximation for (NW-PC-ST) given in [4] can be used with penalties λ to produce the tree T^λ and the dual solution (y^λ, p^λ) such that

$$c(T^\lambda) + 3\lambda(n - |T^\lambda|) \leq 3 \left(\sum_{S \subseteq V \setminus \{r\}} y_S^\lambda + \sum_{v \in V \setminus \{r\}} p_v^\lambda \right), \qquad (2)$$

where T^λ contains all vertices of W. The description of this algorithm is deferred to Subsect. 4.1. Let us now see how we can use it to finish the proof of Lemma 2. We proceed essentially as in [19] and [7]. By subtracting $3\lambda(n - k)$ from both sides of inequality (2) and simplifying the notation so that $\text{DS}_\lambda = \sum_{S \subseteq V \setminus \{r\}} y_S^\lambda + \sum_{v \in V \setminus \{r\}} p_v^\lambda$ denotes the value of a dual solution we have that

$$c(T^\lambda) + 3\lambda(k - |T^\lambda|) \le 3\,(\text{DS}_\lambda - \lambda(n - k))$$
$$\le 3 \cdot \text{DLR}(\lambda) \le 3 \cdot \text{OPT}.$$

Observe that for $\lambda = 0$ the algorithm could output a tree with at least k vertices (because of moats growing around vertices in W, see next subsection). In this case the resulting tree is a 3-approximation so we do not need the merging procedure described in Sect. 3. Otherwise, for some large λ, $e.g.$ the maximum cost of a vertex, the resulting tree would contain all the vertices. Therefore, we do the binary search for λ such that $|T^\lambda|$ is close to k. In a lucky event $|T^\lambda| = k$ and then we don't need the merging procedure described in Sect. 3. Otherwise, we obtain λ_1 and λ_2 such that $|T^{\lambda_1}| < k < |T^{\lambda_2}|$. By making enough steps of the binary search we can ensure that $\lambda_2 - \lambda_1 \le \frac{\varepsilon \cdot \text{OPT}}{3n}$. Let these trees be T_1 and T_2. Now, by setting $\alpha_1 = \frac{|T_2| - k}{|T_2| - |T_1|}$ and $\alpha_2 = \frac{k - |T_1|}{|T_2| - |T_1|}$ and using inequality (2) twice we have that

$$\alpha_1 c(T_1) + \alpha_2 c(T_2) \le 3\,(\alpha_1 \text{DS}_1 + \alpha_2 \text{DS}_2 - \alpha_1 \lambda_1 (n - |T_1|) - \alpha_2 \lambda_2 (n - |T_2|))$$
$$\le 3\,(\alpha_1 \text{DS}_1 + \alpha_2 \text{DS}_2 - \lambda_2 (n - k) + (\lambda_2 - \lambda_1)(n - |T_1|))$$
$$\le 3\,(\text{OPT} + (\lambda_2 - \lambda_1)n)$$
$$\le (3 + \varepsilon)\,\text{OPT},$$

where we used the fact that the convex combination of DS_1 and DS_2 is a feasible solution for $\text{DLR}(\lambda_2)$.

4.1 Moat Growing

In this section we describe the slight technical modification needed in the primal-dual algorithm for NW-PC-ST problem on planar graphs given in [4]. Observe, that there are two differences in the LPs used.

First, we have additional constraints and corresponding dual variables p_v. This is due to the fact, that in our setting all vertices can have both nonzero penalty and cost, while in the previous setting the reduction step was employed so that each vertex is a terminal with some penalty and zero cost or a Steiner vertex with zero penalty. However, this reduction step is equivalent to setting p_v to minimum of cost and penalty and defining the reduced costs and reduced penalties. This does not influence the approximation factor, nor the LMP guarantee. See also Sect. 2.1 of Sadeghian [19] for details.

The second modification comes from the fact that we have to include some guessed vertices W in the solution. However, it is enough to treat these vertices in the same way as terminals.

We now give a description of the resulting LMP primal-dual algorithm. First, we do the reduction of eliminating p_v variables as described above. This makes some vertices terminal and the other Steiner vertices. We also add all the guessed vertices to the set of terminals and set their penalty to infinite.

The algorithm maintains a set of moats, *i.e.*, a family of disjoint sets of vertices. In each step, these moats can be viewed as the components of the graph induced by the so far bought nodes. Each moat has an associated potential equal to the total penalty of vertices inside this moat minus the sum of the dual variables for all the subsets of this moat. The moats with positive potential are active, with an exception that the moat containing the root is always inactive.

The algorithm raises simultaneously the dual variables of all the active moats. For the growth of a moat we pay with its potential. We can have two events.

In the first event, some vertex goes tight, *i.e.*, the inequality for this vertex in the dual program becomes tight. In this case we buy this vertex and merge all the neighboring moats, setting the potential accordingly to the sum of all previous moats' potentials. We declare this new moat inactive whenever it contains a root vertex.

In the second event, some moat goes tight, *i.e.* the inequality in the dual program becomes tight for some set of vertices. This corresponds to the situation when the potential of this moat drops to zero. In this case we declare this moat inactive and we mark all the previously unmarked terminals inside it as marked with the current time. Observe that in the dual we do not have these inequalities for sets containing guessed vertices W. This means, that all the vertices of W will be connected to the root vertex.

We repeat this process until we do not have any active moats. Then we start a pruning phase. We consider all the bought vertices in the reverse order of buying. We delete a vertex v if the removal of v would not disconnect any unmarked terminal or any terminal marked with time greater than the time of buying the vertex v. We return the pruned set of bought vertices as the solution.

A straightforward adaptation of the analysis in [4] implies that the above algorithm run with initial penalty λ for all vertices in V' returns a tree T^λ satisfying inequality (2).

4.2   Generalization to Non-planar Graph Classes

Note that in our algorithm, we use planarity exclusively by exploiting that the LMP algorithm of Byrka et al. [4] for the prize-collecting version has ratio 3 on planar graphs. Their algorithm, however, can be executed on an arbitrary graph class (*e.g.* H-minor-free graphs). Thus all our calculations can be carried through by replacing 3 with any factor $r \geq 1$ thereby obtaining the following generalization.

Corollary 1. *The above algorithm has performance* $(4/3 + \varepsilon)r$ *for any graph class where the algorithm of Byrka et al.* [4] *has a performance ratio of* r.

5   Trying to Beat the Factor of 4: Relation to the Partial Cover

Here we draw connections to the recent work on the partial cover problems. Könemann et al. [13] showed how to obtain a $(4/3 + \varepsilon)r$-approximation algorithm for the partial cover problems using an r-approximate LMP algorithm for the corresponding prize-collecting version as a black-box. Their approach is roughly as follows. First, the most expensive sets from the optimum solution are guessed and all sets which are more expensive are discarded. Further, the black-box algorithm is used together with binary search to find two solutions, one, say S_1, feasible but possibly expensive, and the other, say S_2, infeasible but inexpensive. Then the merging procedure is employed to obtain a solution S_3. Finally, the cheapest solution of the S_1 and S_3 is returned.

5.1   Generalizing the Algorithm of Könemann et al.

Extending a folklore reduction from set cover type problems to node-weighted Steiner tree problems, we argue that our algorithm may be interpreted as a non-trivial generalization of the above-outlined algorithm by Könemann et al. [13].

First of all, the following reduction shows that the partial covering problem can be encoded as the quota node-weighted Steiner tree problem. The reduction creates for each element a vertex with zero cost and profit 1. Then, for each set it creates a node with the same cost and zero profit and connects it to the elements covered by this set. Finally, the root vertex is added and connected to all the set-corresponding nodes. The target quota profit is set to be the same as the requirement for the partial cover problem.

For such a reduced instance, we can run the preprocessing step from Sect. 2 which will remove the expensive sets (we could also employ the Könemann's preprocessing beforehand). Then, we would run any LMP algorithm for the prize-collecting cover problems within the Lagrangian relaxation framework which would indicate two families of sets to merge. Putting it on the reduced instance, these would correspond to two trees to merge. More precisely, take to the tree the set-corresponding nodes, the root vertex and the elements covered by sets. Now, we can apply the merging procedure described in the Lemma 3 with a slight adjustment needed to account for quota variant. In particular we modify the notion of cost-effectiveness to account profits instead of cardinalities and we also redefine the super-cardinality to be the sum of profits. To retrieve the solution from the tree, simply take the sets corresponding to non-zero cost nodes in the tree. Finally, output the cheaper of the two feasible solutions giving a partial cover with the same quality as the one by obtained via the algorithm by Könemann et al.

We remark that the above argument does not work in the reverse direction. The graph instances that are created have a very specific structure with three node layers ensuring that any partial cover solution is automatically connected at no additional cost. Achieving connectivity for *general* graphs, however, is not implied and guaranteeing this structural property without loss in the performance guarantee of the algorithm can be seen as a main contribution of our work.

Analogous arguments as in the result of Mestre [15] can be used to deduce the following.

Corollary 2. *For any $r > 1$ there is an infinite family of graphs where the natural moat growing algorithm for NW-PC-ST [4] has a ratio r but where any feasible solution to the NW-k-MST problem using only the nodes returned by this algorithm has cost at least $4/3 \cdot r$ times that of an optimum solution.*

The proof will be included in the full version of the paper.

Interpretation. In the edge-weighted case of k-MST, Garg [10] was able to carefully exploit the inner workings of the Goemans-Williamson algorithm [11] for the Lagrangian relaxation to match its ratio of 2. Corollary 2 means that our approach is in a certain sense optimal and that we would need to deviate from this framework to improve on the loss of factor $4/3$ in the tree-merging step. This could possibly be achieved by exploiting structural properties of the underlying graph class or using nodes outside the solution returned by the LMP algorithm.

Even when we exploit planarity it seems to be non-trivial to beat factor 4 along the lines of Garg [9,10]. The changes in the solutions by increasing initial potentials of vertices can be much larger than those in the edge-weighted variant. In particular, one can observe situations of node-flips in which two potentially distant vertices exchange their presence in the solution. Also, in contrast to edge-weighted variant, a single node can be adjacent to any number of moats and not only two. This in turn causes the large difference in two trees produced by the algorithm. In particular, the OLD vertices as described by Garg [9] can form any number of connected components which may be expensive to connect even when the graph is planar.

6   Conclusions and Comments

The $4 + \varepsilon$ approximation factor was obtained for the NW-k-MST problem on planar graphs. In the process we used the Lagrangian Relaxation technique. Our work can be interpreted as a generalization of a work on partial cover [13]. The result by Mestre [15] implies that our factor is essentially best possible using the underlying LMP algorithm for the NW-PC-ST as a black-box. It shows that one would have to exploit planarity in the merging process to beat factor 4.

Our ultimate hope would be to match the factor of 3 of the LMP algorithm. We think that the question of whether this is possible is very interesting and challenging.

Acknowledgements. We would like to thank Zachary Friggstad for initial discussions on the problem. The authors were supported by the NCN grant number 2015/18/E/ST6/00456.

References

1. Arora, S., Karakostas, G.: A $(2+\varepsilon)$-approximation algorithm for the k-MST problem. Math. Program. **107**(3), 491–504 (2006). https://doi.org/10.1007/s10107-005-0693-1

2. Bateni, M.H., Hajiaghayi, M.T., Liaghat, V.: Improved approximation algorithms for (budgeted) node-weighted steiner problems. In: Fomin, F.V., Freivalds, R., Kwiatkowska, M., Peleg, D. (eds.) ICALP 2013. LNCS, vol. 7965, pp. 81–92. Springer, Heidelberg (2013). https://doi.org/10.1007/978-3-642-39206-1_8

3. Berman, P., Yaroslavtsev, G.: Primal-dual approximation algorithms for node-weighted network design in planar graphs. In: Gupta, A., Jansen, K., Rolim, J., Servedio, R. (eds.) APPROX/RANDOM -2012. LNCS, vol. 7408, pp. 50–60. Springer, Heidelberg (2012). https://doi.org/10.1007/978-3-642-32512-0_5

4. Byrka, J., Lewandowski, M., Moldenhauer, C.: Approximation algorithms for node-weighted prize-collecting Steiner tree problems on planar graphs. In: Proceedings of the 15th Scandinavian Symposium and Workshops on Algorithm Theory (SWAT 2016), pp. 2:1–2:14 (2016). https://doi.org/10.4230/LIPIcs.SWAT.2016.2

5. Chekuri, C., Ene, A., Vakilian, A.: Node-weighted network design in planar and minor-closed families of graphs. In: Czumaj, A., Mehlhorn, K., Pitts, A., Wattenhofer, R. (eds.) ICALP 2012. LNCS, vol. 7391, pp. 206–217. Springer, Heidelberg (2012). https://doi.org/10.1007/978-3-642-31594-7_18

6. Chekuri, C., Ene, A., Vakilian, A.: Prize-collecting survivable network design in node-weighted graphs. In: Gupta, A., Jansen, K., Rolim, J., Servedio, R. (eds.) APPROX/RANDOM -2012. LNCS, vol. 7408, pp. 98–109. Springer, Heidelberg (2012). https://doi.org/10.1007/978-3-642-32512-0_9

7. Chudak, F.A., Roughgarden, T., Williamson, D.P.: Approximate k-MSTs and k-Steiner trees via the primal-dual method and lagrangean relaxation. Math. Program. **100**(2), 411–421 (2004). https://doi.org/10.1007/s10107-003-0479-2

8. Demaine, E.D., Hajiaghayi, M.T., Klein, P.N.: Node-weighted Steiner tree and group Steiner tree in planar graphs. ACM Trans. Algorithms **10**(3), 13:1–13:20 (2014). https://doi.org/10.1145/2601070

9. Garg, N.: A 3-approximation for the minimum tree spanning k vertices. In: Proceedings of the 37th Annual Symposium on Foundations of Computer Science (FOCS 1996), pp. 302–309 (1996). https://doi.org/10.1109/SFCS.1996.548489

10. Garg, N.: Saving an epsilon: a 2-approximation for the k-MST problem in graphs. In: Proceedings of the 37th Annual ACM Symposium on Theory of Computing (STOC 2005), pp. 396–402 (2005). https://doi.org/10.1145/1060590.1060650

11. Goemans, M.X., Williamson, D.P.: A general approximation technique for constrained forest problems. SIAM J. Comput. **24**(2), 296–317 (1995). https://doi.org/10.1137/S0097539793242618

12. Jain, K., Vazirani, V.V.: Approximation algorithms for metric facility location and k-median problems using the primal-dual schema and lagrangian relaxation. J. ACM **48**(2), 274–296 (2001). https://doi.org/10.1145/375827.375845

13. Könemann, J., Parekh, O., Segev, D.: A unified approach to approximating partial covering problems. Algorithmica **59**(4), 489–509 (2011). https://doi.org/10.1007/s00453-009-9317-0

14. Könemann, J., Sadeghabad, S.S., Sanità, L.: An LMP $O(\log n)$-approximation algorithm for node weighted prize collecting Steiner tree. In: Proceedings of the 54th Annual IEEE Symposium on Foundations of Computer Science (FOCS 2013), pp. 568–577 (2013). https://doi.org/10.1109/FOCS.2013.67

15. Mestre, J.: Lagrangian relaxation and partial cover. In: Proceedings of the 25th Annual Symposium on Theoretical Aspects of Computer Science (STACS 2008), pp. 539–550 (2008). https://doi.org/10.4230/LIPIcs.STACS.2008.1315

16. Moldenhauer, C.: Primal-dual approximation algorithms for node-weighted Steiner forest on planar graphs. Inf. Comput. **222**, 293–306 (2013). https://doi.org/10.1016/j.ic.2012.10.017

17. Moss, A., Rabani, Y.: Approximation algorithms for constrained node weighted Steiner tree problems. SIAM J. Comput. **37**(2), 460–481 (2007). https://doi.org/10.1137/S0097539702420474

18. Ravi, R., Sundaram, R., Marathe, M.V., Rosenkrantz, D.J., Ravi, S.S.: Spanning trees short or small. In: Proceedings of the Fifth Annual ACM-SIAM Symposium on Discrete Algorithms, 23–25 January 1994, Arlington, Virginia, pp. 546–555 (1994)

19. Sadeghabad, S.S.: Node-weighted prize-collecting Steiner tree and applications. Master's thesis (2013)

Exploring Sparse Graphs with Advice
(Extended Abstract)

Hans-Joachim Böckenhauer[1]([⊠]), Janosch Fuchs[2], and Walter Unger[2]

[1] Department of Computer Science, ETH Zürich, Universitätstrasse 6,
8092 Zürich, Switzerland
hjb@inf.ethz.ch

[2] Computer Science 1, RWTH Aachen University, Ahornstr. 55,
52074 Aachen, Germany
{fuchs,quax}@algo.rwth-aachen.de

Abstract. Moving an autonomous agent through an unknown environment is one of the crucial problems for robotics and network analysis. Therefore, it received a lot of attention in the last decades and was analyzed in many different settings. The *graph exploration problem* is a theoretical and abstract model, where an algorithm has to decide how an agent, also called *explorer*, moves through a network with n vertices and m edges such that every point of interest is visited at least once. For its decisions, the knowledge of the algorithm is limited by the perception capacities of the explorer. We look at the *fixed-graph scenario* proposed by Kalyanasundaram and Pruhs (ICALP, 1993), where the explorer starts at a vertex of the network and sees all reachable vertices, their unique names and their distance from the current position.

Because the algorithm only learns the structure of the graph during computation, it cannot deterministically compute an optimal tour that visits every vertex at least once without prior knowledge. Therefore, we are interested in the amount of crucial a-priori information needed to solve the problem optimally, which we measure in terms of the well-studied model of *advice complexity*. Here, a deterministic algorithm can at any time access a binary advice tape written beforehand by an oracle that knows the optimal solution, the graph and the behavior of the algorithm. The number of bits read by the algorithm until the end of computation is called the *advice complexity*.

We look at the graph exploration problem on unknown directed graphs and focus on cyclic solutions. It is known that $\mathcal{O}(n \log n)$ bits of advice are necessary and sufficient to compute an optimal solution, for general graphs. In this work, we present algorithms with an advice complexity of $\mathcal{O}(m)$, thus improving the classical bound for sparse graphs.

Keywords: Graph exploration · Advice complexity
Fixed-graph scenario

This work is supported by the German research council (DFG) Research Training Group 2236 UnRAVeL.

L. Epstein and T. Erlebach (Eds.): WAOA 2018, LNCS 11312, pp. 102–117, 2018.
https://doi.org/10.1007/978-3-030-04693-4_7

1   Introduction

Orientation and navigation in an unknown environment is one of the basic tasks for autonomous agents. The environment can be physical or virtual like in a network of computers, where the connections between the computers are unknown. To send messages as fast as possible, it is helpful to know the structure of the network. Thus an explorer can be used to visit and test the connections for every computer in the network. A more physical example is the field of robotics. There are many applications for robots that explore unknown environments on their own [23, 27, 28]. For example, exploring caves or abandoning mines is often a dangerous task and autonomous robots can be used to create maps of such environments.

Because there are many different applications for exploring, there are also many different models for the environment and for the perception of the explorer. The survey of Berman [3] gives an overview of navigation problems and distinguishes the following main properties: The representation of the environment, the task that should be solved, and the senses of the explorer. The environment can be a geometric space with obstacles [4] or, this is the case that we analyze in this paper, an abstract and discrete space, where the explorer can move from one point to a neighboring one, i.e., a graph consisting of vertices and edges. Starting from a vertex in the given graph, the task can be to find a shortest path to a target vertex or to compute a shortest tour that visits every vertex at least once.

The task to compute a shortest tour that visits every vertex at least once is related to the well-known *Traveling Salesman Problem*. Kalyanasundaram et al. introduced the graph exploration problem as an online version of the TSP [20] with the *fixed-graph scenario* which defines the senses of the explorer. In this model, the vertices have unique labels, and the explorer sees all reachable vertices, their labels and their distances to the current position. Moving onto a new vertex reveals the adjacent vertices and the explorer recognizes if a vertex was already reachable from a previous step.

Obviously, not having complete knowledge of the graph beforehand makes it impossible in general for the explorer to find a tour of optimal length. Thus, algorithms achieving some provable approximation guarantees have been investigated. The best known lower bound on the approximation ratio for exploring general and undirected graphs in the fixed-graph scenario is $\frac{5}{2} - \epsilon$ [11]. For the special case of undirected graphs with bounded genus g, an upper bound of $16(1 + 2g)$ is known [25]. The case of directed graphs in the fixed-graph scenario is also well studied [1, 14, 15]. In [15], the authors give tight bounds for deterministic and randomized graph exploration in directed graphs with weighted or unweighted edges. Moreover, they look at a variation of the problem where the explorer has to search for a specific vertex in the graph. There are many slight variations of the graph exploration problem. The memory of the algorithm [10, 16], the number of explorers in the graph [7–9] or the abilities to set pebbles [2] are well studied variations.

For a more fine-grained analysis of how much information about the unknown graph is really needed by the explorer, we look at a variation of the graph exploration problem where the algorithm has access to some information in the form of a bit string, provided by a helpful oracle that knows the network. The number of bits that the algorithm reads until it finishes its computation is then called its *advice complexity*. We are interested in how many bits are needed to compute an optimal exploration sequence. Dobrev et al. [12] introduced this model in the setting of online algorithms, which was later improved by Hromkovič et al. [19] and Böckenhauer et al. [6] as well as by Emek et al. [13]. The setting and many results are explained in detail in [21]. We use the model of Böckenhauer et al. [6] in this paper.

The first time that the graph exploration problem was analyzed using the advice complexity model was in [17], where Fraigniaud et al. were able to improve the classical upper bound of 2 on the competitive ratio for tree exploration by adding advice. They proved that $\log \log(D) - c$ bits of advice suffice for c-competitiveness, where D is the diameter of the input tree and $c < 2$. Moreover, they showed that every algorithm that uses less advice bits has to be at least 2-competitive.

Since then, there have been many results regarding graph searching problems with advice [11,18,22]. The search for a specific vertex in the graph stands in focus of research in [22]. The authors present an algorithm that uses $\Theta(n/r)$ bits of advice for a competitive ratio r. In [11], Dobrev et al. looked at the trade-off between advice and competitiveness for the cyclic graph exploration problem. Moreover, they show that $\Omega(n \log n)$ bits of advice are necessary for optimality. Gorain et al. [18] show bounds for a weaker oracle model, where the oracle does not know the starting position of the algorithm. Moreover, they show that $\mathcal{O}(n)$ advice bits results in a solution of quadratic size.

In the following, we will prove that $\mathcal{O}(m)$ bits of advice suffice to compute an optimal solution. Note that an upper bound of $\mathcal{O}(n \log n)$ advice bits can be easily achieved by sorting the vertices by their first visits and encoding this order. There also exists a lower bound of $\Omega(n \log n)$ advice bits for general graphs [24]. These (almost) tight bounds motivate the investigation of special graph classes. In this paper, we present an improvement over the general strategy for sparse graphs, i.e., for graphs with $o(n \log n)$ edges. Note that we can assume that $m \geq n - 1$, otherwise the graph would not be connected, making its exploration impossible. We first concentrate on the case where the algorithm should compute a cyclic tour visiting all vertices on a directed graph that is unknown and has bounded in- and outdegree. After that, we show how the problem for unbounded degree graphs can be solved by some modification of the algorithm. Among the different possible settings, the a-priori knowledge of the number of vertices and how they are connected by the edges has the largest impact on the advice complexity. Only the cost values are unknown in the known graph model. Since our algorithm relies on the encoding of an optimal solution within the given advice and does not take the edge costs into account, we formulate our results for the more general case of arbitrary edge weights only.

The paper is organized as follows. In Sect. 2, we give the basic definitions for dealing with the graph exploration problem. Section 3 gives some basic observations and, in Sect. 4, we prove our main result for directed graphs with in- and outdegree bounded by 2. This result is extended to directed graphs of arbitrary degree in Sect. 5. Due to space restrictions, we refer the reader to the technical report [5] for full details.

2   Basic Definitions

We start with a definition of the basic variant of the graph exploration problem that we consider in this paper.

Definition 1. *Let $G = (V, E)$ be a directed graph. Every vertex $v \in V$ has a fixed unique identifier. There is an agent, called* explorer, *initially positioned on some start vertex $v_0 \in V$. The algorithm has to move this explorer along the directed edges of G to visit all vertices and return to v_0. The edges are weighted by a cost function cost: $E \to \mathbb{N}$, and the goal is to minimize the total cost along the cyclic tour traveled by the explorer. In every vertex the explorer is located, it sees the outgoing edges, their costs, and the vertex identifiers at the endpoints of these edges, but not the incoming edges.*

Throughout the paper, we denote, for any graph $G = (V, E)$ with vertex set V and edge set E, the number of vertices by $n = |V|$ and the number of edges by $m = |E|$. We continue with some notations for the solutions or search sequences computed by online and offline algorithms.

Definition 2. *Let $G = (V, E)$ be a directed graph. The* out-neighborhood *of a vertex $v \in V$ is defined as $N_{out}(v) = \{w \mid (v, w) \in E\}$. Analogously, we define the* in-neighborhood *of a vertex v as $N_{in}(v) = \{w \mid (w, v) \in E\}$.*

We describe the tour followed by the explorer in terms of a *search sequence*.

Definition 3. *Let $G = (V, E)$ be a graph. A sequence $S = (v_0, v_1, \ldots, v_s)$ is called a* search sequence *if $(v_{i-1}, v_i) \in E$ for all $1 \leq i \leq s$. If $v_0 = v_s$, we call S a* cyclic search sequence.

For a search sequence $S = (v_0, \ldots, v_k)$ with $e_i = (v_{i-1}, v_i)$, for $1 \leq i \leq k$, we denote by $E(S) = (e_1, e_2, \ldots, e_k)$ the sequence of edges in S and by $V(S) = \{v_0, \ldots, v_k\}$ the set of vertices in S. We also interpret $E(S)$ as a multiset of edges and thus write $e \in E(S)$ if there exists some i with $e = e_i$. The cost of a search sequence $S = (v_0, \ldots, v_k)$ is defined by $cost(S) = \sum_{e \in E(S)} cost(e) \cdot \#_S(e)$.

The search sequence is determined by the algorithm as follows. In each step, the explorer is located at some vertex v and the algorithm chooses one of the vertices from $N_{out}(v)$ as target and moves the explorer towards it. As soon as the explorer arrives at the target vertex, a new round starts and the algorithm again receives the unique identifiers for the out-neighborhood and has to make an unrecoverable decision.[1] The goal is to compute a cyclic search sequence visiting each vertex at least once, we call such a sequence an *exploration sequence*.

Since the algorithm lacks global information about the structure of the graph, there is no deterministic algorithm that finds an optimal exploration sequence for any graph. We employ the model of online algorithms with advice as defined in [6,19] for measuring the amount of missing information, which we can define in the framework of graph exploration as follows.

An *online algorithm with advice* computes for an unknown graph a search sequence $S = (v_0, v_1, \ldots, v_{end})$, where v_i is computed from $N_{out}(v_0), \ldots, N_{out}(v_{i-1})$ (i.e., from the partial knowledge about the graph gathered in the first $i - 1$ rounds) and the content ϕ of the advice tape, i.e., an unbounded binary sequence of *advice bits* computed by an *oracle* that sees the complete input graph together with its edge-cost function. An online algorithm with advice *solves* the graph exploration problem if, for any input $(G, cost)$, there exists a computable advice ϕ such that S is an optimal exploration sequence. The algorithm has sequential access to the bits from the advice tape, and its *advice complexity* is the number of accessed advice bits. As usual, we measure the advice complexity with respect to the input size by considering a worst-case input of the respective size.

We now define how the algorithm makes a decision to extend a search sequence S, based on some advice from the oracle. The basic idea is that the oracle chooses some fixed optimal exploration sequence S^* and communicates a sufficient amount of information about it such that the algorithm can compute this sequence without taking the costs of the edges into consideration. We observe that it is sufficient for the algorithm to know the exact number of traversals for each edge in $E(S^*)$. Explicitly communicating these traversal numbers could be done with $\mathcal{O}(m \log n)$ advice bits in a straightforward way since no edge can be traversed more than n times in an optimal exploration sequence. But this would be too expensive, so the oracle only communicates some partial information from which the algorithm can compute the traversal numbers. As a first step, we partition the edges into three sets according to their number of traversals (none, one, or multiple times) in an optimal exploration sequence.

To this end, we use the following notation. Let $G = (V, E)$ be a graph, let S^* be an optimal exploration sequence and let S be an arbitrary search sequence. Then $\#_S(e)$ is the number of traversals through an edge $e \in E$ in a search sequence S, and E_0, E_1, and E_{Multi} are the sets of edges in E which are visited 0, 1, and multiple times by S^*, respectively. We denote the set of edges in E which are visited at least once by S^* by $E_{used} = E_{Multi} \cup E_1$. If $\#_S(e) = \#_{S^*}(e)$

---

[1] Note that, with each decision, the algorithm influences the new input for the next decision. Thus, strictly speaking, the graph exploration problem is no classical online problem. But the adversary still knows the behavior of the deterministic algorithm and can, with this knowledge, prepare the input graph, the unique identifiers for the vertices, and thus the enumeration of the edges. Hence, we can analyze the graph exploration problem using the same methodology as used for online problems.

holds, e.g., an edge e is as often used in S as in the fixed optimal solution S^*, we say e is *exhausted* in S.

The number of traversals for the edges could differ for different optimal solutions, but the oracle fixes one of the optimal solutions such that the given advice is consistent during the exploration. Figure 1(a) shows a sample graph where the number of traversals for the edges in an optimal solution is non-unique. The five traversals over the vertex x needs to be split up between the two paths (y, v_1, x) and (y, v_2, x).

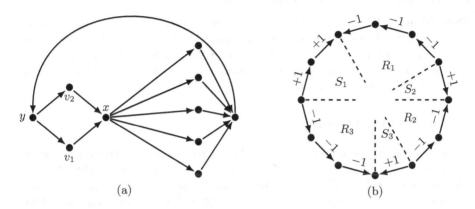

(a)   (b)

Fig. 1. (a) The optimal number of traversals for the edges of this graph is non-unique. The five successors of x and the two possible paths to x require that the algorithm traverses (y, v_1, x) or (y, v_2, x) multiple times. (b) A cycle of multi-edges in an exploration sequence.

Knowing, for every edge e, its membership to the three sets, the traversal number is unknown only for the edges from E_{Multi}. Because the edges from E_{Multi} need additional advice to compute their exact number of traversals, the oracle is interested in fixing a solution that minimizes the number of edges from E_{Multi}. For the graph presented in Fig. 1(a), the oracle would chose a solution where either the edges (y, v_2) and (v_2, x) or the edges (y, v_1) and (v_1, x) are used exactly once. Such a solution sets the number of edges from E_{Multi} to three, instead of five.

To ease the analysis of our algorithm, the oracle constructs a graph G' from the given graph G such that the optimal solution S^* for G' is unique in the sense that, for all other optimal solutions S', $\#_{S^*}(e) = \#_{S'}(e)$. The oracle constructs G' by perturbing the costs on the edges by a small amount such that, for all x and y, only one of two paths $p_1 = (y, \ldots, x)$ and $p_2 = (y, \ldots, x)$ with equal costs is used multiple times. This minimizes the number of edges from E_{Multi} and fixes the traversal number for every edge. The following lemma guarantees that we can choose a sufficiently small perturbation such that S^* is also optimal for G.

Lemma 1. *Let S^* be an optimal exploration sequence for a graph with n vertices. Then $\#_{S^*}(e) \leq n$, for all $e \in E(S^*)$.*

From now on, we assume that the optimal exploration sequence S^* fixed by the oracle is unique.

If the input graph has a solution that visits every vertex at least once, it must be strongly connected. Because the connectivity of the graph alone does not reflect the current position of the explorer and the already exhausted edges, we introduce the term of an *expandable search sequence* to formulate a stronger and more precise connectivity requirement.

Definition 4. *We call a search sequence $S = (v_0, \ldots, v_k)$ expandable with respect to a fixed optimal exploration sequence $S^* = \{v_0, \ldots, v_{end}\}$ if there exists an outgoing edge $e = (v_k, w)$ that is less used than in S^* and there is a search sequence from v_k that uses only non-exhausted edges and reaches the final vertex v_{end} such that $cost(S^*) = cost(S)$ and $V(S) = V$.*

3   Structural Observations

In this section, we focus on the structural properties of the fixed optimal solution S^*. We distinguish between two different graphs that can be induced by S^*. If we remove the edges from E_0 and just look at the graph $G_{S^*} = (V, E_{used})$, we have a graph for which all edges are needed in an optimal exploration sequence. If we are more interested in traversed subsequences and want to distinguish different traversals over the same edge from E_{Multi}, we look at the multigraph $M_{S^*} = (V, E_{used}, \#_{S^*})$. In M_{S^*}, every edge $e = (v, u) \in E_{Multi}$ is replaced by $\#_{S^*}(e)$ many edges that point from v to u. The multigraph $M_{S^*} = (V, E_{used}, \#_{S^*})$ is an Eulerian graph. Let $d_{in}(v)$ denote the indegree of the vertex v in M_{S^*}.

The following lemmas describe structural properties of optimal exploration sequences which will be helpful for deducting the traversal numbers of the edges from a rather small amount of advice.

Lemma 2. *Let $G = (V, E)$ be a connected graph, and let S^* be the optimal exploration sequence for G. A vertex v is part of $d_{in}(v)$ cycles $c_i = (v, \ldots, v) \in S^*$ with $0 \leq i \leq d_{in}(v)$, and at least $d_{in}(v) - 1$ cycles are interchangeable, i.e., changing their order in S^* does not change the cost of S^*.*

Lemma 3. *Let $G = (V, E)$ be a graph with the optimal exploration sequence S^*, then there is no vertex pair v, w with $(v, w) \in E_1$ and $(w, v) \in E_{Multi}$.*

Lemma 4. *Let $G = (V, E)$ be a graph with the optimal exploration sequence S^*. Let $S = (v_0, \ldots, v_{end})$ be a simple cyclic contiguous subsequence of S^*. Then there is an edge $e \in S(E)$ with $\#_{S^*}(e) = 1$.*

Proof. Assume that $\#_{S^*}(e) > 1$ holds for all edges in $E(S)$. We now define a traversal function on $c: E \to \mathbb{N}_0$ with respect to the optimal solution S^*:

$$c(e) = \begin{cases} \#_{S^*}(e) - 1 & \text{for } e \in E(S), \\ \#_{S^*}(e) & \text{otherwise.} \end{cases}$$

Note that $\#_{S^*}(e) - 1 > 0$ for $e \in E(S)$, thus the multigraph $M = (V, E, c)$ is still connected. Furthermore, the condition $d_\Delta(v) = 0$ also holds for all vertices $v \in V$. Thus, the graph $M = (V, E, c)$ is Eulerian and has an exploration sequence S' with lower cost than the exploration sequence S^*. This is a contradiction to the minimality of the cost of S^*. □

Our next observation regarding the edges from E_{Multi} is that they cannot even form a cycle in the underlying undirected graph. For a given optimal exploration sequence S^*, we look at the underlying undirected graph and separate it into cycles. For each hypothetical cycle of multi-edges, we summarize equally directed edges into sequences S_i and R_i, as presented in Fig. 1(b). If we increment the traversal numbers for the sequences S_i and decrease them for the sequences R_i, the resulting graph is still Eulerian. This leads to the following lemma.

Lemma 5. *Let $G = (V, E)$ be a graph with the optimal exploration sequence S^*. Let $S_i = (v^i, \dots, u^i)$ and $R_i = (w^i, \dots, u^i)$, with $1 \le i \le k$, be $2k$ simple search subsequences of S^* with $w^i = v^{i+1}$ ($1 \le i < k$) and $v^1 = w^k$. Then there is an edge $e \in S_i \cup R_i$ with $\#_{S^*}(e) = 1$.*

We showed in Lemmas 3 to 5 that the information about the usage of edges in an unique optimal solution S^* provides some structural information. Because the graph induced by E_{Multi} is cycle-free, it is a forest and we will from now on call it F. This holds even for the underlying undirected graph. As next step, we show that it is possible to derive the exact traversal number for every edge from this induced structure.

Lemma 6. *Let $G = (V, E)$ be a graph with the optimal exploration sequence $S^* = (v_0, \dots, v_{end})$. There is an algorithm $\mathcal{A}(V, E_0, E_1, E_{Multi})$ which computes, for each edge $e \in E$, the number of visits by S^*.*

4   Unknown Directed Graphs with In- and Outdegree Two

While Lemma 6 immediately yields an algorithm with advice for the case that the graph structure is known beforehand, it can, in unknown graphs, only be applied after all edges are explored. Therefore, we now informally explain an extended algorithm that requires that every vertex in the input graph has at most two incoming and also at most two outgoing edges. In Sect. 5, we will show how every graph can be transformed into such a degree-bounded graph. We now give a high-level description of an algorithm with advice that computes the precise number of traversals while traversing such a given degree-bounded graph.

As soon as the explorer is located in a vertex v, the algorithm gets the identifiers of v and of all out-neighbors $N_{out}(v)$ of v. If a neighbor $w \in N_{out}(v)$ was already the out-neighbor of a previously visited vertex, the algorithm recognizes this vertex. The vertices $N_{in}(v)$ that lead to v and their corresponding edges stay hidden as long as the explorer is positioned at v. The algorithm accesses the advice tape to overcome this lack of information.

On the first visit of v, the algorithm reads advice bits to classify the outgoing edges into the three sets E_0, E_1, or E_{Multi} and to know the number of incoming edges adjacent to v and their membership in the sets E_0, E_1, or E_{Multi}. So, the algorithm knows, for every visited vertex, all incoming and outgoing edges, and whether they are used once, more than once, or never in an optimal exploration sequence. The edges from the set E_0 are ignored in the further process, so they are not involved in computing the exploration sequence. Therefore, every vertex in G_{S*} looks like one of the vertices shown in Fig. 2.

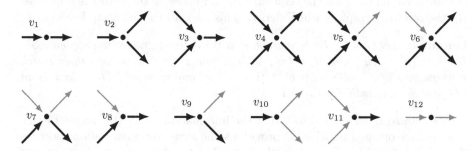

Fig. 2. All possible edge configurations for G_{S*} with in- and outdegree bounded by two. The thick black edges are from E_{Multi}. The gray edges are from E_1.

Our algorithm will prefer a path that is relatively often used and will avoid to exhaust an edge. As long as we know that the explorer does not use an edge for the last time, we know that the incident vertices are visited again. Thus, the algorithm will choose one of many exchangeable cycles and does not make any mistake. If the explorer stands on a vertex with only one outgoing edge, the algorithm does not need to make a decision because the only valid move is to use this edge.

Therefore, we assume that the explorer is positioned at a previously unvisited vertex where the algorithm has to decide between two edges which were not used in some previous step. If there is only one edge of E_{Multi}, the algorithm will prefer it without knowing its number of traversals. But, if both outgoing edges are used multiple times, the lack of precise traversal numbers makes it impossible for the algorithm to choose the more often used edge. Thus, the algorithm accesses the advice tape to compare which edge is more often used in the optimal exploration sequence.

Definition 5. *Let G be a graph where every vertex has at most two outgoing and two incoming edges and let S^* be the optimal exploration sequence. For a vertex v with two outgoing edges e_1 and e_2 from E_{Multi} we call e_1 light if $\#_{S^*}(e_1) < \#_{S^*}(e_2)$. The other edge e_2 will be called heavy. If v has only one outgoing edge from E_{Multi}, it is also called heavy. Analogously, we define for one* *or two incoming multi-edges of v one edge as heavy and, if a second edge exists, the fewer traversed edge as light.*

If both edges are equally often used, the oracle can decide arbitrarily which edge is heavy. The algorithm will ask for the exact number of traversals for the light edges and moves the explorer along the heavy one.

Thus, the algorithm moves the explorer along a path of edges for which the precise number of traversals is unknown, but larger than one. But, the algorithm knows the exact number of traversals for all the *light* edges (which are not yet used for leaving the vertex for the first time). From Lemma 5, we know that the path of multi-edges, which are preferred by the algorithm, are part of a tree T in the forest $F = (V, E_{Multi})$. When the explorer reaches a leaf $w \in T$, the algorithm knows how often w will be visited, because w has only outgoing edges from E_1, like v_{10} or v_{12} in Fig. 2.

Additionally, due to the bounded degree and the Eulerian property, w can have only one incoming edge $e \in E_{Multi}$ and therefore, the algorithm knows the exact number of outgoing traversals for e is exactly 2. This number of traversals can be used to compute the number of traversals for preceding edges, similar to Lemma 6. For every preceding vertex with two outgoing edges, there was at most one incoming and one outgoing edge for which the number of traversals was unknown. Since, the number of traversals for the outgoing heavy edge can be computed, there is only one incident edge with an unknown number of traversals left. Thus, the algorithm can compute the number of traversals for all heavy edges adjacent to the traversed path in a bottom-up approach as soon as the explorer reaches a leaf $w \in T$. So, when a vertex is visited for the second time, the algorithm knows, for all adjacent edges, their precise number of traversals in the optimal solution.

Often, the algorithm just chooses between some interchangeable cycles, but it can happen that the graph has some critical edges whose early traversal would prevent the algorithm from completing an optimal exploration sequence. If the algorithm is in the situation to choose between two edges with only one traversal remaining (or two edges from E_1), it asks for advice to know which edge should be used as *last* edge to leave the vertex. We say that such an edge is important to sustain the expandability (see Definition 4) of the current exploration sequence.

The correctness of this procedure is immediately implied by the following theorem.

Theorem 1. *Let $G = (V, E)$ be a graph with the optimal exploration sequence $S^* = (v_0, \ldots, v_{end})$ and let $S = (v_0, \ldots, v_p)$ be an expandable prefix of it such that, for all $v \in V(S^*)$ the algorithm knows the edge through which v is left for the last time in S^*, and, for all $e \in E(S^*)$, it knows whether $\#_{S^*}(e) - \#_S(e) > 1$ or $\#_{S^*}(e) - \#_S(e)$ or $\#_{S^*}(e) - \#_S(e) = 0$. Then S can be extended to an exploration sequence $S' = (v_0, v_p, v_{p+1}, \ldots, v_{end})$ with $cost(S^*) = cost(S')$.*

It remains to calculate the necessary amount of advice: Every vertex in the input graph has at most two outgoing and two incoming edges. The algorithm asks for advice to classify every adjacent edge, also the invisible incoming ones, into one of the three sets E_0, E_1 or E_{Multi}. For every vertex v in $G_{S^*} = (V, E_{used})$ that has two outgoing or incoming edges from E_{Multi}, the algorithm asks which edge is light. For every light edge, it asks for the number of traversals. Additionally, if v has two outgoing edges from E_{used}, the algorithm asks which edge is *last*. Because the algorithm traverses a path of heavy edges which must end at a vertex with known number of traversals, it can compute the number of traversals for the heavy edges in a bottom-up way.

Lemma 7. *The algorithm can learn, for every edge, its membership to the sets* E_0, E_1, *or* E_{Multi} *on the first visit of an incident vertex using* $n + \log(3)m$ *advice bits overall.*

The n bits are used to learn, for all vertices, whether they have one or two incoming edges. Using standard coding schemes (see, e.g., [26]), the m one-out-of-three decisions for the edges can be communicated by amortized $\log(3)$ bits each.

All edges from the set E_0 are not considered for any further decision of the algorithm and are subsequently ignored. The information about heavy and light edges and the last edges can be communicated using $\mathcal{O}(n)$ bits.

Lemma 8. *For a given graph* $G_{S^*} = (V, E_{used})$ *with in- and outdegree bounded by 2, with an optimal exploration sequence* S^*, *the algorithm needs at most* $3n$ *bits of advice to learn about the light incoming and outgoing edges and which outgoing edge is* last *in* S^*, *for all vertices.*

The next step of the algorithm is to ask, for every light edge, its exact number of traversals in the optimal solution S^*. We claim that this information can be communicated with linearly many bits in the number of edges.

Lemma 9. *For a given graph* $G_{S^*} = (V, E_{used})$ *with in- and out-degree bounded by 2, with an optimal exploration sequence* S^*, *the algorithm needs at most* $5m$ *advice bits to learn the exact traversal numbers for all multi-edges.*

The technically involved proof of Lemma 9 is based on the following ideas. If there exists a light edge $e = (x, y)$ which is traversed many times, this implies that the subtree of multi-edges rooted at y contains many light edges with small traversal numbers.

To illustrate the idea, we assume for the moment that a tree T of outgoing multi-edges is a full binary tree of height h with no additional incoming edges as shown in Fig. 3(a). Then T has $n' = 2^h$ leaves, and, for all inner vertices $v \in T$, both outgoing edges have the same traversal number. Thus, the single light edge at the root is traversed n' times. In the next level, there are two light edges, which are traversed $n'/2$ times each. Adding all up, we need at most $\sum_{i=1}^{h} 2^{i-1} \log(n'/2^{i-1}) \leq 2n'$ advice bits for encoding the traversal numbers of all light edges.

For an arbitrary tree of multi-edges, the incoming number of traversals for a vertex v has to be split between the two outgoing multi-edges as shown in

Fig. 3(b). Analyzing this general structure leads to a more complicated recursive function, which can be shown to be bounded by $5m/2$. The same number of bits can be used to encode the traversal numbers for all incoming light edges.

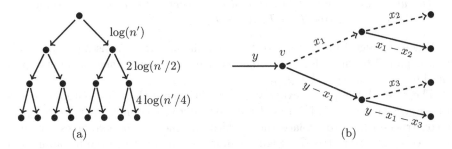

(a)   (b)

Fig. 3. The incoming number of traversals y at the vertex v needs to be split up between the two outgoing multi-edges. This continues recursively downwards in the tree. An edge is dashed if it is light.

Lemmas 7 to 9 imply the following bound on the advice complexity.

Theorem 2. *There exists an online algorithm which solves the graph exploration problem using $4n + (\log(3) + 5)m$ bits of advice on a given unknown directed graph $G = (V, E, \text{cost})$ with in- and outdegree bounded by 2.*

5   General Unknown Directed Graphs

We now explain how our algorithms can be adapted to solve the graph exploration problem on general unknown directed graphs. To count the number of advice bits more easily, we transform the given unbounded graph G into a graph H with in- and out-degree bounded by 2, which has an increased number of vertices and edges. To be more precise, H is implicitly constructed from $G_{S^*} = (V, E_{used})$ during the traversal. The algorithm and the oracle agree on this construction and the oracle provides the advice for H. One step in G_{S^*} will be represented by a sequence of steps in H.

To construct H, the algorithm needs to know the number of incoming edges. The approach from Lemma 7 only works if we know that there are either one or two incoming edges. If their number is between 1 and $n - 1$, the algorithm starts to classify the incoming edges with advice until it reads a delimiter that tells the algorithm that there are no more incoming edges.

Lemma 10. *The algorithm can learn for every edge its membership to the sets E_0, E_1, or E_{Multi} on the first visit of an incident vertex using $2(n + m)$ advice bits overall.*

For constructing the graph H, we introduce the terms *compact out-tree* and *compact in-tree*, which are used to replace a large number of directed edges.

Definition 6. *A* compact out-tree $T_{out}(v)$ *is a directed binary tree, directed from the root v to the leaves, that minimizes the maximum distance between v and some leaf. Analogously, we define a* compact in-tree $T_{in}(v)$, *which is directed from the leaves to the root v.*

The union of a compact in-tree and a compact out-tree, with the same root v, is called an in-out-tree $T_v = T_{out}(v) \cup T_{in}(v)$.

When the algorithm visits a vertex v for the first time, it uses the approach from Lemma 7 to know, for every edge incident to v, about their membership to the three sets E_0, E_1 and E_{Multi}.

The algorithm replaces every vertex v with more than two outgoing edges from E_{used} with $T_{out}(v)$. If v has also more than two incoming edges from E_{used}, it constructs $T_{in}(v)$ to replace these edges and merges the two trees. So, v is replaced by an in-out-tree T_v. Figure 4 shows a vertex with eleven incoming and nine outgoing edges and the resulting in-out-tree $T_v = T_{out}(v) \cup T_{in}(v)$. Note that all newly created edges in an in-out-tree, drawn in gray in Fig. 4, are multi-edges by construction. Thus, the algorithm does not need to read additional advice to know their membership to the three sets.

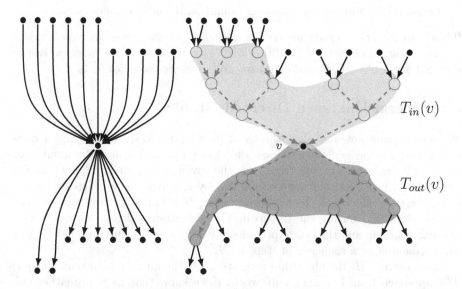

Fig. 4. A vertex of high degree is replaced by an in-out-tree. The gray vertices and dotted edges are virtually added and not part of G_{S^*}. The black edges incident to the gray leaves of the in-out-tree represent the edges from G_{S^*}.

The construction of H itself does not cost advice bits, but the increased number of edges and vertices influences the number of advice bits used overall. Therefore, we now analyze how the number of vertices and edges changes.

Lemma 11. *A graph $G = (V, E)$ with an optimal exploration sequence S^* can be transformed into a graph $H = (V', E')$, with $|V'| = n'$ and $|E'| = m'$, such that every vertex has an outgoing and incoming degree of at most 2 by replacing vertices v with an in-out-tree T_v.*

The graph $H = (V', E')$ is bounded in the number of vertices and edges by $n' \le 2m$ and $m' \le 3m$.

Although the construction increases the number of vertices and edges, the computed solution for H can be easily modified to obtain a solution for G.

Lemma 12. *Let G be an arbitrary directed graph with an optimal exploration sequence S^*. Let H be the degree-bounded graph resulting from the construction described in Lemma 11. An algorithm computing an optimal exploration sequence S for the transformed graph H can be used to compute an optimal exploration sequence S' for G.*

Note that the construction visualizes the binary search approach that the algorithm uses to compute the exact number of traversals for some edges. The important information, like which edge is used for the last traversal or which edge is light, can now be communicated with only one bit per (virtual or real) edge such that we can apply Theorem 2. Obviously, we also increased the number of vertices by transforming G into H and therefore it is also increased how often the explorer asks for such information. Therefore, we need to compute how transforming G into H increases the needed advice, if we apply the same approach as in Theorem 2.

Theorem 3. *There exists an online algorithm which solves the graph exploration problem using $2n + 23m$ bits of advice on a given unknown directed graph $G = (V, E, cost)$.*

6   Conclusion

In this work, we focused on the cyclic graph exploration problem on directed graphs with weighted edges. The presented algorithm can be modified to work on undirected graphs and also to solve the non-cyclic graph exploration problem. The advice complexity decreases if the algorithm already has some a-priori knowledge, e.g., about the graph structure without the edge weights.

References

1. Albers, S., Henzinger, M.R.: Exploring unknown environments. SIAM J. Comput. **29**(4), 1164–1188 (2000)
2. Bender, M.A., Fernández, A., Ron, D., Sahai, A., Vadhan, S.: The power of a pebble: exploring and mapping directed graphs. Inf. Comput. **176**(1), 1–21 (2002)
3. Berman, P.: On-line searching and navigation. In: Fiat, A., Woeginger, G.J. (eds.) Online Algorithms. LNCS, vol. 1442, pp. 232–241. Springer, Heidelberg (1998). https://doi.org/10.1007/BFb0029571

4. Blum, A., Raghavan, P., Schieber, B.: Navigating in unfamiliar geometric terrain. SIAM J. Comput. **26**(1), 110–137 (1997)
5. Böckenhauer, H.J., Fuchs, J., Unger, W.: The graph exploration problem with advice. CoRR abs/1804.06675 (2018)
6. Böckenhauer, H.-J., Komm, D., Královič, R., Královič, R., Mömke, T.: On the advice complexity of online problems. In: Dong, Y., Du, D.-Z., Ibarra, O. (eds.) ISAAC 2009. LNCS, vol. 5878, pp. 331–340. Springer, Heidelberg (2009). https://doi.org/10.1007/978-3-642-10631-6_35
7. Brass, P., Vigan, I., Xu, N.: Improved analysis of a multirobot graph exploration strategy. In: 2014 13th International Conference on Control Automation Robotics & Vision (ICARCV), pp. 1906–1910. IEEE (2014)
8. Chalopin, J., Flocchini, P., Mans, B., Santoro, N.: Network exploration by silent and oblivious robots. In: Thilikos, D.M. (ed.) WG 2010. LNCS, vol. 6410, pp. 208–219. Springer, Heidelberg (2010). https://doi.org/10.1007/978-3-642-16926-7_20
9. Das, S., Flocchini, P., Kutten, S., Nayak, A., Santoro, N.: Map construction of unknown graphs by multiple agents. Theor. Comput. Sci. **385**(1–3), 34–48 (2007)
10. Diks, K., Fraigniaud, P., Kranakis, E., Pelc, A.: Tree exploration with little memory. J. Algorithms **51**(1), 38–63 (2004)
11. Dobrev, S., Královič, R., Markou, E.: Online graph exploration with advice. In: Even, G., Halldórsson, M.M. (eds.) SIROCCO 2012. LNCS, vol. 7355, pp. 267–278. Springer, Heidelberg (2012). https://doi.org/10.1007/978-3-642-31104-8_23
12. Dobrev, S., Královič, R., Pardubská, D.: How much information about the future is needed? In: Geffert, V., Karhumäki, J., Bertoni, A., Preneel, B., Návrat, P., Bieliková, M. (eds.) SOFSEM 2008. LNCS, vol. 4910, pp. 247–258. Springer, Heidelberg (2008). https://doi.org/10.1007/978-3-540-77566-9_21
13. Emek, Y., Fraigniaud, P., Korman, A., Rosén, A.: Online computation with advice. Theor. Comput. Sci. **412**(24), 2642–2656 (2011)
14. Fleischer, R., Trippen, G.: Exploring an unknown graph efficiently. In: Brodal, G.S., Leonardi, S. (eds.) ESA 2005. LNCS, vol. 3669, pp. 11–22. Springer, Heidelberg (2005). https://doi.org/10.1007/11561071_4
15. Foerster, K.T., Wattenhofer, R.: Lower and upper competitive bounds for online directed graph exploration. Theor. Comput. Sci. **655**, 15–29 (2016)
16. Fraigniaud, P., Ilcinkas, D.: Digraphs exploration with little memory. In: Diekert, V., Habib, M. (eds.) STACS 2004. LNCS, vol. 2996, pp. 246–257. Springer, Heidelberg (2004). https://doi.org/10.1007/978-3-540-24749-4_22
17. Fraigniaud, P., Ilcinkas, D., Pelc, A.: Tree exploration with advice. Inf. Comput. **206**(11), 1276–1287 (2008)
18. Gorain, B., Pelc, A.: Deterministic graph exploration with advice. In: 44th International Colloquium on Automata, Languages, and Programming, ICALP 2017, Warsaw, Poland, 10–14 July 2017, pp. 132:1–132:14 (2017). https://doi.org/10.4230/LIPIcs.ICALP.2017.132
19. Hromkovič, J., Královič, R., Královič, R.: Information complexity of online problems. In: Hliněný, P., Kučera, A. (eds.) MFCS 2010. LNCS, vol. 6281, pp. 24–36. Springer, Heidelberg (2010). https://doi.org/10.1007/978-3-642-15155-2_3
20. Kalyanasundaram, B., Pruhs, K.R.: Constructing competitive tours from local information. In: Lingas, A., Karlsson, R., Carlsson, S. (eds.) ICALP 1993. LNCS, vol. 700, pp. 102–113. Springer, Heidelberg (1993). https://doi.org/10.1007/3-540-56939-1_65
21. Komm, D.: An Introduction to Online Computation - Determinism, Randomization, Advice. Texts in Theoretical Computer Science. An EATCS Series. Springer, Heidelberg (2016). https://doi.org/10.1007/978-3-319-42749-2

22. Komm, D., Královič, R., Královič, R., Smula, J.: Treasure hunt with advice. In: Scheideler, C. (ed.) Structural Information and Communication Complexity. LNCS, vol. 9439, pp. 328–341. Springer, Cham (2015). https://doi.org/10.1007/978-3-319-25258-2_23

23. Kortenkamp, D., Bonasso, R.P., Murphy, R.: Artificial Intelligence and Mobile Robots: Case Studies of Successful Robot Systems. MIT Press, Cambridge (1998)

24. Královič, R.: Personal communication (2017)

25. Megow, N., Mehlhorn, K., Schweitzer, P.: Online graph exploration: new results on old and new algorithms. Theor. Comput. Sci. **463**, 62–72 (2012)

26. Seibert, S., Sprock, A., Unger, W.: Advice complexity of the online coloring problem. In: Spirakis, P.G., Serna, M. (eds.) CIAC 2013. LNCS, vol. 7878, pp. 345–357. Springer, Heidelberg (2013). https://doi.org/10.1007/978-3-642-38233-8_29

27. Thrun, S., et al.: Autonomous exploration and mapping of abandoned mines. IEEE Robot. Autom. Mag. **11**(4), 79–91 (2004)

28. Thrun, S., et al.: Robotic mapping: a survey. In: Exploring Artificial Intelligence in the New Millennium, vol. 1, pp. 1–35 (2002)

Call Admission Problems on Grids
with Advice (Extended Abstract)

Hans-Joachim Böckenhauer, Dennis Komm, and Raphael Wegner[⊠]

Department of Computer Science, ETH Zürich, Universitätstrasse 6, 8092 Zürich,
Switzerland
{hjb,dennis.komm}@inf.ethz.ch, rwegner@student.ethz.ch

Abstract. We analyze the call admission problem on grids, thus generalizing previous results for the path graph, where a central authority receives requests that two of the computers in a given network arranged as a two-dimensional grid structure want to communicate. The central authority can then, for every request, either grant it by establishing one of the possible connections in the grid, or reject the request. Thereby, the requests have to be answered in an online fashion, every connection is permanent, and connections have to be edge-disjoint. The goal is to admit as many requests as possible. We are particularly interested to examine how much information about the future requests the central authority needs in order to compute an optimal solution or a solution of some given quality compared to the optimal solution; we quantify this information by studying the advice complexity of the problem.

Our results show that, without additional information, the central authority cannot perform satisfactorily well, and we establish a lower bound linear in $|E|$ for the number of advice bits needed for near-optimal solutions, where $|E|$ denotes the number of edges in the grid. Furthermore, concerning optimality, we are able to prove nearly tight bounds of at least $0.94|E|$ and at most $3|E|$ advice bits. In addition, we state another upper bound in the number of requests k and the number of vertices $|V|$ in the grid of $\lceil \log_2(5) \cdot k + \log_2(3) \cdot |V| \rceil + \lceil 2\log_2(k) \rceil$ advice bits, which is stronger for a small number of requests.

Keywords: Disjoint path allocation · Call admission
Advice complexity · Competitive analysis

1 Introduction

Imagine you are the administrator of a computer network, where each computer can request a connection to any of the other ones from a central authority, which immediately either grants the request or rejects it. Of course the different properties of network topologies are manifold, and the priorities among them are not the same in every case. However, your main concern is to be able to establish connections for the largest possible portion of requests. Being interested in what

L. Epstein and T. Erlebach (Eds.): WAOA 2018, LNCS 11312, pp. 118–133, 2018.
https://doi.org/10.1007/978-3-030-04693-4_8

can be guaranteed, we examine the worst case, i.e., the case where the requests are the most unfavorable. Hence, given a specific topology, the question of how good this topology suits our problem arises.

Now, for simplicity let us assume that once a connection between two computers has been established, the connection is fixed, i.e., it does not get terminated or changed in any way. Further, every wire can only be utilized for one connection, so there are no two connections sharing a wire. In the literature, this informally described setting is known as the *disjoint path allocation problem* or as the *call admission problem* [8].

Throughout this paper, we use the term disjoint path allocation problem when referring to the problem on a path network, although the problem definition actually comprises the complete variety of network topologies. Based on the research on path networks, we aim to analyze the problem on two-dimensional grid networks, and for distinction we specify this as the *call admission problem on grids*. Note that, while some network topologies, e.g., a complete graph, naturally perform well without the central authority being remarkably powerful, this is not the case for path graphs or grids. Therefore, we study the *advice complexity* of algorithms for the problem. Here, the central authority has access to an oracle that knows the future requests and can transmit any information, called *advice*, to the central authority. Deploying this model, we are not longer restricted to just stating that some ratio of satisfied requests cannot be achieved, but can give a precise measure of how much information the central authority has to obtain from the oracle in order to be able to reach exactly this ratio of satisfied requests.

Since the lack of knowledge about the future impeding a reasonably good solution unifies a lot of online problems, suitable complexity models have been invented: The fundamental idea of comparing the solutions of a specific algorithm with the best possible solutions has been introduced by Sleator and Tarjan [18], and the worst case ratio between them is called the *competitive ratio*. Further, the notion of *advice complexity* is due to Dobrev et al. [11]. The model we consider in this paper is a refined version proposed by Hromkovič et al. [15] and Böckenhauer et al. [3,4]. Using this model, various online problems have already been studied, including the ski rental problem, the paging problem [4,11], the k-server problem [5,6,12,14], as well as the problem motivating this paper, the disjoint path allocation problem [1,3,9,10,13].

For the latter, without the help of advice, no deterministic online algorithm can achieve a constant competitive ratio. More precisely, measured in the length l of the path of the network, every online algorithm accepts in the worst case at most $1/l$ of the requests that are satisfied by an optimal algorithm [8]. Even if we allow for a constant number of unsatisfied requests that are ignored for the competitive ratio, Komm [16] proved that every algorithm is at least $\sqrt{l}$-competitive, i.e., accomplishes at most a ratio of $1/\sqrt{l}$ of granted requests. Similarly, with regards to the total number of requests k, Böckenhauer et al. [3] already showed in their initial paper about this model that no algorithm is better than $(k - \mathcal{O}(1))$-competitive, or in case of a randomized online algorithm, $(k/4 - \mathcal{O}(1))$-competitive in expectation.

In the context of deterministic online algorithms with advice, $l - 1$ bits of advice are both necessary and sufficient in order to compute an optimal solution for the disjoint path allocation problem [1]. Digressing from strict optimality, for an algorithm to obtain a competitive ratio of c, again Böckenhauer et al. [3] established that at least $(k + 2)/(2c) - 2 \in \Omega(k/c)$ advice bits have to be read. Later, Boyar et al. [9] proved this to be asymptotically tight, i.e., $\Theta(k/c)$ bits of advice are necessary and sufficient for being c-competitive.

Now, for the call admission problem on grids, we aim to generalize some of the bounds for the disjoint path allocation problem and to establish further ones. Since, in contrast to a path network, on a grid network there are multiple possible paths that can be deliberated to satisfy a request between two computers, the online algorithm has significantly more freedom to build its solution. At the same time, there are far more possible requests that contradict each other, so considering the worst case, the construction of a solution might be remarkably more complicated. Thus, it is not clear whether more or less advice is needed in order to ensure some competitive ratio in comparison to the disjoint path allocation problem. However, this paper will present proofs that also on the grid $G = (V, E)$, advice is needed to achieve a competitive ratio constant in the grid size, and that, for optimality, almost $|E|$ advice bits are necessary. In addition, we show a lower bound on the number of advice bits necessary for non-optimal algorithms with a certain competitive ratio.

Concerning upper bounds, we prove that, for short, horizontally or vertically aligned requests, less than $|E|$ advice bits are sufficient to compute an optimal solution. In general, no more than $3|E|$ bits of advice are needed. Similarly, (partially) measured in the number of requests k, roughly $\log_2(5) \cdot k + \log_2(3) \cdot |V|$ bits of advice suffice.

The paper is organized as follows. In Sect. 2, we introduce our notation and formally define the call admission problem. Section 3 is devoted to lower bounds on the advice complexity, which are then complemented by almost matching upper bounds in Sect. 4. Due to space restrictions, most of the proofs are omitted in this extended abstract; for more detailed discussions, we refer to [7].

2   Preliminaries

Throughout this paper, $\mathbb{N}$ is defined as the set of positive integers, i.e., $\mathbb{N} = \{1, 2, 3, \ldots\}$. For a graph $G = (V, E)$, we denote $V(G) = V$ and $E(G) = E$. If G is a path, then $|E(G)|$ is its length. The degree of a vertex v is $d(v)$, the maximum degree of a graph G is $\Delta(G)$, the chromatic number, i.e., the minimum number of colors needed in a proper vertex-coloring of G, is denoted by $\chi(G)$, and $\omega(G)$ is the maximum clique size.

For $m, n \in \mathbb{N}$, an $(m \times n)$-grid $G = (V, E)$ is the Cartesian product of two paths $p_{\mathrm{ver}} = (V_m, E_m)$, and $p_{\mathrm{hor}} = (V_n, E_n)$, where $V_k = \{1, 2, \ldots, k\}$ and $E_k = \{\{1, 2\}, \{2, 3\}, \ldots, \{k - 1, k\}\}$. Therefore, $V = \{v_{i,j} \mid (i, j) \in V_m \times V_n\}$ and

$E = \{\{v_{a,b}, v_{x,y}\} \mid (a = x \wedge \{b, y\} \in E_m) \vee (b = y \wedge \{a, x\} \in E_n)\}$. Note that p_{ver} is of length $m-1$ and p_{hor} is of length $n-1$. In illustrations and descriptions, $v_{1,1}$ is chosen to be the lower-left corner, whereas $v_{m,n}$ is the upper-right corner. Slightly abusing notation, requests (calls) are denoted by (u, v) instead of $\{u, v\}$, in order to distinguish them from edges, so in this context, exceptionally $(u, v) = (v, u)$.

The essential underlying concept of the online setting is that an online algorithm receives a sequence of requests $r_1, \ldots, r_k$ of an instance of the given online problem, and has to answer each request r_i with an answer a_i before it can receive the next request, and this answer cannot be changed later. At any point in time, the online algorithm is only aware of the requests revealed so far without any knowledge of future requests, so a response a_i may only depend on $r_1, r_2, \ldots, r_i$. For studying online problems, we use the notation from the textbook by Komm [16].

The term *optimal algorithm* will be used for an algorithm that computes an optimal solution on every instance I of some online problem Π; we abbreviate it with OPT. The solution of a concrete deterministic algorithm ALG on some instance I is denoted by ALG(I), and gain(ALG(I)) denotes its gain.

The call admission problem on grids (short CAPG) is an online maximization problem, where the set of instances consists exactly of all possible $I = (r_1, r_2, \ldots, r_k)$, such that the first request r_1 contains two lengths $m, n \in \mathbb{N}$, which define an $(m \times n)$-grid $G = (V, E)$. Similarly to a path, obviously a grid is completely specified through the lengths m and n, and with this observation we have everything to define our primary object of study. Additionally, every request r_i with $1 \leq i \leq k$ consists of a pair of vertices $(u_i, v_i) \in V \times V$, where $u_i \neq v_i$; the pairs of vertices in the requests in I have to be pairwise distinct.

We say a set of paths is *contradicting* if and only if the paths are not pairwise edge-disjoint. A set of requests $\{r'_1, r'_2, \ldots, r'_t\}$ of I is *contradicting* if and only if there is no set of paths $\{p_1, p_2, \ldots, p_t\}$ in G that is not contradicting, and such that p_j is a path between the pair of vertices of request r'_j, for $1 \leq j \leq t$. We define the *length of a request* r to be the length of a shortest path between the pair of vertices in r, and conversely, such path which is not of minimal length is referred to as a *detour*.

A solution $S = (a_1, a_2, \ldots, a_k)$ is an element from $(P \cup \{0\})^k$, where P denotes the set of all paths in G, such that for all $a_i \neq 0$, a_i connects the vertices in r_i and $\{a_i \mid 1 \leq i \leq k \wedge a_i \neq 0\}$ is not contradicting. The set of feasible solutions for an instance I is denoted by $\mathcal{S}(I)$.

The gain of a solution S is the number of satisfied requests, i.e., $|\{a_i \mid 1 \leq i \leq k \wedge a_i \neq 0\}|$, where $S = (a_1, a_2, \ldots, a_k) \in \mathcal{S}(I)$.

If the underlying graph is restricted to be a path graph, i.e., an $(1 \times n)$-grid, we call the problem the *disjoint path allocation problem* (short DPA).

While for the DPA problem it is sufficient to state whether a request should be satisfied or not, because there is only one unique way to satisfy a request on a path, this is not the case for the call admission problem on grids. Here, for each of the respective requests, there may be multiple paths in the grid to grant it. Hence, if a solution would only contain whether a request has been satisfied or not, the algorithm would not need to fix any paths at all, but it only has to ensure the existence of some non-contradicting set of paths connecting the granted requests. However, thinking of the real-world scenario behind CAPG, it seems much more natural that the connection in the network should already be established and therefore be fixed from the time of acceptance. Besides, in some proofs we consider a (partial) solution in which the paths that satisfy some requests are already fixed and use the term *contradicting* to make clear that these fixed paths are not pairwise edge-disjoint.

An *online algorithm* ALG for an online problem Π is an algorithm that, given an instance $I = (r_1, r_2, \ldots, r_k)$ of Π, computes a feasible solution $\text{ALG}(I) = (a_1, a_2, \ldots, a_k)$, where a_i may only depend on $r_1, r_2, \ldots, r_i$.

For $c \in \mathbb{R}$ with $c \geq 1$, ALG is *c-competitive* if there is a constant $\alpha \geq 0$, such that, for all instances I of an online maximization problem Π,

$$\text{gain}(\text{OPT}(I)) \leq c \cdot \text{gain}(\text{ALG}(I)) + \alpha. \tag{1}$$

If this holds for $\alpha = 0$, ALG is called *strictly c-competitive*. The smallest value of c satisfying (1) is called the *competitive ratio* of ALG.

For example, a greedy online algorithm for the DPA problem on an instance I would check, for every request it receives, whether it contradicts any previously satisfied request and, if this is not the case, satisfies it. Note, however, that for CAPG there is no unique greedy online algorithm, since it is not clear which path should be used in order to satisfy a request. However, using any ordering Q of the paths in the grid, we could still define that the algorithm accepts a request r_i using a path p_i between its vertices if and only if p_i is the first path in Q where $\{a_j \mid 1 \leq j < i \wedge a_j \neq 0\} \cup \{p_i\}$ is not contradicting.

Consider an instance of DPA in which first a long request is asked and then this request gets partitioned into requests each of length 1 which are demanded next. The greedy algorithm ALG from above would satisfy the first request and then all the other requests have to be rejected because they contradict (see Fig. 1), leading to a competitive ratio which is as large as the length of the first request.

Fig. 1. Greedy online algorithm for DPA which satisfies the first request (lowest one) and has to reject all subsequent small requests of length 1.

For the ease of explanation, we will at times slightly abuse terminology and also speak about the competitive ratio as the performance of an algorithm on a single, specific instance. Moreover, if there is no constant c, such that some algorithm is c-competitive, in some literature this is expressed as being "not competitive" or that "there is no competitive ratio," or similar [8,16]. However, even if $c = c(\cdot)$ is a function, e.g., of the grid size or the number of requests, the ratio with respect to an optimal solution can still be of interest, so in the respective cases we prefer to formulate this as something in the sense of *the algorithm is $c(\cdot)$-competitive*. Let us remark that in case that the grid size itself is not part of the input, α can be a constant with respect to the instances, and yet depend on the grid size. Then, for a given grid size there are only finitely many instances, and thus, there always would exist a sufficiently large constant α, such that any algorithm is 1-competitive, independently of what the algorithm actually does. This is the reason for including the length of the path or the size of the grid, respectively, into the first request in the definitions of DPA and CAPG.

As the competitive ratio of all deterministic (or even randomized) algorithms for the call admission problem is undesirably high, we use the model of *advice complexity* to analyze how much information a central authority needs to compute a satisfactory solution. This model can be described as follows. The online algorithm has access to an unbounded tape of *advice bits* which it can linearly access during its computation. More formally, an *online algorithm* ALG *with advice* computes, on a request sequence $I = (r_1, r_2, \ldots, r_k)$, the output sequence $\text{ALG}^\phi(I) = (y_1, y_2, \ldots, y_k)$, where y_i is computed from $r_1, r_2, \ldots, r_i, \phi$, and ϕ is the content of the advice tape, i.e. an unbounded binary sequence. The advice tape is beforehand prepared by a computationally unbounded *oracle* that has access to the complete instance. We say that ALG is *c-competitive with advice complexity* $q(k)$ if it is c-competitive (as defined in (1)) and reads at most $q(k)$ bits of ϕ during its computation.

For a better comprehension, let us examine another example for DPA. Consider an algorithm ALG with advice that, on every request r_i, reads one bit b_i of the advice tape and, if $b_i = 1$, it satisfies r_i (recall that for DPA there is only one unique possibility to grant a request), and else ALG rejects r_i. We know that the oracle provides the best possible string of advice, so we are safe to assume that, according to some optimal solution S, it writes a 1 for every request that is granted in S and a 0 for every non-satisfied request on the advice tape. Since then ALG obviously computes the optimal solution S, it is strictly 1-competitive (see Fig. 2).

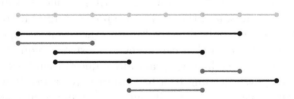

Fig. 2. Example instance for DPA on a path of length 7 (in gray). If 0100101 is the prefix on the advice tape, ALG satisfies the 2nd, 5th and 7th requests from above, which constitute one of the optimal solutions.

For instances of length k, ALG reads exactly k bits of advice and the advice complexity is therefore $q(k) = k$. Although not convenient in this case, we could measure the advice complexity with respect to the length l of the underlying path as well: Since all requests are pairwise distinct, there are at most $\sum_{i=1}^{l} i = \frac{l(l+1)}{2}$ requests in an instance. Hence, with respect to l, the advice complexity is $q(l) \leq \frac{l(l+1)}{2}$.

We observe the following facts which will be helpful in our subsequent proofs.

Fact 1. *In an $(m \times n)$-grid $G = (V, E)$, there are $|V| = mn$ vertices and $|E| = m(n-1) + n(m-1)$ edges, so $|E| = 2|V| - m - n$.*

Fact 2. *For every graph G, we have $\omega(G) \leq \chi(G) \leq \Delta(G) + 1$.*

Remark 2.1. If we want to pass some value $x \in \mathbb{N}$ to an algorithm ALG using parts of the advice tape, the algorithm has to know the beginning and the end of the encoding of x. This can be ensured by using some self-delimiting encoding, e.g., using $2\lceil \log_2(x) \rceil$ bits of advice [16].

3 Lower Bounds

First, we focus on lower bounds and establish some for CAPG that are similar to those for DPA, just for general grid sizes. Of course, since paths are corner cases of grids, DPA is a special case of CAPG, and its known lower bounds immediately carry over to CAPG. However, we are interested in learning which lower bounds hold for general grids, and how they depend on the grid sizes.

We start with a lower bound that is valid for grids where one dimension is logarithmically bounded in the other.

Theorem 3.1. *If $k(l) \in o(\log(l))$ for some $k \colon \mathbb{N} \to \mathbb{N}$, the competitive ratio of every online algorithm without advice for CAPG on a $(k(l) \times (l+1))$-grid is at least $c \in \Omega(l^{\frac{1}{k(l)+1}} \cdot k(l)^{-1})$.*

Proof (idea). The proof is based on the following idea. The adversary constructs a set of instances which consist of $k+1$ phases. In each phase P_j, many consecutive requests of length $c(l, j)$ are presented in row j of the grid, where $c(l, j+1)$ is much smaller than $c(l, j)$. As soon as one of these requests is granted by the algorithm, the phase ends, and the next phase presents all of its requests in the next row of the grid, but only directly "above" the granted request, as shown in Fig. 3. This way, the algorithm can grant at most one request per phase. If there is a phase where ALG does not grant any request, then OPT could take all requests from this phase, leading to a bad competitive ratio. If ALG takes one request from every phase $P_1, \ldots, P_k$, it cannot take any from phase P_{k+1}. The only way would be via a detour, but for each detour, some necessary horizontal edge is blocked. In contrast, OPT can take all requests from phase P_{k+1}.

Note that by the considerations stated priorly, in every row the number of requests has to be increasing in l, we observe that as a result the partitioning of l into the product of $k+1$ increasing functions is actually needed and the bound on the competitive ratio is maximized when these functions are about the same, i.e., $l = c(l, 1)^{k+1}$ and thus $c(l, 1) = l^{\frac{1}{k(l)+1}}$. □

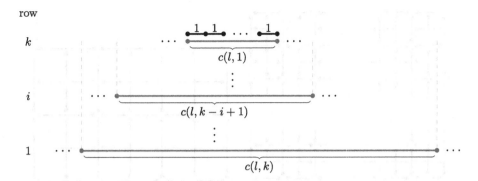

Fig. 3. Schematic illustration on a general grid, where in each row of the grid a request has been satisfied. In every row there are up to $c(l, 1)$ many requests from the first k phases. Then, in row k there are $c(l, 1)$ many additional requests which cannot be accepted anymore.

Also, let us remark that this proof generalizes the case of DPA very smoothly, in the sense that for $k = 1$ it actually boils down to $c(l, 1) = l^{1/(k+1)} = \sqrt{l}$, which is precisely what has been used by Komm [16].

Corollary 3.1. *For any constant $k \in \mathbb{N}$, there is no online algorithm without advice for CAPG which achieves a constant competitive ratio on a $(k \times (l+1))$-grid.*

Returning to the DPA problem, as already mentioned, the partitioning of some former request r into two requests yields requests r'_1 and r'_2, which are mutually exclusive to r. Since, given an instance containing such requests, every algorithm then needs to decide whether to grant r or any requests from $\{r'_1, r'_2\}$ instead, some advice should be necessary. This basic concept can be successively deployed in order to obtain further bounds: First, most obviously one can attempt to make statements about the necessary amount of advice to acquire an optimal solution, which in case of the DPA problem is proved to be at least half the length of the path [3]. Secondly, by reducing the bit guessing problem [2] to DPA, statements about the amount of advice which is needed to achieve some competitive ratio can be established [19].

These proofs extend to grids nicely. Even so, there is one obstacle in the not naturally given mutual exclusivity of these kinds of requests on grids. Since in this case there are plenty of possibilities to satisfy a given request, we will pack the requests very densely to cover the grid and provide some reasoning why then any detours would contradict other requests.

Lemma 3.1. *Every $(m \times n)$-grid G, with m and n odd, can be completely covered with requests of length 2, i.e., $|E(G)|/2 = mn - (m + n)/2$ requests of length 2 can be satisfied.*

Proof (idea). The arrangement of the requests is shown in Fig. 4a.   □

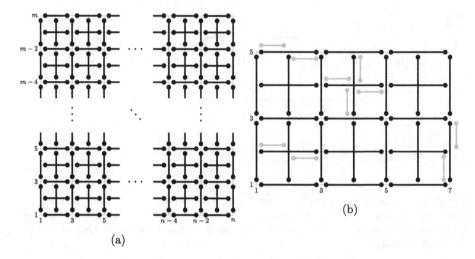

Fig. 4. (a) Grid with odd m and n, which is completely covered by requests of length 2. (b) Some instance with ten requests in phase P_2 (depicted in gray), requests from phase P_1 are depicted in black.

Theorem 3.2. *Every optimal online algorithm with advice for CAPG on a grid G uses at least $|E(G)|/2$ advice bits.*

Proof (idea). It suffices to only consider $(m \times n)$-grids G with m and n odd. We construct a set $\mathcal{I}$ of instances consisting of two phases P_1 and P_2, where phase P_1 is constructed as in the proof of Lemma 3.1, and some of the requests from P_1 are subdivided by two requests from P_2 as shown in Fig. 4b. This leads to $2^{|E(G)|/2}$ possible instances.

Any optimal online algorithm ALG needs to decide already in phase P_1, which requests to reject. Moreover, it can be shown that detours can never be profitable. Thus, ALG needs $|E(G)|/2$ advice bits to distinguish all instances. □

As mentioned beforehand, this construction can be employed to prove lower bounds beyond strict optimality, i.e., to make statements about the number of advice bits that are needed in order to obtain some competitive ratio (larger than 1). To this end, we use the string guessing problem and a known lower bound on its advice complexity. Then, reducing this problem to CAPG, we can transform the lower bound to one that is valid for CAPG.

Definition 3.1 (String Guessing with Unknown History [2]). *The problem of string guessing with unknown history over an alphabet Σ with $|\Sigma| \geq 2$ is an online minimization problem Π_{SGU}. Every instance $I \in \mathcal{I}(\Pi_{SGU})$ consists of some $n \in \mathbb{N}$, followed by $n - 1$ requests "?" containing no additional information, and a string $s = s_1 s_2 \ldots s_n \in \Sigma^n$ of length n, i.e., $I = (n, ?, ?, \ldots, ?, s)$. A corresponding solution $S \in \mathcal{S}(I)$ is of the form $S = (a_1, a_2, \ldots, a_n)$, where $a_i \in \Sigma$, and the last request remains unanswered. The measurement function is the Hamming distance between $s_1 s_2 \ldots s_n$ and $a_1 a_2 \ldots a_n$.*

If $|\Sigma| = 2$, the problem is commonly referred to as the *bit guessing problem with unknown history* [2]. The next theorem states the lower bound for bit guessing we will use in our reduction.

Theorem 3.3 (Böckenhauer et al. [2]). *If $1/2 \leq \gamma \leq 1$ and an online algorithm with advice guesses at least γn bits correctly of every instance of bit guessing with unknown history, then it uses at least $(1+(1-\gamma)\log_2 (1 - \gamma)+\gamma\log_2 \gamma)n$ bits of advice.*

We can now use Theorem 3.3 to prove a lower bound for CAPG.

Theorem 3.4. *Every online algorithm with advice for CAPG which achieves a competitive ratio of $c \leq 12/11$ on a grid G has to read at least*

$$\left(1 + \left(6 - \frac{6}{c}\right)\log_2 \left(6 - \frac{6}{c}\right) + \left(\frac{6}{c} - 5\right)\log_2 \left(\frac{6}{c} - 5\right)\right) \frac{|E(G)|}{2}$$

bits of advice.

Proof (idea). The proof is based on the following idea. Let CAPG be some online algorithm with advice for CAPG. Then, we can devise another algorithm BGUESS for bit guessing with unknown history, which uses CAPG and forwards the advice bits on demand, i.e., we reduce the bit guessing problem to the call admission problem on grids.

Initially, on an instance $I' \in \mathcal{I}'$ for bit guessing, BGUESS obtains the number n' of bits it has to guess. We use the same instances $I \in \mathcal{I}$ for CAPG as in the proof of Theorem 3.2, consisting of phases P_1 and P_2.

BGUESS also operates in two phases: First, for every "?" received, it demands one of the requests from P_1 and guesses 1 for every request that CAPG satisfies, and 0 otherwise, so the ith answer of CAPG specifies the ith guess of BGUESS.

Therefore, in P_2, BGUESS partitions every request of P_1 which corresponds to a 0 in s into two requests, includes them into P_2 and demands them from CAPG. Then, the optimal answer for CAPG would have been to grant exactly the requests for which s contains a 1. Note that this construction does not depend on the answers of CAPG or BGUESS, but only on the last request s for BGUESS, which is transformed into possibly multiple requests for CAPG. So, this way BGUESS penalizes CAPG for every guessed bit that deviates from the corresponding bit of s. However, since for multiple erroneously unsatisfied requests of P_1, CAPG might be able to grant requests from P_2 which are not contained in the optimal solution, the cost of CAPG is not obvious. However, a thorough case distinction shows that this compensation is bounded. □

Using the proof method of *partition trees* [1], we are able to improve the lower bound from Theorem 3.2 by almost a factor of 2.

Theorem 3.5. *Every optimal online algorithm with advice for CAPG on an $(m \times n)$-grid G has to read at least $0.94677 \cdot |E(G)| - m - n$ advice bits.*

Proof (idea). The technically involved proof is based on a generalization of the DPA lower bound by Barhum et al. [1]. For their instances, they use the partition-tree method to prove that there are many different instances requiring different advice strings. Implementing subinstances of their instances on every row and every column of the grid, where the length of all requests is bounded from above by 4, leads to the situation that all bit strings associated to the different instances do not contain more than 3 consecutive zeros. The number of such strings can be counted using tetranacci numbers [17] and yields the claimed result. Bounding the length of the requests by 4 is used for proving that detours are not helpful for any online algorithm.   □

Note that, if we could use bit strings of length $n - 2$ or respectively $m - 2$ for every row or column just as on the path, then this would result in $m \cdot (n - 2) + n \cdot (m - 2) = |E(G)| - m - n$ advice bits, so our result is indeed surprisingly close (see Fig. 5).

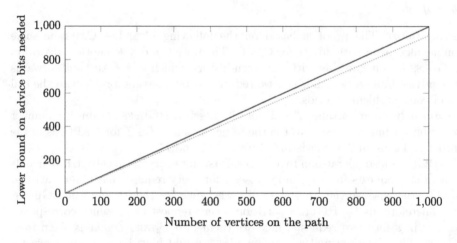

Fig. 5. Illustration of the number of advice bits needed on a path of length n in comparison to the slightly worse lower bound which is obtained by using only requests of length 4 as in the proof of Theorem 3.5.

4   Upper Bounds

Note that, unlike lower bounds, upper bounds for DPA do not inherently carry over to CAPG, since a method to solve DPA does not necessarily have to be applicable for CAPG as well. Let us start with the most obvious upper bound.

Theorem 4.1. *There is an optimal online algorithm with advice for CAPG that reads at most $2|E| \cdot \lceil \log_2(|V|) \rceil \leq 2|E| \cdot \log_2(|E| + m + n)$ bits of advice for every $(m \times n)$-grid $G = (V, E)$.*

Proof (idea). Recall that there are $|V| = mn$ vertices and $|E| = m(n-1) + n(m-1) = 2mn - m - n$ edges in G, so $|V| = (|E| + m + n)/2$. Hence, for an arbitrary but fixed optimal solution, the oracle can encode the two endpoints of every satisfied request using $2 \cdot \lceil \log_2(|V|) \rceil \leq 2 \cdot \log_2(|E| + m + n)$ bits per request. Reading all of this advice at the very beginning of the computation, the algorithm can then solve the resulting offline instance and find a set of non-conflicting paths. Recall that, as usual when dealing with online algorithms, we do not restrict the time complexity of the algorithm.   □

Actually, a further upper bound for a restricted set of instances is already established by the idea of Theorem 3.5, since its proof is (nearly) constructive, just as for the inspiring result regarding DPA [1].

Theorem 4.2. *For instances with either horizontally or vertically aligned requests of length at most 4, there is an optimal online algorithm with advice using at most $|E(G)| - (m+n)/2 + 1$ advice bits for every $(m \times n)$-grid G.*

As observed easily, knowing which edges of the grid are used to satisfy a request and which remain unused for some optimal solution is not sufficient for an algorithm to reconstruct it. Consider, e.g., the instances constructed in the proof of Theorem 3.2, where all edges can be used. Thus, it is necessary to transmit the "membership" of a request in some way as well. For this reason, "neighboring" paths in an optimal solution need to be distinguishable. This leads to the task of vertex-coloring an auxiliary graph $\widehat{G}(S) = (\widehat{V}, \widehat{E})$ whose vertices are the paths of an optimal solution S together with a partition of the unused edges of the grid into connected components. The vertices of this auxiliary graph are connected by an edge if and only if the corresponding paths or components of the grid share a vertex. The partition of the unused edges is chosen in a way that minimizes the chromatic number of $\widehat{G}(S)$.

For reasons of readability, we will hide the dependency on S and simply write $\widehat{G}$ instead of $\widehat{G}(S)$ whenever suitable.

Theorem 4.3. *Let $\mathcal{I}$ denote all possible instances of CAPG on a grid $G = (V, E)$, and let $\mathcal{S}_{opt}(I)$ be the set of optimal solutions for an instance $I \in \mathcal{I}$. Then, there is an optimal online algorithm with advice for CAPG that uses at most*

$$\max_{I \in \mathcal{I}} \min_{S \in \mathcal{S}_{opt}(I)} \lceil |E| \cdot \log_2(\chi(\widehat{G})) \rceil + 2\lceil \log_2(\chi(\widehat{G})) \rceil$$

advice bits.

Corollary 4.1. *If $\max_{I \in \mathcal{I}} \min_{S \in \mathcal{S}_{opt}(I)} \chi(\widehat{G})$ can be bounded by a number c_χ which is known by an online algorithm with advice for CAPG, then the bound of Theorem 4.3 adjusts to $\lceil |E| \cdot \log_2(c_\chi) \rceil$ bits of advice.*

Note that this immediately yields a better constant than Theorem 4.1 since every request has length at least 1, so there are at most $|E| - 1$ requests left which can be neighboring, i.e., $\chi(\widehat{G}) \leq \Delta(\widehat{G})+1 \leq |E|$, so already $\lceil|E|\cdot\log_2(|E|)\rceil$ advice bits are sufficient. However, a short request cannot have many incident paths in a grid, thus this bound is only a rather coarse estimate and can be improved easily as the following corollary shows.

Corollary 4.2. *There is an optimal online algorithm with advice for CAPG that reads at most* $\lceil|E| \cdot \log_2((2|E| + 7)/3)\rceil$ *bits of advice.*

Proof (idea). The idea behind this proof is that there cannot be too many overlaps between different paths from S due to the bounded degree of the grid. This in turn bounds the degree of the vertices in $\widehat{G}$. □

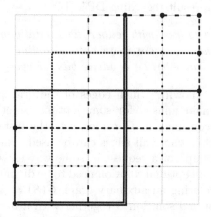

Fig. 6. On the empty edges of the grid the instance contains requests of length 1 and there are no further requests. All requests shown start in the lower left part, go vertically (horizontally) until they reach an empty section or the border and then continue horizontally (vertically). Satisfying the shown requests and all requests of length 1 is the unique optimal solution . Hence, $\widehat{G}$ has a clique of size at least 6, i.e., in G every shown path needs to be colored in a different color.

However, as Fig. 6 illustrates, on an $(n \times n)$-grid with even n there is an instance such that $\widehat{G}$ contains an n-clique. Since $\chi(\widehat{G}) \geq \omega(\widehat{G})$, this implies that, with this approach, the upper bound cannot be enhanced to less than $\lceil|E|\cdot\log_2(n)\rceil = \lceil|E|\cdot\log_2(\sqrt{|V|})\rceil = \lceil\frac{1}{2}|E|\cdot\log_2(|E|+2n)-1\rceil \in \mathcal{O}(|E|\cdot\log(|E|))$ advice bits. Hence, we need an advanced method to prove a more compact bound. Such a bound is given by the next theorem.

Theorem 4.4. *There is an online algorithm with advice for CAPG that computes an optimal solution using at most* $3|E|$ *advice bits.*

Proof (idea). The proof of this claim is based on the following idea. Each path can be partitioned into subpaths using the same row or column. For each edge, 3 advice bits are used to classify the edge into one of eight classes describing whether the edge belongs to the same subpath as its neighboring edges and how the direction of a path possibly changes at the endpoint of the edge. □

We can continue along the same line of reasoning to obtain a slightly different result, bounding the number of advice bits from above in the number of requests: It is sufficient to know for every request whether it has been taken and then to be able to follow the request until its end, in order to reconstruct an optimal solution. Hence, for the first vertex of a request, the oracle can encode whether to reject it or one of the four directions to start, and, for every inner vertex of the path that satisfies the request, the oracle can encode one of the three directions left, straight, or right to continue. Moreover, given that, for a path satisfying a request, the incoming direction for a vertex v is known and already two incident edges of v are used to satisfy another request, there is only one direction left, so there is no additional information needed for v. In other words, for a particular solution, every vertex used as an inner vertex of any path satisfying a request only needs advice once (see Fig. 7). The next theorem formalizes this observation.

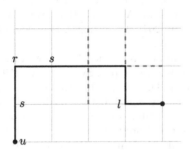

Fig. 7. The dashed lines indicate formerly satisfied requests. For a start vertex of a path, N, l, u, r, d are interpreted as "not satisfied", "left", "upwards", "right", or respectively "downwards", and, for an inner vertex v, the labels l, s, r are interpreted as "left", "straight" and "right" with respect to the incoming edge at v. On the request $((1,1),(2,5))$ the string $usrsl$ translates to the depicted path.

Theorem 4.5. *Let $\mathcal{I}$ denote all possible instances of CAPG on a grid $G = (V, E)$, and let $\mathcal{S}_{opt}(I)$ be the set of optimal solutions for an instance $I \in \mathcal{I}$. Then, there is an optimal online algorithm with advice for CAPG that uses at most*

$$\lceil \log_2(5) \cdot k \rceil + \lceil \log_2(3) \cdot |V| \rceil + \lceil 2 \log_2(k) \rceil$$

advice bits, where k is the number of requests in I.

For fewer than approximately $0.95|E| - 0.34(m + n)$ requests, Theorem 4.5 is indeed an improvement, otherwise (e.g., in the worst case of both bounds) Theorem 4.4 yields the stronger result, since there can be up to $\binom{|V|}{2} = (1/8)(|E| + m + n - 1)^2 - 1/8$ requests in an instance.

References

1. Barhum, K., et al.: On the power of advice and randomization for the disjoint path allocation problem. In: Geffert, V., Preneel, B., Rovan, B., Štuller, J., Tjoa, A.M. (eds.) SOFSEM 2014. LNCS, vol. 8327, pp. 89–101. Springer, Cham (2014). https://doi.org/10.1007/978-3-319-04298-5_9
2. Böckenhauer, H.-J., Hromkovič, J., Komm, D., Krug, S., Smula, J., Sprock, A.: The string guessing problem as a method to prove lower bounds on the advice complexity. Theor. Comput. Sci. **554**, 95–108 (2014)
3. Böckenhauer, H.-J., Komm, D., Královič, R., Královič, R., Mömke, T.: On the advice complexity of online problems. In: Dong, Y., Du, D.-Z., Ibarra, O. (eds.) ISAAC 2009. LNCS, vol. 5878, pp. 331–340. Springer, Heidelberg (2009). https://doi.org/10.1007/978-3-642-10631-6_35
4. Böckenhauer, H.-J., Komm, D., Královič, R., Královič, R., Mömke, T.: Online algorithms with advice: the tape model. Inf. Comput. **254**, 59–83 (2017)
5. Böckenhauer, H.-J., Komm, D., Královič, R., Královič, R.: On the advice complexity of the k-server problem. In: Aceto, L., Henzinger, M., Sgall, J. (eds.) ICALP 2011. LNCS, vol. 6755, pp. 207–218. Springer, Heidelberg (2011). https://doi.org/10.1007/978-3-642-22006-7_18
6. Böckenhauer, H.-J., Komm, D., Královič, R., Královič, R.: On the advice complexity of the k-server problem. J. Comput. Syst. Sci. **86**, 159–170 (2017)
7. Böckenhauer, H.J., Komm, D., Wegner, R.: An analysis of call admission problems in grids. Technical report. www.ita.inf.ethz.ch/gdpa.pdf. Accessed 20 June 2018
8. Borodin, A., El-Yaniv, R.: Online Computation and Competitive Analysis. Cambridge University Press, Cambridge (1998)
9. Boyar, J., Favrholdt, L.M., Kudahl, C., Mikkelsen, J.W.: The advice complexity of a class of hard online problems. CoRR, abs/1408.7033 (2014)
10. Burjons, E., Frei, F., Smula, J., Wehner, D.: Length-weighted disjoint path allocation - advice and parametrization. In: Böckenhauer, H.-J., Komm, D., Unger, W. (eds.) Adventures Between Lower Bounds and Higher Altitudes. LNCS, vol. 11011, pp. 231–256. Springer, Cham (2018). https://doi.org/10.1007/978-3-319-98355-4_14
11. Dobrev, S., Královič, R., Pardubská, D.: How much information about the future is needed? In: Geffert, V., Karhumäki, J., Bertoni, A., Preneel, B., Návrat, P., Bieliková, M. (eds.) SOFSEM 2008. LNCS, vol. 4910, pp. 247–258. Springer, Heidelberg (2008). https://doi.org/10.1007/978-3-540-77566-9_21
12. Emek, Y., Fraigniaud, P., Korman, A., Rosén, A.: Online computation with advice. Theor. Comput. Sci. **412**(24), 2642–2656 (2011)
13. Gebauer, H., Komm, D., Královič, R., Královič, R., Smula, J.: Disjoint path allocation with sublinear advice. In: Xu, D., Du, D., Du, D. (eds.) COCOON 2015. LNCS, vol. 9198, pp. 417–429. Springer, Cham (2015). https://doi.org/10.1007/978-3-319-21398-9_33
14. Gupta, S., Kamali, S., López-Ortiz, A.: On advice complexity of the k-server problem under sparse metrics. Theory Comput. Syst. **59**, 476–499 (2016)
15. Hromkovič, J., Královič, R., Královič, R.: Information complexity of online problems. In: Hliněný, P., Kučera, A. (eds.) MFCS 2010. LNCS, vol. 6281, pp. 24–36. Springer, Heidelberg (2010). https://doi.org/10.1007/978-3-642-15155-2_3
16. Komm, D.: An Introduction to Online Computation - Determinism, Randomization, Advice. Texts in Theoretical Computer Science. An EATCS Series. Springer, Switzerland (2016). https://doi.org/10.1007/978-3-319-42749-2

17. Noe, T., Piezas III, T., Weisstein, E.W.: Fibonacci n-step number. From MathWorld-A Wolfram Web Resource. http://mathworld.wolfram.com/Fibonaccin-StepNumber.html. Accessed 20 June 2018

18. Sleator, D.D., Tarjan, R.E.: Amortized efficiency of list update and paging rules. Commun. ACM **28**(2), 202–208 (1985)

19. Smula, J.: Information content of online problems: advice versus determinism and randomization. Ph.D. thesis, ETH Zurich (2015)

Improved Approximation Algorithms for Minimum Power Covering Problems

Gruia Calinescu[1], Guy Kortsarz[2], and Zeev Nutov[3]($\boxtimes$)

[1] Illinois Institute of Technology, Chicago, IL, USA
calinescu@iit.edu
[2] Rutgers University, Camden, USA
guyk@camden.rutgers.edu
[3] The Open University of Israel, Raanana, Israel
nutov@openu.ac.il

Abstract. Given an undirected graph with edge costs, the **power of a node** is the maximum cost of an edge incident to it, and the **power of a graph** is the sum of the powers of its nodes. Motivated by applications in wireless networks, we consider two network design problems under the power minimization criteria. In both problems we are given a graph $G = (V, E)$ with edge costs and a set $T \subseteq V$ of terminals. The goal is to find a minimum power edge subset $F \subseteq E$ such that the graph $H = (V, F)$ satisfies some prescribed requirements. In the MIN-POWER EDGE-COVER problem, H should contain an edge incident to every terminal. Using the Iterative Randomized Rounding (IRR) method, we give an algorithm with expected approximation ratio 1.41; the ratio is reduced to $73/60 < 1.217$ when T is an independent set in G. In the case of unit costs we also achieve ratio 73/60, and in addition give a simple efficient combinatorial algorithm with ratio 5/4. For all these NP-hard problems the previous best known ratio was 3/2. In the related MIN-POWER TERMINAL BACKUP problem, H should contain a path from every $t \in T$ to some node in $T \setminus \{t\}$. We obtain ratio 3/2 for this NP-hard problem, improving the trivial ratio of 2.

Keywords: Approximation algorithms
Iterative randomized rounding · Minimum power · Edge-cover
Terminal backup

1 Introduction

Wireless networks are studied extensively due to their wide applications. The power consumption of a station determines its transmission range, and thus also the stations it can send messages to. Assigning power levels to the stations (nodes) determines the resulting communication network. Conversely, given a communication network, the power required at v only depends on the farthest

Partially supported by NSF grant number 1540547.

L. Epstein and T. Erlebach (Eds.): WAOA 2018, LNCS 11312, pp. 134–148, 2018.
https://doi.org/10.1007/978-3-030-04693-4_9

node that is reached directly by v. This is in contrast with wired networks, in which every pair of stations that need to communicate directly incurs a cost. Thus the minimal power $p(v)$ of a node v equals the largest cost of an edge incident to v in the communication network. The first work under the minimum power model is from 1989 [7]. For a sample of other works under this model see for example [1,4–6,8,9,14–18,20,21,23–25,27,29–31].

Definition 1. *Let $H = (V, F)$ be a graph with edge-costs $\{c(e) : e \in F\}$. For $v \in V$, the* **power** $p(v) = p_H(v) = p_F(v)$ *of v in H (w.r.t. c) is the maximum cost of an edge in F incident to v (or zero, if no such edge exists), i.e., $p(v) = p_F(v) = \max_{vu \in F} c(vu)$. The power of H is the sum of the powers of its nodes, namely, $p(H) = p(F) = \sum_{v \in V} p_F(v)$.*

All the graphs are assumed to be undirected, unless stated otherwise. In our problems, the input is a graph $G = (V, E)$ with edge costs $\{c(e) : e \in E\}$ and a subset $T \subseteq V$ of terminals; the goal is to find a minimum power subgraph $H = (V, F)$ of G that satisfies some prescribed properties. We refer the reader to a recent survey [28] on such problems. We consider the min-power variant of two classic problems, EDGE-COVER and TERMINAL-BACKUP, defined below.

Definition 2. *For a graph $H = (V, F)$ and a set $T \subseteq V$ of terminals, we say that F (or H) is:*

- *a T-**cover** if every $t \in T$ has some edge in F incident to it (equivalently, no connected component of H is a single terminal);*
- *a T-**backup** if every $t \in T$ has a path to some other node in T (equivalently, no connected component of H contains a single terminal).*

> MIN-POWER EDGE-COVER
> Here F should be a T-cover, namely, every $t \in T$ has some edge in F incident to it.

> MIN-POWER TERMINAL BACKUP
> Here F should be a T-backup, namely, every $t \in T$ has a path to some other node in T.

For illustration, suppose we have two sets A, B of stations. The stations in A can communicate with each other via an existing wired infrastructure, while each station in B should have a wireless communication with some station in A. We want to assign energy levels to the stations while minimizing the total energy. This is modeled as a MIN-POWER EDGE-COVER problem in a bipartite graph with terminal set $T = B$. This bipartite case models many other practical scenarios, that were studied in the (easier) geometric setting, c.f. [22,26].

In the case $T = V$ the problems coincide; the resulting min-power problem is still NP-hard with a standard reduction from SET COVER. The min-cost versions (where one seeks to minimize $c(F) = \sum_{e \in F} c(e)$) of these problems can be solved in polynomial time see [10] and [2], respectively. MIN-COST EDGE COVER is among the most basic problems in combinatorial optimization and theoretical computer science, see for example the book [32]. However, the MIN-POWER EDGE-COVER problem is NP-hard even if T is an independent set in the input graph G and all costs are equal to 1 [15]. The NP-hardness proof in [15] easily extends to the MIN-POWER TERMINAL BACKUP problem.

For each of these problems, any inclusion-minimal solution is a forest, since removing any edge from any cycle keeps the solution feasible and does not increase the objective function. It is known that if F is a forest then $c(F) \leq p(F) \leq 2c(F)$. This implies that both problems admit ratio 2, by simply computing an optimal min-cost solution.

For MIN-POWER EDGE-COVER the trivial ratio 2 was improved to 1.5 in [19]. No better ratio was known even for the case when T is an independent set in G and all costs are equal to 1. We improve this as follows.

Theorem 1. MIN-POWER EDGE-COVER *admits a polynomial time algorithm with expected approximation ratio* 1.41. *If T is an independent set in G then the ratio can be reduced to* $73/60 < 1.217$.

The algorithm in Theorem 1 uses the Iterative Randomized Rounding (IRR) method. We also use a method of analyzing the best of two algorithm using a convex combination of their results; we have seen this technique in [12].

In the case of unit costs we show a simple approximation ratio preserving reduction to the case when T is an independent set, thus obtaining for this case ratio $73/60 < 1.217$. In addition, we use a different method to obtain an efficient combinatorial approximation algorithm with good ratio.

Theorem 2. MIN-POWER EDGE-COVER *with unit costs admits a polynomial time algorithm with expected approximation ratio* $73/60 < 1.217$. *The problem also admits a* 5/4-*approximation algorithm with running time* $O(n^3)$.

We also improve the trivial ratio 2 for MIN-POWER TERMINAL BACKUP.

Theorem 3. MIN-POWER TERMINAL BACKUP *admits a polynomial time algorithm with approximation ratio* 1.5.

The proof of the latter theorem uses the idea of the 1.5-approximation algorithm in [19] for MIN-POWER EDGE-COVER, but the details are more involved.

We now briefly survey some work where the IRR method is used. This method is due to Byrka, Grandoni, Rothvoß and Sanita [3], that gave a $\ln 4 + \epsilon < 1.39$ approximation for the MIN-COST STEINER TREE problem. This is currently the best ratio known for the problem. Goemans, Olver, Rothvoß and Zenklusen [11] gave faster and simpler $\ln 4 + \epsilon$ approximation for the same problem, and also obtained a better ratio $73/60$ for quasi-bipartite graphs. Grandoni [13] used the IRR method to give the currently best known ratio 1.91 for the MIN-POWER STEINER TREE problem. Our paper has similarities with [13] including a Harmonic potential function, and two main differences: (i) it is technically easier (for us) to cover terminals than to cover all cuts separating terminals as in [13]; (ii) we combine iterative randomized rounding with another algorithm, since by itself, iterative randomized rounding fails to improve the ratio 3/2 in some cases.

2   Algorithm for MIN-POWER EDGE-COVER (Theorem 1)

A **star** is a rooted tree R such that only its root r, called the **center**, may have degree ≥ 2. Note that any inclusion-minimal T-cover F is a collection of disjoint stars, as if F has a path of length three, then the middle edge e of this path can be removed and $F \setminus \{e\}$ remains an T-cover.

For $S \subseteq T$ let π_S be the minimum power of a star R_S that contains S ($\pi_S = \infty$ if no such star exists). Note that given S, both R_S and π_S can be computed in polynomial time by "guessing" the center of R_S. For an integer $k \geq 1$ let $\mathcal{T}_k = \{S \subseteq T : |S| \leq k\}$. We say that a subfamily $\mathcal{T} \subseteq \mathcal{T}_k$ is a k-**restricted** T-**cover** if the union of the sets in $\mathcal{T}$ is T; the power of $\mathcal{T}$ is defined to be $p(\mathcal{T}) = \sum_{S \in \mathcal{T}} \pi_S$. In what follows we denote by $t = |T|$ the number of terminals and by $n = |V(G)|$.

Lemma 1. MIN-POWER EDGE-COVER *with* $t = |T|$ *terminals can be solved optimally in time* $O\left(2^{t \log_2 t} poly(n)\right)$.

Proof. For $S \subseteq T$ let R_S be some minimum power star that contains S. For every partition $\mathcal{P}$ of T compute a solution $F_\mathcal{P} = \cup_{S \in \mathcal{P}} R_S$ for this partition, and among the solutions $F_\mathcal{P}$ computed return one of minimum power. Note that $F_\mathcal{P}$ can be computed in polynomial time for any $\mathcal{P}$, since R_S can be computed in polynomial time for each $S \in P$. For the partition $\mathcal{P}$ defined by the stars of some optimal solution, $F_\mathcal{P}$ is an overall optimal solution for the problem. The number of partitions of a set of size t is the Bell number B_t, and it is known that $B_t \leq 2^{t \log_2 t}$. The lemma follows.   □

The "hypergraphic" linear program $LP_k(T)$ below has a variable x_S for every $S \in \mathcal{T}_k$, and it is a relaxation for the problem of finding a k-restricted T-cover of minimum power.

$$\min \quad \sum_{S \in \mathcal{T}_k} \pi_S x_S$$

$$\text{s.t.} \quad \sum_{S \in \mathcal{T}_k, S \ni v} x_S \geq 1 \quad \forall v \in T$$

$$x_S \geq 0 \quad \forall S \in \mathcal{T}_k$$

By Lemma 1, $LP_k(T)$ can be solved in polynomial time for any constant k.

Let us call a feasible solution x to $LP_k(T)$ **irreducible** if no coordinate of x can be lowered while keeping feasibility.

Lemma 2. *Let* x *be an irreducible solution to* $LP_k(T)$. *Then* $\sum_{S \in \mathcal{T}_k} x_S \leq n$, *and* $\mathcal{T}_k$ *with probabilities* $Pr[S] = x_S/n$ *for* $S \neq \emptyset$ *and* $Pr[\emptyset] = 1 - \sum_{S \in \mathcal{T}_k} x_S/n$ *is a sample space, in which* $Pr[\{S \in \mathcal{T}_k : S \ni v\}] \geq 1/n$ *holds for any* $v \in T$.

Proof. Since x is irreducible, for any $S \in \mathcal{T}_k$ with $x_S > 0$ there exists $v \in S$ such that the inequality of v in $LP_k(T)$ is tight. For every $S \in \mathcal{T}_k$ with $x_S > 0$ choose one such node v_S. Let $W = \{v_S : x_S > 0, S \in \mathcal{T}_k\}$ be the set of chosen nodes, and note that $W \subseteq T$. Then

$$\sum_{S \in \mathcal{T}_k} x_S \leq \sum_{v \in W} \sum_{S \in \mathcal{T}_k, S \ni v} x_S \leq \sum_{v \in W} 1 \leq |W| \leq n .$$

This implies that $Pr[\emptyset] = 1 - \sum_{S \in \mathcal{T}_k} x_S/n \geq 0$ and thus we have a sample space. Furthermore, $Pr[\{S \in \mathcal{T}_k : S \ni v\}] = \sum_{S \in \mathcal{T}_k, S \ni v} x_S/n \geq 1/n$, by the constraint of v in $LP_k(T)$. □

The following lemma provides a (tight) bound on the ratio between the power of an optimal T-cover and a k-restricted T-cover.

Lemma 3 ([19]). *For any T-cover F there exists a k-restricted T-cover $\mathcal{T}$ of power $p(\mathcal{T}) \leq (1 + 1/k)p(F)$.*

We run two algorithms and take the best of the two. The first algorithm is the 3/2-approximation algorithm of Kortsarz & Nutov [19]; we call it the KN-Algorithm.

Algorithm 1. KN-ALGORITHM$(G = (V, E), c, T)$

---

1 for all $u, v \in T$ (possibly $u = v$) compute a min-power $\{u, v\}$-cover J_{uv}
2 let (T, E') be a complete graph with all loops and edge costs $c_{uv} = p(J_{uv})$
  for all $u, v \in T$
3 compute a minimum cost T-cover $J' \subseteq E'$
4 return $J = \bigcup_{uv \in J'} J_{uv}$

The second algorithm is an Iterative Randomized Rounding algorithm, abbreviated by IRR-Algorithm. For previous applications of this type of algorithms see [3,11] for the MIN-COST STEINER TREE problem, and [13] for the MIN-POWER STEINER TREE problem.

Algorithm 2. IRR-ALGORITHM$(G = (V, E), c, T, k)$

---

1 **initialize** $J \leftarrow \emptyset$
2 **while** $T \neq \emptyset$ **do**
3 ⎪   compute an irreducible optimal solution x for $LP_k(T)$ and
  ⎪   sample one set $S \in \mathcal{T}_k$ with probabilities as in Lemma 2
4 ⎣ $T \leftarrow T \setminus S$, $J \leftarrow J \cup R_S$
5 **return** J

Note that in every iteration, the set of terminals may change. In such a case, the IRR-Algorithm solves a new LP with respect to the new set of terminals. To ensure polynomial time, after $2n \ln n$ iterations the while-loop is terminated, and we add to J a solution for the residual problem computed by the KN-Algorithm. The following lemma shows that the expected loss in the approximation ratio incurred by such modification is negligible.

Lemma 4. *In every iteration, every $v \in T$ is hit with probability at least $1/n$. The probability that $T \neq \emptyset$ after $2n \ln n$ iterations is at most $1/n$. The expected loss in the approximation ratio incurred by stopping the IRR algorithm after $2n \ln n$ iterations is at most $\frac{3}{2n}$.*

Proof. The first statement follows from Lemma 2. The probability that after $i = 2n \ln n$ iterations a terminal is not hit is at most $(1 - 1/n)^i \leq 1/n^2$. By the union bound the probability that there exists a non hit terminal is at most $1/n$. Finally, in the case that there exists a non hit terminal, the algorithm has an approximation ratio of $3/2$. Thus the loss in the ratio is at most $\frac{3}{2n}$.   □

We now give properties of these algorithms that will enable us to prove the approximation ratio. We say that a star R is a **proper star** if R has at least one terminal and, if R has at least two edges, then all the leaves of R are terminals (a star with one edge may have one terminal, that may be the leaf or the center). Fix some proper star R with center r. Note that if R has a single edge then r can be the unique terminal in R. Denote the leaves of R by $v_1, v_2, \ldots, v_q$ arranged by non-increasing edge costs $c_1 \geq c_2 \geq \ldots \geq c_q$ where $c_j = c(rv_j)$ and assume that $c_1 > 0$. Note that $p(R) = c_1 + c(R) = c_1 + \sum_{j=1}^{q} c_j$. Let $\psi(R)$ be defined by:

$$\psi(R) = \begin{cases} c_3 + c_5 + \cdots + c_q & q \geq 3 \text{ odd} \\ c_3 + c_5 + \ldots + c_{q-1} & q \geq 4 \text{ even, } r \notin T \\ c_3 + c_5 + \ldots + c_{q-1} + c_q & q \geq 4 \text{ even, } r \in T \end{cases}$$

Here $\psi(R) = 0$ if $q \in \{1, 2\}$, except that $\psi(R) = c_2$ if $q = 2$ and $r \in T$.

The following lemma is proved in [19], but we provide a proof-sketch for completeness of exposition.

Lemma 5 ([19]). *Let R be a proper star as above. Then there exists a 2-restricted cover T of the terminals in R such that $p(T) \leq p(R) + \psi(R) \leq \frac{3}{2}p(R)$.*

Proof. It is not hard to verify that the following T is as required:

$$\begin{array}{ll} T = \{\{v_1, v_2\}, \{v_3, v_4\}, \ldots, \{v_{q-2}, v_{q-1}\}, \{v_q\}\} & q \text{ odd, } r \notin T \\ T = \{\{v_1, v_2\}, \{v_3, v_4\}, \ldots, \{v_{q-2}, v_{q-1}\}, \{v_q, r\}\} & q \text{ odd, } r \in T \\ T = \{\{v_1, v_2\}, \{v_3, v_4\}, \ldots, \{v_{q-3}, v_{q-2}\}, \{v_{q-1}, v_q\}\} & q \text{ even, } r \notin T \\ T = \{\{v_1, v_2\}, \{v_3, v_4\}, \ldots, \{v_{q-3}, v_{q-2}\}, \{v_{q-1}\}, \{v_q, r\}\} & q \text{ even, } r \in T \end{array}$$

It is also not hard to see that $\psi(R) \leq \frac{1}{2}p(R)$.   □

Assume for a moment that proper star R as above contains all terminals and is an optimal solution to our problem. Then $p(R) + \psi(R)$ bounds the solution value produced by the KN-Algorithm. We will show later that the expected solution value produced by the IRR-Algorithm is bounded by $p(R) + \phi(R)$ where

$$\phi(R) = \begin{cases} \sum_{j=1}^{q} c_j/j & q \geq 1, V(R) \subseteq T \\ \sum_{j=2}^{q} c_j/j & q \geq 1, r \notin T \\ 0 & \text{otherwise } (q = 1, V(R) \cap T = \{r\}) \end{cases}$$

The function $\phi(R)$ is built so that the proof of Lemma 7 to follow holds; we note that Harmonic functions are also used in [13] and [11].

If we know that our optimal solution is just one star R, then by taking the best outcome of the two algorithms, the (expected) value of the produced solution will be $p(R) + \min\{\psi(R), \phi(R)\}$. In the case of many stars, we take a convex combination of the two algorithms: KN-Algorithm with probability $\theta = 2/3$ and IRR-Algorithm with probability $1 - \theta = 1/3$. Since any inclusion-minimal solution is a collection of node-disjoint proper stars, we conclude that the (expected) approximation ratio of the convex combination algorithm is bounded by the maximum possible value of

$$\frac{\theta(p(R) + \psi(R)) + (1 - \theta)(p(R) + \phi(R))}{p(R)} = 1 + \frac{1}{3} \cdot \frac{2\psi(R) + \phi(R)}{p(R)}$$

over all the stars R (this assumes that, as shown in Lemma 8 below, the expected power of the output of the IRR-Algorithm is $p(R) + \phi(R)$).

For a proper star R as above let us denote (with some abuse of notation) $p(q) = p(R)$, $\psi(q) = \psi(R)$, and $\phi(q) = \phi(R)$. Then the expected approximation ratio of the convex combination algorithm is bounded by $\max_{q \geq 1} \rho(q)$, where

$$\rho(q) = 1 + \frac{1}{3} \max_{c_1 \geq \cdots \geq c_q \geq 0, c_1 > 0} \frac{2\psi(q) + \phi(q)}{p(q)}.$$

We will show later that:

Lemma 6. $\rho(q) \leq 1\frac{73}{180} < 1.4056.$

Let $\Phi(R) = p(R) + \phi(R)$ and $\Psi(R) = p(R) + \psi(R)$. It is convenient to also have $\Phi(R) = 0$ if the star R has only a center and no leaves (this is not a proper star, and has $p(R) = 0$). The next lemma, to be proved in Sect. 3, is the heart of the proof of Theorem 1.

Lemma 7. *Consider an iteration of the IRR-Algorithm. Let R be a proper star at the beginning of the iteration and let R' be a star obtained from R by removing the leaves of R that are terminals covered at the iteration, with one exception: if only the center of R is an uncovered terminal among $V(R)$ after the iteration, we keep in R' the leaf closest to the center (this means that, unless all the terminals of R are covered, R' remains a proper star). Then $\Phi(R) - E[\Phi(R')] \geq p(R)/n$.*

For a collection $\mathcal{R}$ of stars let $\Phi(\mathcal{R}) := \sum_{R \in \mathcal{R}} \Phi(R)$ and $p(\mathcal{R}) := \sum_{R \in \mathcal{R}} p(R)$.

Lemma 8. *Let $\mathcal{R} = \{R_S : S \in T\}$ be a set of stars of a k-restricted (optimal) T-cover T and J a solution produced by the IRR-Algorithm. Then $E[p(J)] \leq \Phi(\mathcal{R})$.*

Proof. Let T_{i-1} be the set of terminals uncovered at the beginning of iteration i and τ_i^* the expected optimal value of $LP_k(T_{i-1})$. Let $\mathcal{R}_0 = \mathcal{R}$ and for $i \geq 1$ obtain $\mathcal{R}_i$ from $\mathcal{R}_{i-1}$ by taking, for each proper star in $R \in \mathcal{R}_{i-1}$, the star R' as in Lemma 7. Now, note the following:

- $E[p(J)] \leq \sum_{i\geq 1} \tau_i^*/n$, since after solving $LP_k(T_{i-1})$ at iteration $i \geq 1$, each star R_S is selected with probability x_S/n.
- $\tau_i^* \leq E[p(\mathcal{R}_{i-1})]$ at iteration $i \geq 1$, since the stars in $\mathcal{R}_{i-1}$ cover T_{i-1} while τ_i^* is the expected optimal value of $LP_k(T_{i-1})$.
- $E[p(\mathcal{R}_{i-1})]/n \leq E[\Phi(\mathcal{R}_{i-1}) - \Phi(\mathcal{R}_i)]$ at iteration $i \geq 1$, by Lemma 7.

Combining we get that the expected power of J is bounded by:

$$E[p(J)] \leq \sum_{i\geq 1} \tau_i^*/n \leq \sum_{i\geq 1} E[p(\mathcal{R}_{i-1})]/n \leq \sum_{i\geq 1} E[\Phi(\mathcal{R}_{i-1}) - \Phi(\mathcal{R}_i)] = \Phi(\mathcal{R})$$

The last equality holds since the sum is telescopic and since $\Phi(\mathcal{R}_0) = \Phi(\mathcal{R})$ is not a random variable. □

Let J_{KN} and J_{IRR} be the outputs of the KN-Algorithm and the IRR-Algorithm, respectively. Let $\mathcal{R}$ and $\mathcal{R}_k$ be optimal and k-restricted optimal set of stars that cover T, respectively. Then $p(\mathcal{R}_k) \leq (1 + 1/k)p(\mathcal{R})$, by lemma 3. As was mentioned, in [19] it is proved that $p(J_{KN}) \leq \Psi(\mathcal{R})$. By Lemma 8, $p(J_{IRR}) \leq \Phi(\mathcal{R})$. Combining we get that the power of the solution produced by the convex combination of the two algorithms is bounded by

$$\theta p(J_{KN}) + (1 - \theta)p(J_{IRR}) \leq \theta\Psi(\mathcal{R}) + (1 - \theta)\Phi(\mathcal{R}_k) \leq \left(1 + \frac{1}{k}\right)(\theta\Psi(\mathcal{R}) + (1 - \theta)\Phi(\mathcal{R}))$$

From Lemma 6 we conclude that $\theta\Psi(\mathcal{R}) + (1-\theta)\Phi(\mathcal{R}) \leq 1.4056 p(\mathcal{R})$ for $\theta = 2/3$. Consequently, we get that for $\theta = 2/3$ and constant k large enough

$$\theta p(J_{KN}) + (1 - \theta)p(J_{IRR}) \leq 1.41 p(\mathcal{R}) = 1.41 \cdot \text{opt} \, .$$

To complete the proof of the 1.41 approximation ratio it only remains to prove Lemmas 6 and 7; Lemma 6 is proved below, while Lemma 7 is proved in the next section.

For the proof of Lemma 6 we bound the function $h(q) = 3(\rho(q) - 1)$, so $\rho(q) = 1 + \frac{1}{3}h(q)$. For simplicity of notation let us write

$$h(q) = \frac{2\psi(q) + \phi(q)}{p(q)} \quad \text{meaning} \quad h(q) = \max_{c_1 \geq c_2 \geq \cdots \geq c_q, c_1 > 0} \frac{2\psi(q) + \phi(q)}{p(q)} \, .$$

Note that Lemma 6 follows immediately from the following lemma:

Lemma 9. $h(q) \leq \frac{73}{60}$.

Proof. Let us consider the cases $q = 1, 2, 3, 4, 5$.

1. $\psi(1) = 0$, $\phi(1) \leq c_1$ and $p(1) = 2c_1 > 0$, hence $h(1) = 1/2$.
2. $\psi(2) = c_2$, $\phi(2) = c_1 + c_2/2$, and $p(2) = 2c_1 + c_2$, hence $h(2) \leq \frac{c_1 + \frac{5}{2}c_2}{2c_1 + c_2} \leq \frac{7}{6}$.
3. We have $h(3) = \frac{2c_3 + c_1 + c_2/2 + c_3/3}{2c_1 + c_2 + c_3} \leq \frac{23}{24}$.

4. We have $h(4) = \frac{2c_3 + 2c_4 + c_1 + c_2/2 + c_3/3 + c_4/4}{2c_1 + c_2 + c_3 + c_4}$, and this can be verified to be at most $\frac{73}{60}$ by expanding and using $c_1 \geq c_2 \geq c_3 \geq c_4$ (we get equality when $c_1 = c_2 = c_3 = c_4$).

5. We have $h(5) = \frac{2c_3 + 2c_5 + c_1 + c_2/2 + c_3/3 + c_4/4 + c_5/5}{2c_1 + c_2 + c_3 + c_4 + c_5}$, and this can be verified to be at most $\frac{73}{60}$ by expanding and using $c_1 \geq c_2 \geq c_3 \geq c_4 \geq c_5$.

For $q > 5$, we use induction on q. For even $q > 4$, we must prove: $60(2(c_3 + c_5 + \cdots c_{q-1} + c_q) + \sum_{j=1}^{q} c_j/j) \leq 73(c_1 + \sum_{j=1}^{q} c_j)$, which follows from summing up the inductive hypothesis: $60(2(c_3 + c_5 + \cdots c_{q-3} + c_{q-2}) + \sum_{j=1}^{q-2} c_j/j) \leq 73(c_1 + \sum_{j=1}^{q-2} c_j)$ and the inequalities $60c_q \leq 60c_{q-2}$, $60c_{q-1} \leq 60c_{q-2}$, $60c_q(1 + 1/q) \leq 73c_q$, and $60c_{q-1}(1 + 1/(q-1)) \leq 73c_{q-1}$.

For odd $q > 5$, we must prove: $60(2(c_3 + c_5 + \cdots + c_q) + \sum_{j=1}^{q} c_j/j) \leq 73(c_1 + \sum_{j=1}^{q} c_j)$, which follows from summing up the inductive hypothesis: $60(2(c_3 + c_5 + \cdots + c_{q-2}) + \sum_{j=1}^{q-2} c_j/j) \leq 73(c_1 + \sum_{j=1}^{q-2} c_j)$ and $(120 + 60/q)c_q + c_{q-1}60/(q-1) \leq 73(c_{q-1} + c_q)$. $\quad\square$

In the case when T is an independent set in G, no star has center in T. In this case, we simply run the IRR-Algorithm (the improvement one gets from using a convex combination is minor). The approximation ratio stated for this case in Theorem 1 follows from the following lemma.

Lemma 10. *If T is an independent set in G then $\Phi(q)/p(q) \leq \frac{73}{60}$.*

Proof. In this case, we have $\Phi(q) = \sum_{j=1}^{q} c_j(1 + 1/j)$. One obtains that $60\Phi(q) \leq 73p(q)$ for $q = 1, 2, 3, 4$ by inspection, using $c_1 \geq c_2 \geq c_3 \geq c_4$. The bound is tight for $q = 4$ and $c_1 = c_2 = c_3 = c_4$. For $q \geq 5$, the bound follows from the fact that $60(1 + 1/j) \leq 73$ for any $j \geq 5$. $\quad\square$

3   Proof of Lemma 7

Let us write explicitly the function Φ:

$$\Phi(R) = p(R) + \phi(R) = \begin{cases} c_1 + \sum_{j=1}^{q} c_j(1 + 1/j) & q \geq 1, V(R) \subseteq T \\ \sum_{j=1}^{q} c_j(1 + 1/j) & q \geq 1, r \notin T \\ 2c_1 & \text{otherwise } (q = 1, V(R) \cap T = \{r\}) \end{cases}$$

We split the proof into two cases: $r \notin T$ and $r \in T$.

3.1   The Case $r \notin T$

Recall that a set-function f on a groundset U is submodular if for any $A \subseteq U$ and $a_j, a_k \in U \setminus A$ we have:

$$\Delta_f(A, \{a_j, a_k\}) := f(A \cup \{a_j\}) + f(A \cup \{a_k\}) - f(A) - f(A \cup \{a_j, a_k\}) \geq 0.$$

We will need the following lemma. We believe this lemma is known, but we failed to find its proof in the literature.

Lemma 11. *Let U be a set of items with non-negative weights $\{w(u) : u \in U\}$ and let $z_1 \geq z_2 \geq \cdots \geq z_{|U|}$ be reals. Let $f(\emptyset) := 0$ and for $\emptyset \neq A \subseteq U$ define $f(A) := \sum_{i=1}^{|A|} z_i w(a_i)$, where $a_1, \ldots, a_{|A|}$ is an ordering of A such that $w(a_1) \geq \cdots \geq w\left(a_{|A|}\right)$. Then f is submodular and non-decreasing.*

Proof. Let $A \subseteq U$ and $a_j, a_k \in U \setminus A$. Order the elements in $A \cup \{a_j, a_k\}$ in non-increasing order $a_1, \ldots, a_{|A|+2}$ by the weights $w_1 \geq \cdots \geq w_{|A|+2}$, and suppose w.l.o.g. that this order is $a_1, \ldots, a_{j-1}, a_j, a_{j+1}, \ldots, a_{k-1}, a_k, a_{k+1}, \ldots, a_{|A|+2}$. Note that the terms in the sums defining $f(A \cup \{a_k\})$ and $f(A)$ coincide up to the kth term, and this so also for $f(A \cup \{a_j\})$ and $f(A \cup \{a_j, a_k\})$. Then we have:

$$f(A \cup \{a_k\}) - f(A) = \sum_{i=k}^{|A|+2} w_i z_{i-1} - \sum_{i=k+1}^{|A|+2} w_i z_{i-2} = \sum_{i=k}^{|A|+2} w_i z_{i-1} - \sum_{i=k}^{|A|+1} w_{i+1} z_{i-1}$$

$$f(A \cup \{a_j\}) - f(A \cup \{a_j, a_k\}) = \sum_{i=k+1}^{|A|+2} w_i z_{i-1} - \sum_{i=k}^{|A|+2} w_i z_i = \sum_{i=k}^{|A|+1} w_{i+1} z_i - \sum_{i=k}^{|A|+2} w_i z_i$$

Consequently,

$$\Delta_f(A, \{a_j, a_k\}) = \sum_{i=k}^{|A|+2} w_i z_{i-1} - \sum_{i=k}^{|A|+1} w_{i+1} z_{i-1} + \sum_{i=k}^{|A|+1} w_{i+1} z_i - \sum_{i=k}^{|A|+2} w_i z_i$$

$$= \sum_{i=k}^{|A|+2} w_i (z_{i-1} - z_i) - \sum_{i=k}^{|A|+1} w_{i+1}(z_{i-1} - z_i)$$

$$\geq \sum_{i=k}^{|A|+1} (w_i - w_{i+1})(z_{i-1} - z_i) \geq 0$$

This shows that f is submodular. It is easy to see that f is non-decreasing. $\square$

We want to show that $\Phi(R) - E[\Phi(R')] \geq p(R)/n$. Let $\bar{R}$ be the set of leaves of R. The case $\bar{R} = \emptyset$ is obvious hence we assume that $\bar{R} \neq \emptyset$.

In this case, by definition, $\Phi(R) = \sum_{j=1}^{q}(1 + 1/j)c_j$. Therefore if we set in Lemma 11 $w_i = c_i$ for every i and and $z_i = 1 + 1/i$ for $1 \leq i \leq q$, then by definition $f(\bar{R}) = \Phi(R)$.

Definition 3. *Let $\tilde{H}$ be the random variable of the set of terminals hit in iteration i. For $H \subseteq T$ we denote the probability that $\tilde{H} \cap \bar{R} = H$ by $Pr[H]$, namely, that H is exactly the set of hit terminals among the vertices of R.*

Denote $\Delta(H) = \Phi(R) - \Phi(R')$; in this case $(r \notin T)$, we have $\Delta(H) = f(\bar{R}) - f(\bar{R} \setminus H)$. Consider some arbitrary set $H \subseteq \bar{R}$ of possible terminals that could be hit. The following lemma is a standard consequence of submodularity:

Lemma 12. $\Delta(H) \geq \sum_{v \in H} \Delta(\{v\})$.

Proof. We have

$$\Delta(H) = f(\bar{R}) - f(\bar{R} \setminus H) = \qquad \text{(As the sum is telescopic)}$$

$$= \sum_{\ell=1}^{p} f\left(\bar{R} \setminus \{v_1, \ldots v_{\ell-1}\}\right) - f(\bar{R} \setminus \{v_1, \ldots v_\ell\}) \geq \qquad \text{(As } f \text{ is submodular)}$$

$$\geq \sum_{\ell=1}^{p} f(\bar{R}) - f(\bar{R} \setminus \{v_\ell\}) = \sum_{\ell=1}^{p} \Delta(\{v_\ell\})$$

$$\square$$

Therefore

$$E[\Delta(H)] = \sum_{H \subseteq \bar{R}} Pr[H]\Delta(H) \geq \qquad \text{(As } \Delta(H) \geq \sum_\ell \Delta(v_\ell))$$

$$\geq \sum_{H \subseteq \bar{R}} \left(Pr[H] \sum_{v \in H} \Delta(\{v\}) \right) = \qquad \text{(By changing summation order)}$$

$$= \sum_{v \in \bar{R}} \left(\Delta(\{v\}) \sum_{H \subseteq \bar{R} \mid v \in H} Pr[H] \right)$$

$$= \sum_{v \in \bar{R}} \Delta(v) Pr[v \text{ is hit}] \geq \sum_{v \in \bar{R}} \Delta(v) \frac{1}{n}$$

To justify the last equality, note that $\sum_{H \subseteq \bar{R} \mid v \in H} Pr[H] = Pr[v \text{ is hit}]$ because we sum the probabilities of all sets H that contain v. The last inequality follows from Lemma 2.

What remains to be proved is that

$$\sum_{v \in \bar{R}} \Delta(v) \geq p(R) \qquad (1)$$

We need to measure the change in the potential $\Delta(v_\ell)$ (recall that v_ℓ is the ℓth child of the star R). Also recall that in the potential function $\Phi(R)$, c_ℓ is multiplied by $(1 + 1/\ell)$. The addition of v_1 (and its most expensive edge) shifts all indexes by 1. This means that $v_{\ell-1}$ becomes v_ℓ. In the new star with v_1 the coefficient of the edge number ℓ is $1 + 1/\ell$ and in the star without this edges it was $1/(\ell - 1)$. Thus the difference between the coefficients is $-(1/(\ell-1) - 1/\ell)$.

Suppose that we add an edge rv_p, $p \geq 2$. Then the coefficients are shifted only for edges that are $p+1$ smallest or later. This means that the sum will start with $\ell = p + 1$. Indeed adding edge number p does not change the location of the $p - 1$ first edges. Thus the changes are as follows:

$$\Delta(\{v_1\}) \geq 2c_1 - \sum_{\ell=2}^{q} c_\ell \left(\frac{1}{\ell - 1} - \frac{1}{\ell} \right)$$

$$\Delta(\{v_2\}) = c_2 \left(1 + \frac{1}{2} \right) - \sum_{\ell=3}^{q} c_\ell \left(\frac{1}{\ell - 1} - \frac{1}{\ell} \right)$$

$$\cdots$$

$$\Delta(\{v_k\}) = c_k \left(1 + \frac{1}{k} \right) - \sum_{\ell=k+1}^{q} c_\ell \left(\frac{1}{\ell - 1} - \frac{1}{\ell} \right)$$

$$\cdots$$

$$\Delta(\{v_q\}) = c_q \left(1 + \frac{1}{q}\right)$$

Note that the coefficient of edge k is counted $k - 1$ times and thus we get by summing up these equations that:

$$\sum_{k=1}^{q} \Delta(v_k) \geq 2c_1 + \sum_{k=2}^{q} c_k \left(1 + \frac{1}{k} - (k-1)\left(\frac{1}{k-1} - \frac{1}{k}\right)\right) = 2c_1 + \sum_{k=2}^{q} c_k = p(R),$$

ending the proof for the case $r \notin T$.

3.2   The Case $r \in T$

Note that the equality $\Phi(R) - \Phi(R') = f(\bar{R}) - f(\bar{R}')$ no longer holds in all the cases, because $\Phi(R) = f(\bar{R}) + c_1$, but this may not hold for $\Phi(R')$. Precisely, the bound $\Delta(H)$ is by definition:

$$\Delta(H) = f(\bar{R}) - f(\bar{R}') + c_1 \qquad \text{if } r \text{ is hit } (H \ni r)$$
$$\Delta(H) = f(\bar{R}) - f(\bar{R}') + c_1 - c_1' \qquad \text{if } r \text{ is not hit } (H \not\ni r) \text{ and } \bar{R} \neq H$$
$$\Delta(H) = f(\bar{R}) + c_1 - 2c_q \qquad \text{if } \bar{R} = H$$

Indeed, if r is hit, then c_1 does not appear anymore in $\phi(R')$, since r is no longer a terminal. If r is not hit, its power goes from c_1 to c_1'. In the case $\bar{R} = H$ we get that $\Phi(R) - \Phi(R') = (c_1 - 2c_q) + \sum_{j \geq 1}(1 + 1/j) \cdot c_j$. This is because R' is defined to keep from R only the leaf closest to the center, and therefore $\Phi(R') = 2c_q$.

Corollary 1. If $\bar{R} \neq H$ then $\Phi(R) - \Phi(R') \geq f(\bar{R}) - f(\bar{R}')$. If $\bar{R} = H$ then $\Delta(H) = f(\bar{R}) + c_1 - 2c_q \geq f(\bar{R}) - c_1$.

We continue with the proof of Lemma 7 for the case $r \in T$. We first assume that $Pr[\bar{R}] \leq 1/n$. Then we have

$$E[\Delta(H)] = Pr[\bar{R}] \cdot \Delta(\bar{R}) + \sum_{H \neq \bar{R}} Pr[H] \cdot \Delta(H) \qquad \text{(Corollary 1 and } Pr[\bar{R}] \leq 1/n)$$

$$\geq -\frac{1}{n} c_1 + Pr[\bar{R}] f(\bar{R}) + \sum_{H \neq \bar{R}} Pr[H] \Delta(H) \qquad \text{(By separating } r \text{ from the sum)}$$

$$\geq -\frac{1}{n} c_1 + Pr[\bar{R}] \Delta(\bar{R}) + \sum_{H \not\ni r, H \neq \bar{R}} Pr[H] \Delta(H) + \sum_{H \ni r} Pr[H] \Delta(H)$$

By the definition of Δ we get that $E[\Delta(H)] + \frac{1}{n} c_1$ is at least

$$Pr[\bar{R}] f(\bar{R}) + \sum_{H \not\ni r, H \neq \bar{R}} Pr[H](f(\bar{R}) - f(\bar{R} \setminus H)) + \sum_{H \ni r} Pr[H](c_1 + f(\bar{R}) - f(\bar{R} \setminus H))$$

$$= \sum_{H \not\ni r} Pr[H](f(\bar{R}) - f(\bar{R} \setminus H)) + \sum_{H \ni r} Pr[H](c_1 + f(\bar{R}) - f(\bar{R} \setminus H))$$

Lemma 12 submodularity implies that the last expression is at least

$$\sum_{H\ni r}\sum_{v\in H} Pr[H]\sum_{v\in H}(f(\bar{R})-f(\bar{R}\backslash\{v\}))+\sum_{H\ni r} Pr[H](c_1+\sum_{v\in H\backslash\{r\}}(f(\bar{R})-f(\bar{R}\backslash\{v\})))$$

By rearranging terms and applying Lemma 4 we get

$$E[\Delta(H)] \geq -\frac{1}{n}c_1 + \left(\sum_{v\in\bar{R}}(f(\bar{R}) - f(\bar{R}\setminus\{v\}))\cdot\sum_{H\ni v} Pr[H]\right) + c_1\cdot\sum_{H\ni r} Pr[H]$$

$$\geq -\frac{1}{n}c_1 + \left(\sum_{v\in\bar{R}}(f(\bar{R}) - f(\bar{R}\setminus\{v\}))\frac{1}{n}\right) + c_1\cdot\frac{1}{n}$$

$$= \frac{1}{n}\sum_{v\in\bar{R}}(f(\bar{R}) - f(\bar{R}\setminus\{v\}))$$

$$\geq p(R)/n,$$

where the last inequality is as in the case $r \notin T$.

The second case is if $Pr[\bar{R}] > 1/n$. In this case only the contribution of disjoint events $H = \bar{R}$ and $r \in H$ is taken into account:

$$E[\Delta(R)] \geq Pr[\bar{R}]\Delta(\bar{R}) + Pr[r \text{ is hit}]\cdot\Delta(\{r\}) \qquad \text{(Corollary 1)}$$

$$\geq Pr[\bar{R}]\left(f(\bar{R}) - c_1\right) + Pr[r \text{ is hit}]\cdot\Delta(\{r\}) \qquad \text{(We assume } Pr[\bar{R}] \geq 1/n)$$

$$\geq \frac{1}{n}\cdot\left(f(\bar{R}) - c_1\right) + Pr[r \text{ is hit}]\cdot\Delta(\{r\}) \qquad \text{(Lemma 4)}$$

$$\geq \frac{1}{n}\cdot\left(f(\bar{R}) - c_1\right) + \Delta(\{r\})/n \qquad \text{(Definition of } \Delta)$$

$$= \frac{1}{n}\cdot f(\bar{R}) \qquad \text{(Definition of } f)$$

$$\geq \frac{1}{n}p(R).$$

This finishes the proof of Lemma 7 and thus the proof of Theorem 1 is complete. Theorems 2 and 3 will be proved in the full version, due to space limitation.

Acknowledgment. Gruia and Zeev thank Neil Olver for many useful discussions. Also, Gruia and Zeev acknowledge the support of the Hausdorff Trimester Program for Combinatorial Optimization (held at the Hausdorff Research Institute for Mathematics, University of Bonn).

References

1. Althaus, E., Calinescu, G., Mandoiu, I., Prasad, S., Tchervenski, N., Zelikovksy, A.: Power efficient range assignment for symmetric connectivity in static ad hoc wireless networks. Wirel. Netw. **12**(3), 287–299 (2006)
2. Anshelevich, E., Karagiozova, A.: Terminal backup, 3D matching, and covering cubic graphs. SIAM J. Comput. **40**(3), 678–708 (2011)

3. Byrka, J., Grandoni, F., Rothvoß, T., Sanità, L.: Steiner tree approximation via iterative randomized rounding. J. ACM **60**(1), 6:1–6:33 (2013)

4. Calinescu, G., Kapoor, S., Olshevsky, A., Zelikovsky, A.: Network lifetime and power assignment in ad hoc wireless networks. In: Di Battista, G., Zwick, U. (eds.) ESA 2003. LNCS, vol. 2832, pp. 114–126. Springer, Heidelberg (2003). https://doi.org/10.1007/978-3-540-39658-1_13

5. Calinescu, G., Wan, P.J.: Range assignment for biconnectivity and k-edge connectivity in wireless ad hoc networks. Mob. Netw. Appl. **11**(2), 121–128 (2006)

6. Caragiannis, I., Kaklamanis, C., Kanellopoulos, P.: Energy-efficient wireless network design. Theory Comput. Syst. **39**(5), 593–617 (2006)

7. Chen, W., Huang, N.: The strongly connecting problem on multi-hop packet radio networks. IEEE Trans. Commun. **37**(3), 293–295 (1989)

8. Clementi, A.E.F., Penna, P., Silvestri, R.: Hardness results for the power range assignment problem in packet radio networks. In: Hochbaum, D.S., Jansen, K., Rolim, J.D.P., Sinclair, A. (eds.) APPROX/RANDOM -1999. LNCS, vol. 1671, pp. 197–208. Springer, Heidelberg (1999). https://doi.org/10.1007/978-3-540-48413-4_21

9. Clementi, A.E.F., Penna, P., Silvestri, R.: The power range assignment problem in radio networks on the plane. In: Reichel, H., Tison, S. (eds.) STACS 2000. LNCS, vol. 1770, pp. 651–660. Springer, Heidelberg (2000). https://doi.org/10.1007/3-540-46541-3_54

10. Edmonds, J.: Paths, trees, and flowers. Can. J. Math. **17**, 449–467 (1965)

11. Goemans, M.X., Olver, N., Rothvoß, T., Zenklusen, R.: Matroids and integrality gaps for hypergraphic Steiner tree relaxations. In: STOC, pp. 1161–1176 (2012)

12. Goemans, M.X., Williamson, D.P.: New 3/4-approximation algorithms for the maximum satisfiability problem. SIAM J. Discrete Math. **7**, 656–666 (1994)

13. Grandoni, F.: On min-power Steiner tree. In: Epstein, L., Ferragina, P. (eds.) ESA 2012. LNCS, vol. 7501, pp. 527–538. Springer, Heidelberg (2012). https://doi.org/10.1007/978-3-642-33090-2_46

14. Hajiaghayi, M.T., Immorlica, N., Mirrokni, V.: Power optimization in fault-tolerant topology control algorithms for wireless multi-hop networks. IEEE-ACM Trans. Netw. **15**(6), 1345–1358 (2007)

15. Hajiaghayi, M.T., Kortsarz, G., Mirrokni, V.S., Nutov, Z.: Power optimization for connectivity problems. Math. Program. **110**(1), 195–208 (2007)

16. Hoffmann, S., Wanke, E.: Minimum power range assignment for symmetric connectivity in sensor networks with two power levels. Report no. 1605.01752 on arXiv, May 2016

17. Jia, X., Kim, D., Makki, S., Wan, P., Yi, C.: Power assignment for k-connectivity in wireless ad hoc networks. J. Comb. Optim. **9**(2), 213–222 (2005)

18. Kirousis, L.M., Kranakis, E., Krizanc, D., Pelc, A.: Power consumption in packet radio networks. Theor. Comput. Sci. **243**(1–2), 289–305 (2000)

19. Kortsarz, G., Nutov, Z.: Approximating minimum-power edge-covers and 2, 3-connectivity. Discrete Appl. Math. **157**(8), 1840–1847 (2009)

20. Krumke, S.O., Liu, R., Lloyd, E.L., Marathe, M.V., Ramanathan, R., Ravi, S.S.: Topology control problems under symmetric and asymmetric power thresholds. In: Pierre, S., Barbeau, M., Kranakis, E. (eds.) ADHOC-NOW 2003. LNCS, vol. 2865, pp. 187–198. Springer, Heidelberg (2003). https://doi.org/10.1007/978-3-540-39611-6_17

21. Lando, Y., Nutov, Z.: On minimum power connectivity problems. J. Discrete Algorithms **8**(2), 164–173 (2010)

22. Lev-Tov, N., Peleg, D.: Polynomial time approximation schemes for base station coverage with minimum total radii. Comput. Netw. **47**(4), 489–501 (2005)
23. Li, M., Li, Z., Vasilakos, A.V.: A survey on topology control in wireless sensor networks: taxonomy, comparative studies and open issues. Proc. IEEE **101**(12), 1.1–1.20 (2013)
24. Lloyd, E.L., Liu, R., Marathe, M.V., Ramanathan, R., Ravi, S.S.: Algorithmic aspects of topology control problems for ad hoc networks. Mob. Netw. Appl. (MONET) **10**, 19–34 (2005)
25. Lloyd, E.L., Liu, R., Ravi, S.S.: Approximating the minimum number of maximum power users in ad hoc networks. Mob. Netw. Appl. (MONET) **11**(2), 129–142 (2006)
26. Mustafa, N.H., Ray, S.: Improved results on geometric hitting set problems. Discrete Comput. Geom. **44**(4), 883–895 (2010)
27. Nutov, Z.: Approximating minimum power covers of intersecting families and directed edge-connectivity problems. Theor. Comput. Sci. **411**(26–28), 2502–2512 (2010)
28. Nutov, Z.: Activation network design problems. In: Gonzalez, T.F. (ed.) Handbook on Approximation Algorithms and Metaheuristics, chap. 15, 2nd edn, vol. 2. Chapman & Hall/CRC, London (2018)
29. Rajaraman, R.: Topology control and routing in ad hoc networks: a survey. SIGACT News **33**(2), 60–73 (2002)
30. Ramanathan, R., Rosales-Hain, R.: Topology control of multi-hop wireless networks using transmit power adjustment. In: INFOCOM, pp. 404–413 (2000)
31. Santi, P.: Topology control in wireless ad hoc and sensor networks. ACM Comput. Surv. **37**(2), 164–194 (2005)
32. Schrijver, A.: Combinatorial Optimization: Polyhedra and Efficiency. Springer, Heidelberg, New York (2004)

DISPATCH: An Optimally-Competitive Algorithm for Maximum Online Perfect Bipartite Matching with i.i.d. Arrivals

Minjun Chang[ID], Dorit S. Hochbaum[ID], Quico Spaen$^{(\boxtimes)}$[ID], and Mark Velednitsky$^{(\boxtimes)}$[ID]

University of California, Berkeley, USA
{minjun.lynn,dhochbaum,qspaen,marvel}@berkeley.edu

Abstract. This work presents an optimally-competitive algorithm for the problem of maximum weighted online perfect bipartite matching with i.i.d. arrivals. In this problem, we are given a known set of workers, a distribution over job types, and non-negative utility weights for each pair of worker and job types. At each time step, a job is drawn i.i.d. from the distribution over job types. Upon arrival, the job must be irrevocably assigned to a worker and cannot be dropped. The goal is to maximize the expected sum of utilities after all jobs are assigned.

We introduce DISPATCH, a 0.5-competitive, randomized algorithm. We also prove that 0.5-competitive is the best possible. DISPATCH first selects a "preferred worker" and assigns the job to this worker if it is available. The preferred worker is determined based on an optimal solution to a fractional transportation problem. If the preferred worker is not available, DISPATCH randomly selects a worker from the available workers. We show that DISPATCH maintains a uniform distribution over the workers even when the distribution over the job types is non-uniform.

Keywords: Perfect matching · i.i.d. arrivals · Competitive ratio

1 Introduction

We consider the problem of *maximum online perfect bipartite matching*. Suppose that we have a set of jobs and a set of workers. At every time step, a single job arrives to be served by one of the workers. Upon a job's arrival, we observe the utility of assigning the job to each of the workers. We must immediately decide which worker will serve the job. Once a worker is assigned a job, it is busy and cannot be assigned to another job. Jobs continue to arrive until all workers are busy.

In the natural bipartite graph that arises, there is an edge between each worker and job with a non-negative utility of assigning that worker to that job. The assignment of workers to jobs will form a perfect matching in this bipartite graph. Our goal is to design a dispatching algorithm that maximizes the expected sum of utilities of the perfect matching.

© Springer Nature Switzerland AG 2018
L. Epstein and T. Erlebach (Eds.): WAOA 2018, LNCS 11312, pp. 149–164, 2018.
https://doi.org/10.1007/978-3-030-04693-4_10

In this work, we consider the maximum online perfect bipartite matching problem with *independent and identically distributed (i.i.d.) arrivals*. This means that, at each time step, a job is drawn i.i.d. from a known distribution over job types.

Examples of online bipartite matching include matching doctors to patients in hospitals, matching operators to callers in call centers, matching drivers to passengers in ride-sharing, and matching impressions to customers in online ad auctions [17].

We introduce the randomized algorithm DISPATCH for the problem of online weighted perfect bipartite matching with i.i.d. arrivals. DISPATCH is 0.5-competitive algorithm: the total expected utility of the perfect matching produced by DISPATCH is at least half of the total expected utility of an optimal algorithm that knows the job arrival sequence in advance. We also describe a family of problem instances for which 0.5 is the best-possible competitive ratio. The DISPATCH algorithm, thus, achieves the best-possible competitive ratio. In contrast, the same problem with adversarial job arrivals cannot be bounded, as observed by Feldman et al. [6].

To assign workers to jobs, DISPATCH first selects a *preferred worker*. This preferred worker is determined based on an optimal solution to a fractional transportation problem. If the preferred worker is available, then job is assigned to this worker. Otherwise, DISPATCH randomly selects a worker from the available workers.

1.1 Related Work

Our work resides in the space of online matching problems, including the Maximum (Imperfect) Bipartite Matching problem and the Minimum (Perfect) Bipartite Matching problem. Another closely related problem is the k-Server problem. For each of these problems, several arrival models are considered. Arrival models including adversarial, where the adversary chooses jobs and their arrival order; random order, where the adversary chooses jobs but not their arrival order; and i.i.d., where the adversary specifies a probability distribution over job types and each arrival is sampled independently from the distribution. We briefly describe each of these problems and present best-known results, contrasting it to the setting considered here. A summary is in Table 1.

Maximum Online (Imperfect) Bipartite Matching. The maximum online (imperfect) bipartite matching problem is defined on a bipartite graph with n known workers and n jobs that arrive one at a time. Jobs either get assigned to a worker or are discarded. The goal is to maximize the cardinality (or sum of weights) of the resulting matching. In contrast to our problem, jobs may be the discarded and the resulting matching may be *imperfect*.

For the unweighted problem with adversarial arrivals, Karp, Vazirani, and Vazirani [10] showed a best-possible algorithm that achieves a competitive ratio of $1 - \frac{1}{e} \approx 0.632$. Variations of the problem have been proposed: addition of edge or vertex weights, the use of budgets, different arrival models, etc. Mehta [17] provides an excellent overview of this literature. When the arrivals are in a random order, it is possible to do better than $1 - \frac{1}{e}$. Mahdian and Yan [14], in 2011, achieved a competitive ratio of 0.696. Manshadi et al. [16] showed that you cannot do better than 0.823. If the problem also has weights, then the best-possible competitive ratio is 0.368 by a reduction from the secretary problem as shown by Kesselheim et al. [11]. They also give an algorithm that attains this competitive ratio.

The problem has also been studied when the jobs are drawn i.i.d. from a known distribution. This problem is also referred to as *Online Stochastic Matching*. The first result to break the $1 - \frac{1}{e}$ barrier for the unweighted case was the 0.67-competitive algorithm of Feldman et. al. [7] in 2009. To date, the best-known competitive ratio of 0.730 is due to Brubach et al. [2]. This is close the best-known bound of 0.745 by Correa et al. [4].

Online Minimum (Perfect) Bipartite Matching. The online minimum (perfect) bipartite matching addresses the question of finding a minimum cost perfect matching on a bipartite graph with n workers and n jobs. Given any arbitrary sequence of jobs arriving one by one, each job needs to be irrevocably assigned to worker on arrival. This problem is the minimization version of the problem considered in this work. However, the obtained competitive ratios do not transfer.

The problem was first considered by Khuller, Mitchell, and Vazirani [12] and independently by Kalyanasundaram and Pruhs [9]. If the weights are arbitrary, then the competitive ratio cannot be bounded. To address this, both papers considered the restriction where the edge weights are distances in some metric on the set of vertices. They give a $2n-1$ competitive algorithm, which is the best-possible for deterministic algorithms. When randomized algorithms are allowed, the best-known competitive ratio is $O(\log^2(n))$ by Bansal et al. [1]. If the arrival order is also randomized, then Raghvendra [19] shows that $2\log(n)$ is attainable. He also shows that this is the best possible.

k-**Server Problem.** In the k-server problem, k workers are distributed at initial positions in a metric space. Jobs are elements of the same metric space and arrive one at a time. When a job arrives, it must be assigned to a worker which moves to the job's location. The goal in the k-server problem is to minimize the total distance traveled by all workers to serve the sequence of jobs. After an assignment, the worker remains available for assignment to new jobs. This *reassignment* distinguishes the k-server problem from ours, where workers are fixed to a job once assigned.

The k-server problem was introduced by Manasse, McGeoch, and Sleater [15]. A review of the k-server problem literature was written by Koutsoupias [13]. For randomized algorithms in discrete metrics, the competitive ratio $O(\log^2(k)\log(n))$ was attained by Bubeck et. al. [3], where n is the number of points in the discrete metric space. On the other hand, $\Omega(\log(k))$ is a known lower bound. In the i.i.d. setting, Dehghani et. al. [5] consider a different kind of competitive ratio: they give an online algorithm with a cost no worse than $O(\log(n))$ times the cost of the optimal *online* algorithm.

Table 1. Best-known competitive ratios and impossibility bounds for various online bipartite matching problems. ★: Results presented in this paper.

Sense	Matching	Arrivals	Restrictions	Best known	Best possible
Max	Imperfect	Advers	0/1	0.632 [10]	0.632 [10]
Max	Imperfect	Rand. Ord	0/1	0.696 [14]	0.823 [16]
Max	Imperfect	Rand. Ord	None	0.368 [11]	0.368 [11]
Max	Imperfect	i.i.d	None	0.730 [2]	0.745 [4]
Min	Perfect	Advers	Metric	$O(\log^2(n))$ [1]	$\Omega(\log(n))$ [18]
Min	Perfect	Rand. Ord	Metric	$2\log(n)$ [19]	$2\log(n)$ [19]
Max	Perfect	Adversarial	None	-	0 [6]
Max	**Perfect**	**i.i.d.**	**None**	$\frac{1}{2}$★	$\frac{1}{2}$★

1.2 Structure of This Work

This paper is organized as follows. Section 2 formally introduces the problem of online perfect bipartite matching with i.i.d. arrivals and defines the concept of competitive ratio. Section 3 describes DISPATCH, presents an example to demonstrate the algorithm, and provides the proof that DISPATCH is 0.5-competitive. Section 4 introduces a family of instances of the online perfect bipartite matching problem for which no online algorithm performs better than $\frac{1}{2}$ in terms of competitive ratio. Finally, Sect. 5 summarizes the results and suggests directions for future research.

2 Preliminaries

The set of workers is denoted by W with size $n = |W|$. The set J denotes the set of job types with size $k = |J|$. For every worker $w \in W$ and job type $j \in J$ there is a utility of $u_{wj} \geq 0$ for assigning a job of type j to worker w. Let $\mathcal{D}(J)$ be a known probability distribution over the job types.

At every time step $t = 1, \ldots, n$, a single job is drawn i.i.d. from J according to $\mathcal{D}$. The job must be irrevocably assigned to a worker before the next job arrives. Workers are no longer available after they have been assigned a job. Let r_j denote the expected number of jobs of type j that arrive. After n steps, each worker is assigned to one job and the resulting assignment forms a perfect matching. Our goal is to design a procedure such that the expected sum of the utilities of the resulting perfect matching is as high as possible.

Throughout this work, we will repeatedly use two bipartite graphs; the *expectation graph* G and the *realization graph* $\widehat{G}$. The expectation graph $G = (W, J, E)$

is a complete bipartite graph defined over the set of workers W and the set of job types J. An edge $[w, j] \in E$ has associated utility $u_{wj} \geq 0$, for $w \in W$ and $j \in J$. The realization graph $\widehat{G} = (W, \widehat{J}, \widehat{E})$ is the random bipartite graph obtained after all n jobs have arrived. $\widehat{J}$ denotes the set of n jobs that arrived. We use $\hat{j}_t \in \widehat{J}$ to denote the job that arrives at time t and $j_t \in J$ to denote its job type. The edge set $\widehat{E}$ consists of all worker-job pairs, such that $\widehat{G}$ is a complete bipartite graph defined over W and $\widehat{J}$. Every edge $[w, \hat{j}] \in \widehat{E}$ has utility u_{wj}, where j is the job type of job $\hat{j}$. It is important to remember that the expectation graph G is deterministic and known in advance whereas the realization graph $\widehat{G}$ is a random graph representing a realization of the job arrival process and is revealed over time.

An instance of the online perfect bipartite matching problem with i.i.d. arrivals is defined by the set of workers W, the job types J, non-negative utilities u_{wj}, and a distribution over the job types $\mathcal{D}(J)$. Equivalently, the expectation graph G and the distribution $\mathcal{D}(J)$ defines an instance of this problem. Here we analyze the family of potentially randomized algorithms that return a perfect matching $\hat{M}$ on $\widehat{G}$. The performance of an algorithm ALG for a single realization $\widehat{G}$ is given by:

$$ALG(\widehat{G}) = \mathbb{E}\left[\sum_{[w,j] \in E} u_{wj} I_{wj} \right],$$

where I_{wj} is a random indicator variable that equals 1 if ALG assigned a job of type j to worker w and equals 0 otherwise. For a given problem instance defined by expectation graph G and distribution $\mathcal{D}(J)$, $\mathbb{E}\left[ALG(\widehat{G}) \right]$ measures the algorithm's expected performance over samples of $\widehat{G}$ from G according to $\mathcal{D}(J)$.

The worst-case performance across instances is measured by the *competitive ratio*. Let $OPT(\widehat{G})$ be the maximum weight perfect matching in the realization graph $\widehat{G}$ and let $\mathbb{E}\left[OPT(\widehat{G}) \right]$ be its expectation across different realizations for a given expectation graph G and distribution $\mathcal{D}(J)$. $\mathbb{E}\left[OPT(\widehat{G}) \right]$ measures the performance of an optimal algorithm that has full information about the arrival sequence. This is known as an adaptive online adversary. The ratio $\frac{\mathbb{E}[ALG(\widehat{G})]}{\mathbb{E}[OPT(\widehat{G})]}$ measures the performance of ALG relative to the optimal algorithm for a given instance of the problem. The competitive ratio is the worst-case, i.e. lowest, ratio among all possible instances of the expectation graph G and distributions $\mathcal{D}(J)$:

Definition 1 (Competitive Ratio). *An algorithm ALG is said to have a competitive ratio of α when for all instances of the expectation graph G and distribution $\mathcal{D}(J)$:*

$$\alpha \leq \frac{\mathbb{E}\left[ALG(\widehat{G}) \right]}{\mathbb{E}\left[OPT(\widehat{G}) \right]}.$$

2.1   Bounding the Performance of OPT

It is difficult to compute $\mathbb{E}\left[OPT(\widehat{G})\right]$ directly. We show that the randomness in $\widehat{G}$ reduces the expected value of the optimal perfect matching compared to the value of the optimal transportation problem where the number of jobs of each type is equal to its expectation. This offline transportation problem is then used to guide the online assignment.

A similar approach was used in the context of unweighted online imperfect bipartite matching by Feldman et al. [7] and Haepler et al. [8]. Here, we use a transportation problem instead of a maximum weight matching. We also bound the performance of OPT differently.

Recall that, in expectation, r_j jobs of job type $j \in J$ will arrive in $\widehat{G}$. An optimal fractional matching of these jobs is obtained by solving a fractional transportation problem on the expectation graph G, where each job type has a demand of r_j and each worker has a supply of 1 and the sum of utilities is maximized.

Formally, let $f_{wj} \geq 0$ be the flow from worker $w \in W$ to job type $j \in J$. This can be interpreted as a fractional assignment of worker w to jobs of job type j. We define the transportation problem TPP:

$$TPP(G) = \max_{f_{wj} \geq 0} \sum_{w \in W} \sum_{j \in J} u_{wj} f_{wj},$$

$$\sum_{w \in W} f_{wj} = r_j \quad \forall j \in J,$$

$$\sum_{j \in J} f_{wj} = 1 \quad \forall w \in W.$$

Let f^*_{wj} be an optimal flow on edge $[w, j] \in E$.

We claim that $\mathbb{E}\left[OPT(\widehat{G})\right] \leq TPP(G)$. The reason is that the weighted average of perfect matchings $OPT(\widehat{G})$ forms a feasible solution to the transportation problem above.

Lemma 1. *Given any expectation graph G and distribution over job types $\mathcal{D}(J)$,*

$$\mathbb{E}\left[OPT(\widehat{G})\right] \leq TPP(G).$$

Proof. Assign each edge in G an indicator variable I_{wj}, which takes on the value 1 if OPT assigns worker w to a job of type j in $\widehat{G}$ and 0 otherwise. We claim that $f_{wj} = \mathbb{E}\left[I_{wj}\right]$ forms a feasible solution to the transportation problem in G. Indeed,

$$\sum_{w \in W} \mathbb{E}\left[I_{wj}\right] = \mathbb{E}\left[\sum_{w \in J} I_{wj}\right] = r_j, \qquad \sum_{j \in J} \mathbb{E}\left[I_{wj}\right] = \mathbb{E}\left[\sum_{j \in J} I_{wj}\right] = 1.$$

Since $\mathbb{E}[I_{wj}]$ is feasible for the transportation problem, it must have objective smaller than $TPP(G)$:

$$\mathbb{E}\left[OPT(\widehat{G})\right] = \mathbb{E}\left[\sum_{[w,j]\in E} u_{wj}I_{wj}\right] = \sum_{[w,j]\in E} u_{ij}\mathbb{E}[I_{wj}] \leq TPP(G).$$

$\square$

This implies that we can bound the performance of an algorithm with respect to $TPP(G)$. We apply this technique in Sect. 3.3.

3   A 1/2-Competitive Algorithm

3.1   The Dispatch Algorithm

Before any jobs arrive, DISPATCH solves the offline transportation problem TPP on the expectation graph G. We find an optimal flow f^*_{wj} from workers to jobs. Throughout the online stage, the algorithm reconstruct this flow between job types and workers as much as possible. For each arriving job, a *preferred worker* w^P is randomly selected with a probability proportional to the optimal flow f^* between the corresponding job type and the worker in the transportation problem. If the preferred worker is no longer available, then the job is assigned to a worker selected randomly from the set of available workers AW. We refer to this worker as the *assigned worker* w^A. The resulting assignment forms a perfect matching on $\widehat{G}$ since each worker is assigned at most once and each job is assigned to a worker.

In the context of online bipartite matching, the idea of using an offline solution to guide the online algorithm was used in the "Suggested Matching" algorithm [7] and subsequent work, e.g. [8]. Our algorithm differs in two ways. First, the offline solution is a transportation problem instead of a maximum weight matching problem. Second, the job is randomly assigned instead of discarded when the preferred worker is no longer available. This random selection ensures that we obtain a perfect matching and is crucial for Lemma 3. The analysis of the competitive performance of DISPATCH is also novel except for Lemma 2.

The algorithm is formally defined in Algorithm 1. We prove the following result:

Theorem 1. DISPATCH *achieves a competitive ratio of* $\frac{1}{2}$ *for the online perfect bipartite matching problem with i.i.d. arrivals.*

3.2   Example

To illustrate DISPATCH, we consider the example shown in Fig. 1. The example has five workers ($n = 5$) and three job types ($k = 3$). The expectation graph is shown in Fig. 1a. Note that the distribution over job types, $\mathcal{D}(J)$, is fully specified by r_j. An instance of the realization graph is shown in Fig. 1c.

Algorithm 1. DISPATCH

Input: Expectation graph G.
Output: Perfect matching $\hat{M}$ on $\hat{G}$.

Initialization:
Solve the transportation problem TTP on G to obtain the optimal flow f^*.
$\hat{M} \leftarrow \emptyset$
$AW \leftarrow W$

Online stage:
for $t = 1, \ldots, n$ **do**
   # Job $\hat{j}_t$ arrives with job type j_t.
   Randomly draw **preferred** worker w^P with probability $p(w) = \frac{f^*_{wj_t}}{r_{j_t}}$ for $w \in W$.
   # Use preferred worker (w^P) as assigned worker (w^A) if possible.
   if $w^P \in AW$ **then**
     $w^A \leftarrow w^P$
   else
     Randomly draw $w^A \in AW$ with equal probability.
   end if
   $\hat{M} \leftarrow \hat{M} \cup [w^A, \hat{j}_t]$
   $AW \leftarrow AW - \{w^A\}$
end for

Figure 1b shows f^*, the solution to the transportation problem on G that is used by DISPATCH. The corresponding objective value is $TPP(G) = 8$. Figures 1d to h show the arrival of the jobs and the corresponding assignment made by DISPATCH. Figure 1h illustrates an instance where the preferred worker selected by DISPATCH is not available, and a different worker is assigned. For this particular realization $\hat{G}$, the perfect matching constructed by DISPATCH has a total utility 6, while the optimal perfect matching on $\hat{G}$ has a total utility 8. Note that these values are for this particular realization of $\hat{G}$. The performance guarantee is with respect to the expectation over all realizations of $\hat{G}$.

3.3 Proof of $\frac{1}{2}$-Competitiveness

To prove that the perfect matching produced by DISPATCH has a competitive ratio of a $\frac{1}{2}$, we rely on a key feature of DISPATCH: It maintains the invariant, Lemma 4, that workers are equally likely to be available even though the distribution over job types may not be uniform. To prove this invariant, we first show that both the preferred and the assigned worker are selected uniformly across workers. Recall that the preferred worker may be different than the assigned worker. In fact, the preferred worker does not have to be available and could have been assigned to another job already. Lemma 2 states this formally for the selection of the preferred worker. The observation underlying this lemma is that each worker is selected with a probability proportional to the total flow f^* originating at the worker, which is equal to one for each worker.

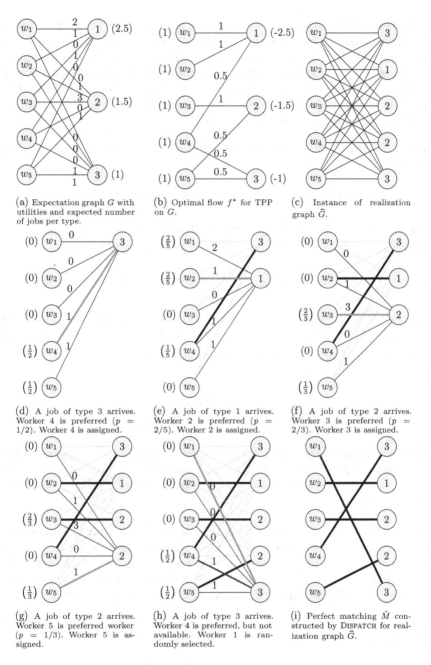

(a) Expectation graph G with utilities and expected number of jobs per type.

(b) Optimal flow f^* for TPP on G.

(c) Instance of realization graph $\widehat{G}$.

(d) A job of type 3 arrives. Worker 4 is preferred ($p = 1/2$). Worker 4 is assigned.

(e) A job of type 1 arrives. Worker 2 is preferred ($p = 2/5$). Worker 2 is assigned.

(f) A job of type 2 arrives. Worker 3 is preferred ($p = 2/3$). Worker 3 is assigned.

(g) A job of type 2 arrives. Worker 5 is preferred worker ($p = 1/3$). Worker 5 is assigned.

(h) A job of type 3 arrives. Worker 4 is preferred, but not available. Worker 1 is randomly selected.

(i) Perfect matching $\widehat{M}$ constructed by DISPATCH for realization graph $\widehat{G}$.

Fig. 1. An example of the DISPATCH algorithm on the realization graph shown in Fig. 1c. The underlying expectation graph G with $n = 5$ and $k = 3$ is shown in Fig. 1a. In Figs. 1d up to 1h, the numbers in parenthesis denote the probability of selecting that worker as the preferred worker. Red edges represent the assignment made by the algorithm, thick black edges are previous assignments, and blue edges mark unavailable preferred workers. Figure 1h shows an instance where the preferred worker is busy. (Color figure online)

Throughout this section we use additional notation. Let the random variable W_t^P represent the preferred worker for the job arriving at time t, and let the random variable W_t^A be the assigned worker. Furthermore, let the random set AW_t consist of the available workers when the job at time t arrives. We make no further assumptions on the expectation graph G and/or distribution $\mathcal{D}(J)$ other than those outlined in Sect. 2. Lemmas and theorems in this section are therefore applicable to all problem instances.

Lemma 2. *At each time t, the **preferred** worker W_t^P is drawn uniformly from all workers:*

$$\mathbb{P}\left(W_t^P = w\right) = \frac{1}{n} \quad \text{for all } w \in W \text{ and } t = 1, \dots, n.$$

Proof. By conditioning on the job type j_t at stage t and using the law of total probability, we can rewrite the probability of selecting worker w as:

$$\mathbb{P}\left(W_t^P = w\right) = \sum_{j \in J} \mathbb{P}\left(W_t^P = w | j_t = j\right) \mathbb{P}\left(j_t = j\right).$$

Since the jobs are drawn i.i.d., a job of type j is selected with probability $\mathbb{P}\left(j_t = j\right) = \frac{r_j}{n}$, by definition of r_j. Given a job of type j, the algorithm selects a worker w as the preferred worker with probability $\mathbb{P}\left(W_t^P = w | j_t = j\right) = \frac{f_{wj}^*}{r_j}$. Thus,

$$\mathbb{P}\left(W_t^P = w\right) = \sum_{j \in J} \frac{f_{wj}^*}{r_j} \frac{r_j}{n} = \sum_{j \in J} \frac{f_{wj}^*}{n}.$$

Finally, recall that every worker supplies a unit of flow in the offline transportation problem, equivalent to the expected number of jobs it serves. The edges adjacent to worker w must thus transport a unit of flow, so $\sum_j f_{wj}^* = 1$. Thus, $\mathbb{P}\left(W_t^P = w\right) = \frac{1}{n}$. □

Next we show that the assigned worker is selected uniformly at random from the set of available workers. For this lemma to hold, it is crucial that the draw of the assigned worker is done uniformly at random when the preferred worker is not available. Recall that W_t^A is the assigned worker for the job arriving at time t and that AW_t are the available workers before the job arrives.

Lemma 3. *At each time step t, the **assigned** worker W_t^P is drawn uniformly from the available workers:*

$$\mathbb{P}\left(W_t^A = w | w \in AW_t\right) = \frac{1}{n - (t-1)}.$$

Proof. Assume that w is fixed and that $w \in AW_t$. There are two ways for w to be the assigned worker. Either w is the preferred worker or the preferred worker is not available and w is randomly selected. We express this as:

$$\mathbb{P}\left(W_t^A = w | w \in AW_t\right) = \mathbb{P}\left(W_t^P = w | w \in AW_t\right)$$
$$+ \mathbb{P}\left(W_t^A = w | W_t^P \notin AW_t, w \in AW_t\right) \times$$
$$\mathbb{P}\left(W_t^P \notin AW_t | w \in AW_t\right)$$

The selection of W_t^P is independent of whether $w \in AW_t$. Therefore,

$$\mathbb{P}\left(W_t^A = w | w \in AW_t\right) = \mathbb{P}\left(W_t^P = w\right)$$
$$+ \mathbb{P}\left(W_t^A = w | W_t^P \notin AW_t, w \in AW_t\right) \mathbb{P}\left(W_t^P \notin AW_t\right)$$

Now we use three observations to complete the proof. First, Lemma 2 implies that $\mathbb{P}\left(W_t^P = w\right) = \frac{1}{n}$. Second, since there are $t - 1$ busy workers, Lemma 2 implies that $\mathbb{P}\left(W_t^P \notin AW_t\right) = \frac{(t-1)}{n}$. Third, the fact that the assigned worker is drawn uniformly at random when the preferred worker is not available implies that $\mathbb{P}\left(W_t^A = w | W_t^P \notin AW_t, w \in AW_t\right) = \frac{1}{n-(t-1)}$. Thus,

$$\mathbb{P}\left(W_t^A = w | w \in AW_t\right) = \frac{1}{n} + \frac{1}{n-(t-1)} \frac{(t-1)}{n} = \frac{1}{n-(t-1)}.$$

$\square$

Lemma 3 specifies each available worker is equally likely to be assigned to the next job. As a consequence, we can derive the probability that a worker is still available after $t - 1$ jobs have arrived:

Lemma 4. DISPATCH *maintains the following invariant throughout the online stage:*

$$\mathbb{P}\left(w \in AW_t\right) = \frac{n - (t-1)}{n} \quad \text{for all } w \in W \text{ and } t = 1, \ldots, n.$$

Proof. At every time step, a worker is chosen randomly from the remaining available workers, as shown in Lemma 3. The probability that an available worker in time step t is still available in time step $t + 1$ is:

$$\mathbb{P}\left(w \in AW_{t+1} | w \in AW_t\right) = 1 - \mathbb{P}\left(W_t^A = w | w \in AW_t\right)$$
$$= 1 - \frac{1}{n - (t-1)} = \frac{n-t}{n-(t-1)}.$$

Thus, the probability of being available for the tth job is equal to:

$$\mathbb{P}\left(w \in AW_t\right) = \prod_{i=1}^{t} \mathbb{P}\left(w \in AW_t | w \in AW_{t-1}\right)$$
$$= \frac{n - (t-1)}{n - (t-2)} \frac{n - (t-2)}{n - (t-3)} \cdots \frac{n-1}{n} = \frac{n - (t-1)}{n}.$$

$\square$

From Lemma 4, we know the probability that a worker is available at each time step. We use this to bound the probability that a worker w is assigned to a job with job type j by DISPATCH. We use the indicator random variable I_{wj}. $I_{wj} = 1$ when the DISPATCH assigns worker w to a job with job type j, and $I_{wj} = 0$ otherwise. We bound the probability with respect to f^*_{wj} in $TPP(G)$. By bounding the algorithm's performance with respect to $TPP(G)$ we can bound the competitive ratio of DISPATCH. See Sect. 2.1 for more details.

Lemma 5. *Given a perfect matching $\hat{M}$ constructed by DISPATCH, the probability that worker w is assigned to a job of type j is bounded by:*

$$\mathbb{P}\left(I_{wj} = 1\right) \geq \frac{1}{2} f^*_{wj}.$$

Proof. If $I_{wj} = 1$, then worker w must have been assigned to a job of type j in one of the time steps. Thus, $I_{wj} = \sum_{t=1}^{n} I^t_{wj}$ where I^t_{wj} is indicator for whether worker w is assigned to a job of type j at time step t:

$$\mathbb{P}\left(I_{wj} = 1\right) = \sum_{t=1}^{n} \mathbb{P}\left(I^t_{wj} = 1\right).$$

Let us bound the probability $\mathbb{P}\left(I^t_{wj} = 1\right)$ for all $t = 1, \ldots, n$. First, we condition on the job type arriving at time t. Note that j_t must equal j:

$$\mathbb{P}\left(I^t_{wj} = 1\right) = \mathbb{P}\left(I^t_{wj} = 1 | j_t = j\right) \mathbb{P}\left(j_t = j\right).$$

Recall that there are two ways for worker w to be assigned after a job of type j arrives. Either w is the preferred worker and is assigned the job, or another worker w' is selected as the preferred worker but is not available. w is then selected as the assigned worker. We lower bound the probability that worker w is assigned for the job of type j by considering only the case where w is the preferred worker.

$$\begin{aligned}
\mathbb{P}\left(I^t_{wj} = 1\right) &\geq \mathbb{P}\left(w \in AW_t, W^P_t = w | j_t = j\right) \mathbb{P}\left(j_t = j\right) \\
&= \mathbb{P}\left(w \in AW_t\right) \mathbb{P}\left(W^P_t = w | j_t = j\right) \mathbb{P}\left(j_t = j\right) \\
&= \frac{n - (t-1)}{n} \frac{f^*_{wj}}{r_j} \frac{r_j}{n} \\
&= \frac{1}{n} \frac{n - (t-1)}{n} f^*_{wj}.
\end{aligned}$$

For the first equality, we use that the job type at time t and the selection of the preferred worker are independent from whether w is available at time t. The second equality follows from Lemma 4, the weighted random selection of the preferred worker, and the job arrival process.

We use $\mathbb{P}\left(I^t_{wj} = 1\right) = \frac{1}{n} \frac{n-(t-1)}{n} f^*_{wj}$ to bound the total probability of assigning worker w for a job of type j:

$$\mathbb{P}\left(I_{wj} = 1\right) = \sum_{t=1}^{n} \mathbb{P}\left(I^t_{wj} = 1\right) \geq \sum_{t=1}^{n} \frac{1}{n} \frac{n-(t-1)}{n} f^*_{wj} = \frac{1}{2} \frac{n+1}{n} f^*_{wj} \geq \frac{1}{2} f^*_{wj}.$$

$\square$

Lemma 5 bounds the probability that worker w is matched to a job of type j. By linearity of expectation, Theorem 1 and the $\frac{1}{2}$ competitive ratio follow almost immediately from Lemma 5.

Proof (Proof of Theorem 1). The expected utility returned by the algorithm is a weighted sum of indicators whether worker w is assigned to a job of type j. Note that each worker is assigned to at most one job (type). We can then apply Lemma 5 to bound the probability $P(I_{wj} = 1)$ and the expected utility of the algorithm:

$$\mathbb{E}\left[\text{DISPATCH}(\widehat{G})\right] = \mathbb{E}\left[\sum_{w \in W, j \in J} u_{wj} I_{wj}\right]$$

$$= \sum_{w \in W, j \in J} u_{wj} \mathbb{E}[I_{wj}]$$

$$= \sum_{w \in W, j \in J} u_{wj} \mathbb{P}(I_{wj} = 1)$$

$$\geq \frac{1}{2} \sum_{w \in W, j \in J} u_{wj} f_{wj}^* = \frac{1}{2} TPP(G).$$

Note that the inequality requires that the utility weights are non-negative.

Finally, we apply Lemma 1 to obtain a bound on the competitive ratio attained by DISPATCH for any expectation graph G and distribution $\mathcal{D}(J)$:

$$\mathbb{E}\left[\text{DISPATCH}(\widehat{G})\right] \geq \frac{1}{2} TPP(G) \geq \frac{1}{2} \mathbb{E}\left[OPT(\widehat{G})\right]$$

$\square$

4   Best-Possible Competitive Ratio

We present here a family of instances for which any online algorithm attains a competitive ratio of at most $\frac{1}{2}$. The DISPATCH algorithm guarantees a competitive ratio of $\frac{1}{2}$ and is thus optimal with respect to competitive ratio.

Theorem 2. *For the online perfect bipartite matching problem with an i.i.d. arrival process, no online algorithm can achieve a competitive ratio* $\frac{\mathbb{E}[ALG(\widehat{G})]}{\mathbb{E}[OPT(\widehat{G})]}$ *better than* $\frac{1}{2}$.

Proof. Consider an instance G with the number of job types $k = n + 1$. Let the job types be indexed from 1 to $n + 1$ and the workers from 1 to n. Job types 1 to n each arrive with probability p/n and job type $n + 1$ arrives with probability $1 - p$. For this graph, we set $u_{wj} = 1$ if $w = j$ and to 0 otherwise. This implies $u_{w,n+1} = 0$ for all $w \in W$.

Note that OPT gains a utility of one per unique job type in $\{1, \ldots, n\}$ that arrives. The expected number of unique job types is computed by considering

each job type as a geometric random variable with a success probability of $\frac{p}{n}$.
Thus, $\mathbb{E}\left[OPT(\widehat{G})\right] = n\left(1 - \left(1 - \frac{p}{n}\right)^n\right)$.

For any online algorithm ALG^*, $t-1$ workers are no longer available at time step t regardless of the strategy. Thus, with probability $(1-p)+p\frac{t-1}{n}$ the increase in utility is zero. Thus, the total expected utility increases by at most $p\frac{n-(t-1)}{n}$ in time step t. The total expected utility obtained by ALG^* is then:

$$\mathbb{E}\left[ALG^*(\widehat{G})\right] \le p\frac{n}{n} + p\frac{n-1}{n} + p\frac{n-2}{n} + \cdots + p\frac{1}{n} = \frac{1}{2}p(n+1)$$

We compute the relevant ratio and then take the limit as n goes to infinity:

$$\lim_{n\to\infty} \frac{\mathbb{E}\left[ALG^*(\widehat{G})\right]}{\mathbb{E}\left[OPT(\widehat{G})\right]} = \lim_{n\to\infty} \frac{\frac{1}{2}p(n+1)}{n\left(1 - \left(1 - \frac{p}{n}\right)^n\right)} = \frac{1/2 \cdot p}{1 - e^{-p}}$$

Since p can take on any value in the interval $(0, 1)$, we consider the limit as p goes to zero:

$$\lim_{p\to 0^+} \frac{1/2 \cdot p}{1 - e^{-p}} = \lim_{p\to 0^+} \frac{1/2}{e^{-p}} = \frac{1}{2}.$$

$\square$

Corollary 1. DISPATCH *achieves the best-possible competitive ratio of $\frac{1}{2}$ for the Online Perfect Bipartite Matching problem.*

5   Conclusion

In this paper, we examine the problem of online perfect bipartite matching with i.i.d. arrivals from a known distribution. We present the DISPATCH algorithm. It attains a competitive ratio of $\frac{1}{2}$. We show that this is the best possible. Thus, the algorithm DISPATCH is optimal in terms of competitive ratio.

There is an intriguing difference between online perfect bipartite matching algorithms for minimization and the DISPATCH algorithm for maximization. Whereas the competitive ratio for minimization is bounded logarithmically, a constant bound was obtained for maximization with i.i.d. arrivals. This raises the question of whether a constant competitive ratio is possible for minimization with i.i.d. arrivals.

It may be possible to translate the analysis in this work to other contexts. Our analysis relied on two key ideas; the use of the expectation graph and proving that, regardless of how the jobs arrive, the DISPATCH algorithm effectively translates the non-uniform sampling over jobs to a uniform sampling over workers.

References

1. Bansal, N., Buchbinder, N., Gupta, A., Naor, J.S.: An $O(\log^2 k)$-competitive algorithm for metric bipartite matching. In: Arge, L., Hoffmann, M., Welzl, E. (eds.) ESA 2007. LNCS, vol. 4698, pp. 522–533. Springer, Heidelberg (2007). https://doi.org/10.1007/978-3-540-75520-3_47

2. Brubach, B., Sankararaman, K.A., Srinivasan, A., Xu, P.: New algorithms, better bounds, and a novel model for online stochastic matching. In: 24th Annual European Symposium on Algorithms, vol. 57, pp. 24:1–24:16. Schloss Dagstuhl-Leibniz-Zentrum fuer Informatik (2016). https://doi.org/10.4230/LIPIcs.ESA.2016.24

3. Bubeck, S., Cohen, M.B., Lee, Y.T., Lee, J.R., Madry, A.: K-server via multiscale entropic regularization. In: Proceedings of the 50th Annual ACM SIGACT Symposium on Theory of Computing, pp. 3–16. ACM (2018). https://doi.org/10.1145/3188745.3188798

4. Correa, J., Foncea, P., Hoeksma, R., Oosterwijk, T., Vredeveld, T.: Posted price mechanisms for a random stream of customers. In: Proceedings of the 2017 ACM Conference on Economics and Computation, pp. 169–186. ACM (2017). https://doi.org/10.1145/3033274.3085137

5. Dehghani, S., Ehsani, S., Hajiaghayi, M., Liaghat, V., Seddighin, S.: Stochastic k-server: how should uber work? In: 44th International Colloquium on Automata, Languages, and Programming, vol. 80, pp. 126:1–126:14. Schloss Dagstuhl-Leibniz-Zentrum fuer Informatik, Dagstuhl, Germany (2017)

6. Feldman, J., Korula, N., Mirrokni, V., Muthukrishnan, S., Pál, M.: Online ad assignment with free disposal. In: Leonardi, S. (ed.) WINE 2009. LNCS, vol. 5929, pp. 374–385. Springer, Heidelberg (2009). https://doi.org/10.1007/978-3-642-10841-9_34

7. Feldman, J., Mehta, A., Mirrokni, V., Muthukrishnan, S.: Online stochastic matching: Beating 1−1/e. In: 50th Annual IEEE Symposium on Foundations of Computer Science, pp. 117–126. IEEE (2009). https://doi.org/10.1109/FOCS.2009.72

8. Haeupler, B., Mirrokni, V.S., Zadimoghaddam, M.: Online stochastic weighted matching: improved approximation algorithms. In: Chen, N., Elkind, E., Koutsoupias, E. (eds.) WINE 2011. LNCS, vol. 7090, pp. 170–181. Springer, Heidelberg (2011). https://doi.org/10.1007/978-3-642-25510-6_15

9. Kalyanasundaram, B., Pruhs, K.: Online weighted matching. J. Algorithms **14**(3), 478–488 (1993). https://doi.org/10.1006/jagm.1993.1026

10. Karp, R.M., Vazirani, U.V., Vazirani, V.V.: An optimal algorithm for on-line bipartite matching. In: Proceedings of the 22nd Annual ACM Symposium on Theory of Computing, pp. 352–358. ACM (1990). https://doi.org/10.1145/100216.100262

11. Kesselheim, T., Radke, K., Tönnis, A., Vöcking, B.: An optimal online algorithm for weighted bipartite matching and extensions to combinatorial auctions. In: Bodlaender, H.L., Italiano, G.F. (eds.) ESA 2013. LNCS, vol. 8125, pp. 589–600. Springer, Heidelberg (2013). https://doi.org/10.1007/978-3-642-40450-4_50

12. Khuller, S., Mitchell, S.G., Vazirani, V.V.: On-line algorithms for weighted bipartite matching and stable marriages. Theor. Comput. Sci. **127**(2), 255–267 (1994). https://doi.org/10.1016/0304-3975(94)90042-6

13. Koutsoupias, E.: The k-server problem. Compu. Sci. Rev. **3**(2), 105–118 (2009). https://doi.org/10.1016/j.cosrev.2009.04.002

14. Mahdian, M., Yan, Q.: Online bipartite matching with random arrivals: an approach based on strongly factor-revealing LPs. In: Proceedings of the 43rd Annual ACM Symposium on Theory of Computing, pp. 597–606. ACM (2011). https://doi.org/10.1145/1993636.1993716

15. Manasse, M.S., McGeoch, L.A., Sleator, D.D.: Competitive algorithms for server problems. J. Algorithms **11**(2), 208–230 (1990). https://doi.org/10.1016/0196-6774(90)90003-W
16. Manshadi, V.H., Gharan, S.O., Saberi, A.: Online stochastic matching: online actions based on offline statistics. Math. Oper. Res. **37**(4), 559–573 (2012). https://doi.org/10.1287/moor.1120.0551
17. Mehta, A., et al.: Online matching and ad allocation. Found. Trends Theor. Comput. Sci. **8**(4), 265–368 (2013). https://doi.org/10.1561/0400000057
18. Meyerson, A., Nanavati, A., Poplawski, L.: Randomized online algorithms for minimum metric bipartite matching. In: Proceedings of the 17th Annual ACM-SIAM Symposium on Discrete Algorithms, pp. 954–959. Society for Industrial and Applied Mathematics (2006)
19. Raghvendra, S.: A robust and optimal online algorithm for minimum metric bipartite matching. In: Approximation, Randomization, and Combinatorial Optimization. Algorithms and Techniques, vol. 60, pp. 18:1–18:16. Schloss Dagstuhl-Leibniz-Zentrum fuer Informatik (2016). https://doi.org/10.4230/LIPIcs.APPROX-RANDOM.2016.18

Strategic Contention Resolution in Multiple Channels

George Christodoulou[1], Themistoklis Melissourgos[1(✉)], and Paul G. Spirakis[1,2]

[1] Department of Computer Science, University of Liverpool, Liverpool, UK
{G.Christodoulou,T.Melissourgos,P.Spirakis}@liverpool.ac.uk
[2] Computer Engineering and Informatics Department, University of Patras, Patras, Greece

Abstract. We consider the problem of resolving contention in communication networks with selfish users. In a *contention game* each of $n \geq 2$ identical players has a single information packet that she wants to transmit using one of $k \geq 1$ multiple-access channels. To do that, a player chooses a slotted-time protocol that prescribes the probabilities with which at a given time-step she will attempt transmission at each channel. If more than one players try to transmit over the same channel (collision) then no transmission happens on that channel. Each player tries to minimize her own expected *latency*, i.e. her expected time until successful transmission, by choosing her protocol. The natural problem that arises in such a setting is, given n and k, to provide the players with a common, anonymous protocol (if it exists) such that no one would unilaterally deviate from it (equilibrium protocol).

All previous theoretical results on strategic contention resolution examine only the case of a single channel and show that the equilibrium protocols depend on the feedback that the communication system gives to the players. Here we present multi-channel equilibrium protocols in two main feedback classes, namely *acknowledgement-based* and *ternary*. In particular, we provide equilibrium characterizations for more than one channels, and give specific anonymous, equilibrium protocols with finite and infinite expected latency. In the equilibrium protocols with infinite expected latency, all players transmit successfully in optimal time, i.e. $\Theta(n/k)$, with probability tending to 1 as $n/k \to \infty$.

Keywords: Contention resolution · Multiple channels
Acknowledgement-based protocol · Ternary feedback · Game theory

1 Introduction and Motivation

In the last fifteen years a great number of works in the Electrical and Electronics Engineering community has been devoted to designing medium access

P. G. Spirakis—The work of this author was partially supported by the ERC Project ALGAME.
For a full version with detailed comments and proofs see [11].

L. Epstein and T. Erlebach (Eds.): WAOA 2018, LNCS 11312, pp. 165–180, 2018.
https://doi.org/10.1007/978-3-030-04693-4_11

control (MAC) protocols that achieve high throughput. Their main approach is to consider, instead of the initial single-channel scheme, multi-channel schemes (*multi-channel* MAC protocols) which resolve contention caused by packet collisions (e.g. [6,20,24–26,29]). Apart from high throughput, an additional benefit of introducing more channels in such a system is robustness, meaning no great dependence on a single node's functionality. However, to the authors' knowledge, *strategic* behaviour in multi-channel systems is limited to the Aloha protocol ([19]), contrary to the case of single-channel systems (e.g. [2,7–9,12]). In this paper, we examine the problem of *strategic contention resolution* in multi-channel systems, where obedience to a suggested protocol is not required. We seek only *anonymous*, equilibrium protocols, that is, protocols which do not use player IDs. If a players' protocol depended on her ID, then equilibria are simple, but can be unfair as well; scheduling each player's transmission through a priority queue according to her ID is an equilibrium.

We provide two types of equilibrium protocols. The first type, called *FIN-EQ*, describes an anonymous, equilibrium protocol that yields finite expected time of successful transmission (*latency*) to a player. Similarly, the second type, called *IN-EQ*, describes an anonymous, equilibrium protocol which yields infinite expected latency to a player but is also *efficient*, i.e, all players transmit successfully within $\Theta(\frac{\#players}{\#channels})$ time with high probability. We study equilibria for two classes of feedback protocols: (a) acknowledgement-based protocols, where the user gets just the information of whether she had a successful transmission or not, only when she tries to transmit her packet, and (b) protocols with ternary feedback, where the user is informed about the number of pending players in each time-step regardless of whether she attempted transmission or not. Previous results on these classes of protocols have been produced only for the case of a single transmission channel ([7,12]). Here we investigate the multiple-channels case.

In the last part of the paper we seek efficient protocols for both feedback classes. Due to an impossibility result that we show (Theorem 7), the technique used in [12] by Fiat et al. for the single-channel setting in order to provide a FIN-EQ that is also efficient, cannot be applied when there are more than one channels. This fact discourages us from searching for efficient FIN-EQ protocols and, instead, points to the search for efficient IN-EQ protocols, which indeed we find. One could argue that an anonymous protocol with infinite expected time until successful transmission, such as the IN-EQ protocols we provide, does not incentivize a player to participate in such a communication system. To this we reply that exponential waiting-time for a large amount of players (see protocol in Subsect. 4.2) is equally bad for a player, since waiting for e.g. e^{10} msec is like waiting forever in Real-Time-Communications.

1.1   Our Results

Most of the proofs are omitted due to lack of space, and can be found in the full version of the paper.

The main contributions of this work are the characterizations of FIN-EQ and IN-EQ protocols in the two aforementioned feedback classes. Note that in the current bibliography regarding the single-channel setting, there are no characterizations of equilibrium in acknowledgement-based protocols. Also, in the single-channel setting the existence of a symmetric equilibrium with finite expected latency in the class of acknowledgement-based protocols remains an open problem, even for three players. However, for the settings with 2 and 3 transmission channels, in a short companion paper [10] we presented simple anonymous FIN-EQ protocols for up to 4 and 5 players respectively. Furthermore, these protocols are memoryless, while the only known FIN-EQ protocol in the single-channel setting ([7]) is not.

The paper is organized in three main parts. Section 3 deals with FIN-EQ protocols in the acknowledgement-based feedback setting. In that section we give two characterizations of equilibrium and also provide FIN-EQ protocols for specific numbers of players and channels. These results have also appeared recently in a complementary short paper ([10]) with no proofs; for detailed proofs of this section we refer the reader to the full version of the paper. Section 4 deals with FIN-EQ protocols in the ternary feedback setting and extends the corresponding results for the single-channel setting by Fiat et al. [12]. Finally, in Sect. 5, IN-EQ protocols with deadline are provided with the property that the time until all n players transmit successfully is $\Theta(n/k)$ with high probability, when there are k channels. The latter result makes clear the advantage (with respect to time efficiency) that multiple channels bring to a system with strategic users, which is that the time until all players transmit successfully with high probability is inversely proportional to the number of available channels.

1.2   Related Work

Contention in telecommunications is a major problem that results to poor throughput due to packet collisions. Motivated mainly by this problem, many works studying conflict-resolution protocols emerged in the late 70's ([4,5,15, 23,28]). Their approach is to resolve a collision when it occurs, and only then allow further transmissions on the channel. In those works the user's packets are assumed either to be generated by some stochastic process, or to appear at the same time in a worst-case scenario. Here, we consider the latter setting, i.e. a worst-case model of slotted time, where at any time-step all users have a packet ready to be transmitted (for an example of a similar bursty-input case, see [3]). As stated in [13], even though real implementations of multiple-access channels do not fit precisely within the slotted-time model, it can be shown (e.g. [14,17]) that results obtained in this model do apply to realistic multiple-access channels.

Also, many works have examined multiple-channel communication protocols. In the data link layer, a Medium Access Control (MAC) protocol is responsible for the flow of data through a multiple-access medium. Our multiple-channels model is motivated by theoretical and experimental results which have shown that higher throughput and lower delay is achieved by using "multi-channel" MAC protocols (see [20,21,25,26]). In [26], the *multi-channel hidden terminal problem* is raised which, additionally to increased packet collisions, results to incapability of the users to "sense" more than one channels at a time (possibly none); therefore a user might not know whether another user transmitted successfully or not (see also [27] for the classical "hidden terminal problem"). This motivates us for the consideration of feedback protocols with minimum feedback, i.e. "acknowledgement-based" protocols (see par.2, Sect. 1). Also, settings with stronger feedback have been studied (e.g. the Aloha protocol in [19]) in which a user is informed about the number of users that have not transmitted successfully yet. This is why we consider "ternary feedback" protocols (see par.2, Sect. 1).

Apart from the latter, all of the aforementioned works assume that the users blindly follow the given protocol, i.e. the users are not strategic. Contention resolution with strategic users has been studied only in single-channel settings or in the special case of the multiple-channel Aloha protocol. Some interesting cooperative and noncooperative models of slotted Aloha have been analysed in [1,18,19]. Aiming to understand the properties of contention resolution under selfishness, apart from various feedback settings, many cost functions have also been studied. One of the most meaningful cost functions is the one that models non-zero transmission costs as in [9] (and also [2,19]).

The theoretical works that relate the most to the current paper are the seminal paper by Fiat, Mansour and Nadav [12] and two by Christodoulou et al. [7,8] which study protocols for strategic contention resolution with zero transmission costs. These works examine the case of a single transmission channel only. In [12] the feedback is ternary. In that work, a characterization of symmetric equilibrium is provided, along with an efficient FIN-EQ protocol that puts an extremely costly equilibrium after a deadline in order to force users to be obedient. The feedback model of [7] and [8] is the acknowledgement-based. Among other results, [7] provides the unique FIN-EQ protocol for the case of two players and a deadline IN-EQ protocol for at least three players.

2   The Model and Definitions

Game Structure. We define a *contention game* as follows. Let $N = \{1, 2, \ldots, n\}$ be the set of players, also denoted by $[n]$, and $K = \{1, 2, \ldots, k\}$ the set of channels. Each player has a single packet that she wants to send through a channel in K, without caring about the identity of the channel. All players know n and K. We assume synchronous communications with discretized time, i.e. time slots $t = 1, 2, \ldots$. The players that have not yet successfully transmitted their packet are called *pending* and initially all n players are pending. At any given time slot t, a pending player i has a set $A = \{0, 1, 2, \ldots, k\}$ of *pure strategies*: a pure strategy $a \in A$ is the action of choosing channel $a \in K$ to transmit her packet on, or no transmission ($a = 0$). At time t, a *(mixed) strategy* of a player i is a probability distribution over A that potentially depends on information that i has gained from the process based on previous transmission attempts. If exactly one player transmits on a channel in a given slot t, then her transmission is *successful*, the successful player exits the game (i.e. she is no longer pending), and the game continues with the rest of the players. On the other hand, whenever two or more players try to access the same channel (i.e. transmit) at the same time slot, a *collision* occurs and their transmissions fail, in which case the players remain in the game. The game continues until all players have successfully transmitted their packets.

Transmission Protocols. Let $X_{i,t} \in A$ be the channel-indicator variable that keeps track of the identity of the channel where player i attempted transmission at time t; value 0 indicates no transmission attempt. For any $t \geq 1$, we denote by $\vec{X}_t$ the transmission vector at time t, i.e. $\vec{X}_t = (X_{1,t}, X_{2,t}, \ldots, X_{n,t})$.

An *acknowledgement-based* protocol uses very limited channel feedback. After each time step t, only players that attempted a transmission receive feedback, and the rest get no information. In fact, the information received by a player i who transmitted during t is whether her transmission was successful (in which case she gets an acknowledgement and exits the game) or whether there was a collision.

In a protocol with *ternary feedback* every pending player in every round is informed about the number of remaining players $m \leq n$. This information is given to the players regardless of their transmission history.

Let $\vec{h}_{i,t}$ be the vector of the *personal transmission history* of player i up to time t, i.e. $\vec{h}_{i,t} = (X_{i,1}, X_{i,2}, \ldots, X_{i,t})$. We also denote by $\vec{h}_t$ the transmission history of all players up to time t, i.e. $\vec{h}_t = (\vec{h}_{1,t}, \vec{h}_{2,t}, \ldots \vec{h}_{n,t})$. A *decision rule* $f_{i,t}$ for a pending player i at time t, is a function that maps $\vec{h}_{i,t-1}$ to a strategy $\vec{P}_{i,t}$, with elements $\Pr(X_{i,t} = a | \vec{h}_{i,t-1})$ for all $a \in A$. When the transmission probability on some $a' \in A$ is not stated in a decision rule it is because it can be deduced from the stated ones.

For a player $i \in N$, a *(transmission) protocol* f_i is a sequence of decision rules $f_i = \{f_{i,t}\}_{t\geq 1} = f_{i,1}, f_{i,2}, \ldots$ Given a protocol f_i for player i, when her decision rules depend on the number of pending players and the personal history of i, then we describe them by the player's probability distribution on the action set A. In this case, we denote by $p_{m,t}^{i,a}$ the probability of player i choosing action a at time t given her personal history h_{t-1} when m players are pending right before t. When the context is clear enough we will drop some of the indices accordingly.

When we state that the players use an *anonymous* protocol f, we will mean that they follow a common protocol $f(= f_1 = \cdots = f_n)$ whose decision rules do not depend on any ID of the player (in our setting players do not have IDs), i.e. the decision rule assigns the same strategy to all players with the same personal history. In particular, for any two players $i \neq j$ and any $t \geq 0$, if $\vec{h}_{i,t-1} = \vec{h}_{j,t-1}$, it holds that $f_{i,t}(\vec{h}_{i,t-1}) = f_{j,t}(\vec{h}_{j,t-1})$. In this case, we drop the subscript i in the notation and write f instead of f_i.

A protocol f_i for player i is a *deadline protocol with deadline* t_0 if and only if there exists a finite $t_0 \geq 1$ such that a particular channel $a_i \in K$ is assigned (deterministically or stochastically) to player i at some time $t \leq t_0$ and $\Pr(X_{i,t} = a_i | \vec{h}_{i,t-1}) = 1$ for every time slot $t \geq t_0$ and any history $\vec{h}_{i,t-1}$.

Efficiency. Assume that all n players follow an anonymous protocol f. We will call f *efficient* if and only if all players will have successfully transmitted by time $\Theta(n/k)$ with high probability (i.e. with probability tending to 1, as $n/k \to \infty$).

Individual Utility. By *protocol profile* $\vec{f} = (f_1, f_2, \ldots, f_n)$ we will call the n-tuple of the players' protocols. For a given transmission sequence $\vec{X}_1, \vec{X}_2, \ldots$, which is consistent with $\vec{f}$, define the *latency* of agent i as $T_i \triangleq \inf\{t : X_{i,t} = a, X_{j,t} \neq a, \text{ for some } a \in K, \forall j \neq i\}$. That is, T_i is the time at which i successfully transmits. Also, define the *finishing time* of $\vec{f}$ as $T \triangleq \sup_i\{T_i\}$, i.e., the least time at which all players have successfully transmitted. Given a transmission history $\vec{h}_t$, the n-tuple of protocols $\vec{f}$ induces a probability distribution over sequences of further transmissions. In that case, we write $C_i^{\vec{f}}(\vec{h}_t) \triangleq \mathbb{E}[T_i|\vec{h}_t, \vec{f}] = \mathbb{E}[T_i|\vec{h}_{i,t}, \vec{f}]$ for the expected latency of a pending agent i given that her current history is $\vec{h}_{i,t}$ and from $t+1$ on she follows f_i. For anonymous protocols, i.e. when $f_1 = f_2 = \cdots = f_n = f$, we will simply write $C_i^f(\vec{h}_t)$ instead. Abusing notation slightly, we will also write $C_i^{\vec{f}}(\vec{h}_0)$ for the *unconditional* expected latency of player i induced by $\vec{f}$. We also define the expected future latency $F_i^{\vec{f}}(\vec{h}_t) \triangleq C_i^{\vec{f}}(\vec{h}_t) - t$ and again, whenever clear from the context, we omit redundant indices or vectors from the notation.

Equilibria. The objective of every player is to minimize her expected latency. We call a protocol g_i a *best response* of player i to the *partial protocol profile* $\vec{f}_{-i}$ if for any transmission history $\vec{h}_t$, player i cannot decrease her expected latency by unilaterally deviating from g_i after t. That is, for all time slots t, and for all protocols f_i' for player i, we have

$$C_i^{(\vec{f}_{-i}, g_i)}(\vec{h}_t) \leq C_i^{(\vec{f}_{-i}, f_i')}(\vec{h}_t),$$

where $(\vec{f}_{-i}, g_i)$ (respectively, $(\vec{f}_{-i}, f_i')$) denotes the *protocol profile* where every player $j \neq i$ uses protocol f_j and player i uses protocol g_i (respectively f_i'). For an anonymous protocol f, we denote by (f_{-i}, g_i) the profile where player $j \neq i$ uses protocol f and player i uses protocol g_i.

We say that $\vec{f} = (f_1, f_2, \ldots f_n)$ is an *equilibrium* if for any transmission history $\vec{h}_t$ the players cannot decrease their expected latency by unilaterally deviating after t; that is, for every player i, f_i is a best response to $\vec{f}_{-i}$.

FIN-EQ and IN-EQ Protocols. We call an anonymous protocol *FIN-EQ* if it is an equilibrium protocol and yields finite expected latency to a player. Similarly, we call an anonymous protocol *IN-EQ* if it is an equilibrium protocol, yields infinite expected latency to a player, and is also efficient.

3   Equilibrium for Acknowledgement-Based Protocols

For the sake of completeness, in this section we state the definitions and results that have also appeared as a companion short paper in the 11th International

Symposium on Algorithmic Game Theory (SAGT 2018) [10], together with some complementary ones.

3.1 Nash Equilibrium Characterizations

The following equilibrium characterizations for the class of acknowledgement-based protocols help us check whether the protocols we subsequently guess are equilibrium protocols. The characterizations are for symmetric and asymmetric equilibria, arbitrary number of channels $k \geq 1$ and number of players $n \geq 2$.

In an acknowledgement-based protocol, the actions of player i at time t depend only (a) on her personal history $\vec{h}_{i,t-1}$ and (b) on whether she is pending or not at t. Let $\vec{f} = (f_1, f_2, \ldots, f_n)$ be a tuple of acknowledgement-based protocols (not necessarily anonymous) for the n players. For a (finite) positive integer τ^*, and a given history $h_{i,\tau^*} = (a_{i,1}, a_{i,2}, \ldots, a_{i,\tau^*})$, define for player i the protocol

$$
g_i = g_i(h_{i,\tau^*}) \triangleq \begin{cases} (\Pr\{X_{i,t} = a_{i,t}\} = 1, \ \Pr\{X_{i,t} \neq a_{i,t}\} = 0) & , \text{ for } 1 \leq t \leq \tau^* \\ f_{i,t}, \quad \text{ for } t > \tau^*. \end{cases}
$$

A personal history $\vec{h}_{i,\tau^*}$ is *consistent with* the protocol profile $\vec{f}$ if and only if there is a non-zero probability that $\vec{h}_{i,\tau^*}$ will occur for player i under $\vec{f}$. Protocol $g_i(h_{i,\tau^*})$ is *consistent with* $\vec{f}$ if and only if h_{i,τ^*} is consistent with $\vec{f}$, and when clear from the context we write g_i instead. We denote the set of all g_i's, that is, all $g_i(h_{i,t})$'s for all $t \geq 1$, which are consistent with $\vec{f}$, by $\mathcal{G}_i^{\vec{f}}$. If $f_i = f \ \forall i$ (i.e. f is anonymous), then instead of g_i and $\mathcal{G}_i^{\vec{f}}$ we write g and $\mathcal{G}^f$ respectively.

Lemma 1 (Equilibrium Characterization 1). *Consider a profile* $\vec{f} = (f_1, f_2, \ldots f_n)$ *of acknowledgement-based protocols and a protocol* $g_i = g_i(h_{i,\tau^*})$ *for some* $\tau^* \geq 1$. *The following statements are equivalent:*

(i) $\vec{f}$ *is an equilibrium.*
(ii) For every player $i \in [n]$,
$$\text{if } g_i \in \mathcal{G}_i^{\vec{f}} \text{ then } C_i^{(\vec{f}_{-i}, g_i)}(\vec{h}_0) = \min_{f_i'} C_i^{(\vec{f}_{-i}, f_i')}(\vec{h}_0) = C_i^{\vec{f}}(\vec{h}_0).$$

Lemma 2 (Equilibrium Characterization 2 [10]). *Consider a profile* $\vec{f} = (f_1, f_2, \ldots f_n)$ *of acknowledgement-based protocols. The following statements are equivalent:*

(i) $\vec{f}$ *is an equilibrium.*
(ii) For every player $i \in [n]$,

$$
\begin{cases} (a) & C_i^{(\vec{f}_{-i}, g_i)}(\vec{h}_0) = C_i^{(\vec{f}_{-i}, r_i)}(\vec{h}_0) = C_i^{\vec{f}}(\vec{h}_0), \quad \forall g_i, r_i \in \mathcal{G}_i^{\vec{f}}, \text{ and} \\ (b) & C_i^{(\vec{f}_{-i}, g_i)}(\vec{h}_0) \leq C_i^{(\vec{f}_{-i}, r_i)}(\vec{h}_0), \quad \forall g_i \in \mathcal{G}_i^{\vec{f}}, r_i \notin \mathcal{G}_i^{\vec{f}}. \end{cases}
$$

3.2   Acknowledgment-Based FIN-EQ Protocols

Regarding the search for FIN-EQ protocols, there is no straight-forward way for our equilibrium characterizations (previous subsection) to be used in order to *find* an equilibrium protocol. However, they allow us to *check* whether the protocols discussed in this subsection are equilibrium protocols. In this subsection we give FIN-EQ protocols for $k = 2$ and $k = 3$. For the detailed proofs see the full version of the paper.

We define the following anonymous, memoryless protocol for $k \geq 2$ channels.

Protocol f^k: For player i, every $t \geq 1$ and any history $\vec{h}_{i,t-1}$,

$$f^k_{i,t} = \left(\Pr\{X_{i,t} = 0\} = 0, \quad \Pr\{X_{i,t} = a\} = \frac{1}{k}, \quad \forall a \in K \right). \qquad (1)$$

---

n Players - 2 Transmission Channels. Here, we first give an example of a method for checking equilibria (Theorem 1). Then, with a better approach, by employing our characterizations of the previous subsection, we prove that f^2 is an equilibrium protocol for $n \in \{2, 3, 4\}$ players and $k = 2$ channels (Theorem 3).

Lemma 3 ([10]). *When all $n \geq 2$ players use protocol f^2 the expected latency of any player is $2^n/n$.*

In the next theorem we will give an example of a method for checking whether a given protocol profile is an equilibrium, which however could be inconclusive in some cases. Suppose you we want to check whether an arbitrary protocol profile $\vec{f}$ is an equilibrium. By definition of the equilibrium, we can fix all protocols except player i's, i.e. $\vec{f}_{-i}$ and check if f_i is a best response to them, and repeat this for every player i. By fixing $\vec{f}_{-i}$ we create a stochastic environment for player i who can be considered to be free to take sequential decisions through time. These decisions correspond to decision rules of f_i. Since, due to the feedback limitations, i has no information about the number of pending players, this situation from her point of view is modeled as an infinite state Partially Observable Markov Decision Process (POMDP). f_i is a best response to $\vec{f}_{-i}$ if and only if f_i is an optimal policy of the POMDP, that is, a set of decisions through time that minimize her expected latency.

However for this kind of POMDPs there are no known techniques to find an optimal policy. In order to circumvent this problem, we can assume that player i is an advantageous player that always knows how many players are pending. This turns the infinite state POMDP into a finite state Markov Decision Process (MDP), whose optimal policy we can find through known techniques (e.g. [22]). One can see that the optimal policy in the MDP of the advantageous player i yields at most the expected latency of the optimal policy in the POMDP of the initial player i. Thus, if the best policy in the MDP yields the same expected latency as what $\vec{f}$ gives to i, then we know that f_i is a best response; however, if the best policy of the MDP yields smaller expected latency, then we get no information about whether f_i is a best response in the POMDP or not. The proof of the next theorem (see full version of the paper) demonstrates the method and shows that protocol f^2 of (1) is an equilibrium protocol for 3 players.

Theorem 1. *For 3 players and 2 channels, f^2 is an equilibrium protocol with expected latency 8/3.*

We subsequently exploit the lack of memory and the anonymity of our protocol f defined in equation (1) and show more general results on equilibria (Theorem 3), using the characterizations from Subsect. 3.1.

Theorem 2. *In a contention game with $k = 2$ channels, consider an anonymous, memoryless protocol of player i with the property: $Pr\{X_{i,t} = 0\} = 0$, for every $t \geq 1$. For more than 4 players any such protocol is not an equilibrium protocol.*

Since protocol f^2 belongs to the class of protocols defined in the statement of Theorem 2, the following corollary is immediate.

Corollary 1 ([10]). *For $n \geq 5$ players and $k = 2$ channels, f^2 is not an equilibrium protocol. In fact, a better response for any player is to not transmit in $t = 1$ and then follow f^2.*

Theorem 3 ([10]). *For $n \in \{2, 3, 4\}$ players and $k = 2$ channels, f^2 is an equilibrium protocol with expected latencies 2, 8/3 and 4, respectively.*

n Players - 3 Transmission Channels. Here, by employing our characterizations from Subsect. 3.1, we give an acknowledgement-based, equilibrium protocol for $n \in \{2, 3, 4, 5\}$ players and $k = 3$ channels.

Theorem 4. *For $n \in \{2, 3, 4, 5\}$ players and $k = 3$ channels, f^3 defined in (1) is an equilibrium protocol with expected latencies 3/2, 15/8, 189/80 and 597/200, respectively. (See also [10].)*

4   Equilibria for Ternary Feedback Protocols

In this section we consider anonymous protocols with *ternary feedback*, that is, a pending player knows at every time t the number $m \leq n$ of pending players. This knowledge is given to each player regardless of her transmission history.

4.1   Nash Equilibrium Characterization

Here we give a characterization of FIN-EQ protocols for $n \geq 1$ players and $k = 2$ channels in the general history-dependent case for ternary feedback. (For details on the characterization, see the full version of the paper.)

Theorem 5. *There exists an anonymous, history-dependent, equilibrium protocol with ternary feedback for n players and 2 transmission channels.*

The equilibrium probability that defines the equilibrium protocol, although guaranteed to exist when expected (future) latencies are finite, is difficult to be expressed in closed form, if possible at all. For more comments, see the full version of the paper.

4.2   History-Independent FIN-EQ Protocols

Let us now consider anonymous, history-independent protocols, that is, protocols whose decision rules depend only on the number $1 \leq m \leq n$ of pending players. Now, the decision rule p_m of the players does not depend on their transmission history (and therefore on time as well), hence a player's expected future latency F_m does not depend on her transmission history. In this class of protocols the following theorem fully characterizes the equilibria.

Theorem 6. *There exists a unique, anonymous, history-independent, equilibrium protocol with ternary feedback for n players and 2 transmission channels, which is: any player among $2 \leq m \leq n$ remaining players, for every $t \geq 1$ attempts transmission to each channel with equal probability p_m. This probability is $\Theta(\frac{1}{\sqrt{m}})$ and yields expected future latency $e^{\Theta(\sqrt{m})}$ for every player.*

The latter result is analogous to the one in [12] that characterizes anonymous, history-independent, equilibrium protocols with ternary feedback for the case of a single channel. However here, the proof methodology is different due to the fact that the transmission probabilities in equilibrium cannot be expressed in closed-form, therefore their asymptotic behaviour can only be extracted from a recurrence relation, which, contrary to the one in [12], is quite complex. Using dynamic programming, we can compute the equilibrium probabilities in linear time (for more details see the full version of the paper).

5   IN-EQ Protocols for Both Feedback Classes

Ideally, we would like to find an anonymous, equilibrium protocol that is efficient (i.e. the time until all players transmit successfully is $\Theta(n/k)$ with high probability) and also has finite expected latency. For the case of ternary feedback and a single channel such a protocol was found in [12]. That protocol set a deadline $t_0 \propto n$ after which it prescribed to the players to transmit with probability 1 on the channel at every time. It is easy to see that transmitting surely at every time is an equilibrium for more than 2 players. The idea of that protocol was to employ that "bad equilibrium" by putting it after the deadline so as to keep the players that were unsuccessful until t_0 for a very long (exponential in n) time. This works as a threat to the players, which they try to avoid by adopting a cooperative behaviour; using a history-dependent, equilibrium protocol they attempt transmission with probability low enough so that all of them are successful before the deadline with high probability. After the long part of the protocol, there is a last part that prescribes to the players to use a history-independent, equilibrium protocol (similar to the one we find for the 2-channel case) which has finite expected future latency. Since all three parts of the protocol are in equilibrium, the whole protocol is in equilibrium as well.

However, for the case of multiple channels in both the ternary feedback and acknowledgement-based feedback classes, a protocol with the above structure cannot be constructed as the following theorem shows (for the proof see the full version of the paper). First, let us define the following notion of equilibrium protocol: By *equilibrium with blocking step (EBS)* we call an anonymous, equilibrium protocol with the property that there exists a time-step of the protocol in which every pending player has probability of successful transmission equal to 0.

Theorem 7. *In the classes of acknowledgement-based and ternary feedback protocols with $k \geq 2$ channels and $n \geq 2$ players, there exists no equilibrium protocol with blocking step (EBS) and finite expected latency.*

This impossibility result discourages the search for multiple-channel protocols with the additional property of finite expected latency, since the only candidate that guarantees efficiency seems to be a deadline protocol. Whether no anonymous, efficient, equilibrium protocol with finite expected latency can be found for multiple channels is one of the most interesting open problems that stem from this work.

In the rest of this section we present IN-EQ protocols within the classes of acknowledgement-based and ternary feedback for the general case of $k \geq 1$ channels and any number of $n \geq 2k + 1$ players. For this, we employ the deadline idea introduced in [12] and consequently used in [7,8]. Our protocols are efficient, even though the expected latency is infinite.

5.1   Acknowledgement-Based Feedback

In the aforementioned companion short paper [10] we provided an efficient deadline protocol with infinite expected latency for $k \geq 1$ channels and $n \geq 2k + 1$ players. This protocol generalizes the efficient protocol of [7] which deals with a single channel and at least 3 players. The general protocol we present uses their idea, that is, estimating the number of pending players (since it is not known in the acknowledgement-based environment) and adjusting the transmission probabilities of the players accordingly, in order to simulate a socially optimal protocol (see protocol SOP below) that allows all transmission to be successful by time $\Theta(n/k)$ with high probability. Our modification is that, instead of prescribing to the players to always transmit to the single channel once they reach the deadline (so that with some positive probability they get blocked forever), we block all channels with positive probability by prescribing a random assignment of each player to a channel. The protocol and proof of equilibrium and efficiency can be found in the full version of the paper.

5.2   Ternary Feedback

Since in the ternary feedback setting the only history-independent equilibrium from Subsect. 4.2 yields exponential expected latency in the number of players n, even one player's latency being any polynomial in n happens with exponentially small probability. This fact points to history-dependent protocols as candidates for efficient equilibria. Here, we construct a protocol (Theorem 8) which imposes a heavy cost on any player that does not manage to transmit successfully until a certain deadline-round. This forces any potential deviator to play "fairly" until the deadline and follow an anonymous, socially optimal protocol, named SOP (guarantees expected time $\Theta(n/k)$ for all players to pass). To prove the main theorem of this subsection we need a series of technical results, namely the following three lemma 4, 5, and 6. Lemma 5 shows that protocol SOP is socially optimal.

Lemma 4. *Consider a single round with $k \geq 1$ channels and $n \geq 1$ players. Assume that for every player the probability of transmission attempt is $z \in [0,1]$ which she splits equally to all k channels. Then, the expected number[1] of players that transmit successfully is $zn \left(1 - \frac{z}{k}\right)^{n-1}$.*

We define the following anonymous, history-independent protocol which we prove to be efficient. However, we remark that it is not in equilibrium, due to Theorem 6 which characterizes the unique, anonymous, equilibrium protocol that is history-independent.

Protocol SOP:
Every player among $1 \leq m \leq n$ pending players, in each round $t \geq 1$ assigns transmission probability $1/\max\{m,k\}$ to each channel.

Lemma 5. *Protocol SOP for $k \geq 1$ channels and $n > k$ players has expected finishing time $O((n-k)/k)$.*

In the sequel, by e we denote the constant named "Euler's number", i.e. $e = 2.7182\ldots$. Using the above lemma we are able to prove the following.

Lemma 6. *(a) If at $t = 0$ there are n pending players, the probability that more than k players are pending at time $t_1 = 2e(n-k)/k$ is at most $exp\left(-\frac{n-k}{2ek}\right)$.*
(b) If at $t = 0$ there are k pending players, the probability that not all players have transmitted successfully at time $t_2 = 2e(n-k)/k$ is at most $exp\left(-\frac{n-k}{2ek}\right)$.

Proof. Let $\{Y_t\}_{t=1}^{t_1}$ be random variables which indicate the number of successful transmissions that occur in each time-step from $t = 1$ up to $t_1 \triangleq 2e(n-k)/k$, given that there are n pending players at time $t = 0$. For the events for which $Y \triangleq \sum_{t=1}^{t_1} Y_t > n - k$ we have the desired outcome. For the rest, since the pending players in each round $1 \leq t \leq t_1$ are $m > k$, the protocol prescribes to each player probability $1/m$ on each channel. Therefore, by Lemma 4, we have $\mathbb{E}[Y_t] = k \left(1 - 1/m\right)^{m-1}$. In the next claim we show that Y_t stochastically dominates a random variable $Z_t \in \{0, 1, \ldots, k\}$ that indicates the number of successful transmissions in round $1 \leq t \leq t_1$ but, in this process, the players that transmit successfully are placed back to the group of pending players.

Claim. $\Pr\{Y_t \geq x\} \geq \Pr\{Z_t \geq x\}$, for all $x \in \{0, 1, \cdots, k\}$.

---
[1] We define $0^0 = 1$..

Proof. We will prove the above claim by showing the stronger fact that, for any fixed number $1 \leq m \leq n - 1$ of pending players at time t,

$$\Pr\{Y_t \geq x \mid m \text{ pending players}\} \geq \Pr\{Y_t \geq x \mid m + 1 \text{ pending players}\},$$

for all $x \in \{0, 1, \cdots, k\}$.

Indeed, by substituting the probabilities of the above inequality we get,

$$\binom{m}{x} x! \left(\frac{1}{m}\right)^x \left(1 - \frac{x}{m}\right)^{m-x} \geq \binom{m+1}{x} x! \left(\frac{1}{m+1}\right)^x \left(1 - \frac{x}{m+1}\right)^{m+1-x},$$

or equivalently, $(m+1)^m (m-x)^{m-x} \geq m^m (m - x + 1)^{m-x}$,

and finally, $\left(1 + \frac{1}{m}\right)^m \geq \left(1 + \frac{1}{m-x}\right)^{m-x}$,

which is true, since the function $f(w) = (1 + 1/w)^w$ is strictly increasing. The claim follows from the fact that for any fixed $x \in \{0, 1, \cdots, k\}$,

$$\Pr\{Z_t \geq x\} = \Pr\{Y_t \geq x \mid n \text{ pending players}\}.$$

□

Clearly $\{Z_t\}_{t=1}^{t_1}$ are independent random variables bounded in $[0, k]$. Let $Z \triangleq \sum_{t=1}^{t_1} Z_t$ and $\mu_1 \triangleq \mathbb{E}[Z] = \sum_{t=1}^{t_1} \mathbb{E}[Z_t] = t_1 k (1 - 1/n)^{n-1}$. Then by Hoeffding's inequality [16] and the stochastic domination we have,

$$\Pr(Y \leq n - k) \leq \Pr(Z \leq n - k) = \Pr\left(Z \leq \frac{\mu_1}{2e(1 - 1/n)^{n-1}}\right) \leq \Pr\left(Z \leq \frac{\mu_1}{2}\right)$$

$$\leq exp\left(-\frac{(1 - 1/2)^2 \mu_1^2}{t_1(k-0)^2}\right) \leq exp\left(-\frac{1}{4}\frac{t_1}{e^2}\right) = exp\left(-\frac{n-k}{2ek}\right),$$

where in the last three inequalities we used the fact that $(1 - 1/n)^{n-1} \geq 1/e$.

For the second part of the proof, suppose the process is at round $t = 0$ with k pending players. Let $\{X_t\}_{t=1}^{t_2}$ be random variables which indicate the number of successful transmissions that occur in each time-step from $t = 1$ up to $t_2 \triangleq 2e(n - k)/k$, given that there are k pending players at time $t = 0$. The pending players in each round $1 \leq t \leq t_2$ are $m \leq k$, hence the protocol prescribes to each player probability $1/k$ on each channel. By Lemma 4, we have $\mathbb{E}[X_t] = m (1 - 1/k)^{m-1}$. Now, observe that X_t stochastically dominates a random variable $W_t \in \{0, 1, \ldots, k\}$ that indicates the number of successful transmissions in round $1 \leq t \leq t_2$ but, in this process, the players that transmit successfully are placed back to the group of pending players. The latter observation is easy to see since an argument similar to the Claim that was stated earlier holds in this case.

Clearly, $\{W_t\}_{t=1}^{t_2}$ are independent random variables bounded in $[0, k]$. Let $W \triangleq \sum_{t=1}^{t_2} W_t$ and $\mu_2 \triangleq \mathbb{E}[W] = \sum_{t=1}^{t_2} \mathbb{E}[W_t] = t_2 k (1 - 1/k)^{k-1}$. Then by Hoeffding's inequality [16] and the stochastic domination we have,

$$\Pr(X \leq k - 1) \leq \Pr(W \leq k) = \Pr\left(W \leq \frac{\mu_2 k}{2e(n - k)(1 - 1/k)^{k-1}} \right)$$

$$\leq \Pr\left(W \leq \frac{\mu_2}{2} \right) \leq exp\left(-\frac{(1 - 1/2)^2 \mu_2^2}{t_2(k - 0)^2} \right) \leq exp\left(-\frac{1}{4}\frac{t_2}{e^2} \right)$$

$$= exp\left(-\frac{n - k}{2ek} \right),$$

where in the last three inequalities we used the fact that $(1 - 1/k)^{k-1} \geq 1/e$, and $n \geq 2k + 1$. This completes the proof of the lemma. $\square$

We define the following anonymous protocol. In the next theorem we show that it is an equilibrium protocol and also that it is efficient.

Protocol r:
Let the deadline be $t_0 = 4e(n - k)/k$. Every player among $1 \leq m \leq n$ pending players for $1 \leq t \leq t_0 - 1$ assigns transmission probability $1/\max\{m, k\}$ to each channel. Right before t_0 each pending player is assigned to a random channel equiprobably, and for $t \geq t_0$ always attempts transmission to that channel.

Theorem 8. *Protocol r for $n \geq 2k+1$ players and $k \geq 1$ channels is an efficient, equilibrium protocol.*

Proof. First, we show that it is an equilibrium protocol when $n \geq 2k + 1$. The expected latency of a player using protocol r is ∞. That is because there is an event with positive probability in which some player i finds herself in an equilibrium where at least 2 of the other players have been assigned to each and all of the k channels and transmit there in every time slot. In particular, with probability at least $k(\frac{1}{n})^{t_0-1} > 0$ all players will be pending right after $t_0 - 1$. Given this, with probability $\binom{n-1}{2,2,\ldots,2,n-1-2k}(\frac{1}{k})^{n-1} > 0$ exactly 2 out of $n - 1$ players will be assigned to each of the $k - 1$ channels and the remaining players (including player i), which are at least 3, are assigned to the remaining channel. Therefore, the aforementioned two events occur with positive probability, and then for player i all channels are blocked for every $t \geq t_0$, resulting to infinite latency. Hence, the expected latency of a player using protocol r is ∞.

Now suppose that player i unilaterally deviates to some protocol r'. The event that all players are pending right before t_0 remains non-empty, since the event that all players transmit on the same channel as i for every $1 \leq t \leq t_0 - 1$ happens with positive probability. Given that, the event that at least 2 of the players other than i will be assigned to each channel happens with positive probability. Therefore, the deviator's expected latency remains ∞ and r is an equilibrium protocol.

Finally, we will show that, when $n \in \omega(k)$, this protocol is also efficient, i.e. the time until all n players transmit successfully is linear in n/k with probability tending to 1 as $\frac{n}{k} \to \infty$. By Lemma 6, the probability that not all players have successfully transmitted by time $t_1 + t_2 = 4e(n - k)/k$ is at most $\exp\left(-\frac{n-k}{2ek}\right) + \exp\left(-\frac{n-k}{2ek}\right) = 2\exp\left(-\frac{n-k}{2ek}\right)$. Therefore, when $n \in \omega(k)$, no player is pending after $4e(n - k)/k$ rounds with high probability. $\square$

References

1. Altman, E., Barman, D., Benslimane, A., El Azouzi, R.: Slotted aloha with priorities and random power. In: Boutaba, R., Almeroth, K., Puigjaner, R., Shen, S., Black, J.P. (eds.) NETWORKING 2005. LNCS, vol. 3462, pp. 610–622. Springer, Heidelberg (2005). https://doi.org/10.1007/11422778_49

2. Altman, E., El Azouzi, R., Jiménez, T.: Slotted aloha as a game with partial information. Comput. Netw. **45**(6), 701–713 (2004)

3. Bender, M.A., Farach-Colton, M., He, S., Kuszmaul, B.C., Leiserson, C.E.: Adversarial contention resolution for simple channels. In: Proceedings of the seventeenth annual ACM symposium on Parallelism in algorithms and architectures, pp. 325–332. ACM (2005)

4. Capetanakis, J.: Generalized TDMA: the multi-accessing tree protocol. IEEE Trans. Commun. **27**(10), 1476–1484 (1979)

5. Capetanakis, J.: Tree algorithms for packet broadcast channels. IEEE Trans. Inf. Theory **25**(5), 505–515 (1979)

6. Chen, J., Sheu, S.T., Yang, C.A.: A new multichannel access protocol for IEEE 802.11 ad hoc wireless LANs. In: 14th IEEE Proceedings on Personal, Indoor and Mobile Radio Communications, PIMRC 2003, vol. 3, pp. 2291–2296. IEEE (2003)

7. Christodoulou, G., Gairing, M., Nikoletseas, S.E., Raptopoulos, C., Spirakis, P.G.: Strategic contention resolution with limited feedback. In: 24th Annual European Symposium on Algorithms, ESA 2016, 22–24 August 2016, Aarhus, Denmark, pp. 30:1–30:16 (2016)

8. Christodoulou, G., Gairing, M., Nikoletseas, S., Raptopoulos, C., Spirakis, P.: A 3-player protocol preventing persistence in strategic contention with limited feedback. In: Bilò, V., Flammini, M. (eds.) SAGT 2017. LNCS, vol. 10504, pp. 240–251. Springer, Cham (2017). https://doi.org/10.1007/978-3-319-66700-3_19

9. Christodoulou, G., Ligett, K., Pyrga, E.: Contention resolution under selfishness. Algorithmica **70**(4), 675–693 (2014)

10. Christodoulou, G., Melissourgos, T., Spirakis, P.G.: Short paper: strategic contention resolution in multiple channels with limited feedback. In: Deng, X. (ed.) SAGT 2018. LNCS, vol. 11059, pp. 245–250. Springer, Cham (2018). https://doi.org/10.1007/978-3-319-99660-8_22

11. Christodoulou, G., Melissourgos, T., Spirakis, P.G.: Strategic Contention Resolution in Multiple Channels. Comput. Sci. Game Theory (2018). arXive-prints arxiv:1810.04565

12. Fiat, A., Mansour, Y., Nadav, U.: Efficient contention resolution protocols for selfish agents. In: Proceedings of the Eighteenth Annual ACM-SIAM Symposium on Discrete Algorithms, SODA 2007, New Orleans, Louisiana, USA, 7–9 January 2007, pp. 179–188 (2007). http://dl.acm.org/citation.cfm?id=1283383.1283403

13. Goldberg, L.A.: Notes on contention resolution (2002). http://www.cs.ox.ac.uk/people/leslieann.goldberg/contention.html
14. Håstad, J., Leighton, T., Rogoff, B.: Analysis of backoff protocols for multiple access channels. SIAM J. Comput. **25**(4), 740–774 (1996)
15. Hayes, J.: An adaptive technique for local distribution. IEEE Trans. Commun. **26**(8), 1178–1186 (1978)
16. Hoeffding, W.: Probability inequalities for sums of bounded random variables. J. Am. Stat. Assoc. **58**(301), 13–30 (1963). http://www.jstor.org/stable/2282952
17. Kelly, F.P., MacPhee, I.M.: The number of packets transmitted by collision detect random access schemes. Ann. Probab. **15**(4), 1557–1568 (1987)
18. Ma, R.T., Misra, V., Rubenstein, D.: Modeling and analysis of generalized slotted-aloha MAC protocols in cooperative, competitive and adversarial environments. In: 26th IEEE International Conference on Distributed Computing Systems, ICDCS 2006, pp. 62–62. IEEE (2006)
19. MacKenzie, A.B., Wicker, S.B.: Stability of multipacket slotted aloha with selfish users and perfect information. In: INFOCOM 2003. Twenty-Second Annual Joint Conference of the IEEE Computer and Communications, IEEE Societies, vol. 3, pp. 1583–1590. IEEE (2003)
20. Mo, J., So, H.S.W., Walrand, J.: Comparison of multichannel MAC protocols. IEEE Trans. Mob. Comput. **7**(1), 50–65 (2008)
21. Nasipuri, A., Zhuang, J., Das, S.R.: A multichannel CSMA MAC protocol for multihop wireless networks. In: Wireless Communications and Networking Conference, WCNC 1999, vol. 3, pp. 1402–1406. IEEE (1999)
22. Norris, J.: Markov Chains. Cambridge Series in Statistical and Probabilistic Mathematics. Cambridge University Press, Cambridge (1998). https://books.google.co.uk/books?id=qM65VRmOJZAC
23. Roberts, L.G.: Aloha packet system with and without slots and capture. ACM SIGCOMM Comput. Commun. Rev. **5**(2), 28–42 (1975)
24. Shih, M.J., Lin, G.Y., Wei, H.Y.: A distributed multi-channel feedbackless mac protocol for D2D broadcast communications. IEEE Wirel. Commun. Lett. **4**(1), 102–105 (2015)
25. So, H.W., Walrand, J., Mo, J.: McMAC: a multi-channel MAC proposal for ad-hoc wireless networks. In: Proceedings of IEEE WCNC, pp. 334–339 (2007)
26. So, J., Vaidya, N.H.: Multi-channel MAC for ad hoc networks: handling multi-channel hidden terminals using a single transceiver. In: Proceedings of the 5th ACM International Symposium on Mobile Ad Hoc Networking and Computing, pp. 222–233. ACM (2004)
27. Tobagi, F., Kleinrock, L.: Packet switching in radio channels: part II-the hidden terminal problem in carrier sense multiple-access and the busy-tone solution. IEEE Trans. Commun. **23**(12), 1417–1433 (1975)
28. Tsybakov, B.S., Mikhailov, V.A.: Free synchronous packet access in a broadcast channel with feedback. Problemy Peredachi Informatsii **14**(4), 32–59 (1978)
29. Zhang, J., Zhou, G., Huang, C., Son, S.H., Stankovic, J.A.: Tmmac: an energy efficient multi-channel MAC protocol for ad hoc networks. In: IEEE International Conference on Communications, ICC 2007, pp. 3554–3561. IEEE (2007)

Sublinear Graph Augmentation for Fast Query Implementation

Artur Czumaj[1], Yishay Mansour[2], and Shai Vardi[3(✉)]

[1] University of Warwick, Coventry, UK
A.Czumaj@warwick.ac.uk
[2] Tel Aviv University, Tel Aviv, Israel
mansour@tau.ac.il
[3] Purdue University, West Lafayette, IN, USA
svardi@purdue.edu

Abstract. We introduce the problem of augmenting graphs with sublinear memory in order to speed up replies to queries. As a concrete example, we focus on the following problem: the input is an (unpartitioned) bipartite graph $G = (V, E)$. Given a query $q \in V$, the algorithm's goal is to output q's color in some legal 2-coloring of G, using few probes to the graph. All replies have to be *consistent with the same 2-coloring*. We show that if a linear amount of preprocessing is allowed, there is a randomized algorithm that, for any α, uses $O\left(\frac{m}{\alpha}\right)$ probes and $\tilde{O}(\alpha)$ memory, where m is the number of edges in the graph. On the negative side, we show that for a natural family of algorithms that we call probe-first local computation algorithms, this trade-off is optimal even with unbounded preprocessing.

We describe a randomized algorithm that replies to queries using $\tilde{O}\left(\frac{\sqrt{n}}{\Phi^2}\right)$ probes with no additional memory on regular graphs with conductance Φ (n is the number of vertices in G). In contrast, we show that any deterministic algorithm for regular graphs that uses no memory augmentation requires a linear (in n) number of probes, even if the conductance is the largest possible. We give an algorithm for grids and tori that uses a sublinear number of probes and no memory. Last, we give an algorithm for trees that errs on a sublinear number of edges (i.e., a sublinear number of edges are monochromatic under this coloring) that uses sublinear preprocessing, memory and probes per query.

Keywords: Sublinear algorithms · Graph augmentation
Local computation algorithms

Part of this work was done while the authors were visiting the Simons Institute for the Theory of Computing. Artur Czumaj was supported by the Centre for Discrete Mathematics and its Applications (DIMAP) and by EPSRC grant EP/N011163/1. Yishay Mansour was supported in part by the Israel Science Foundation (ISF). Shai Vardi was supported in part by the Linde Foundation and NSF grants CNS-1254169 and CNS-1518941.

L. Epstein and T. Erlebach (Eds.): WAOA 2018, LNCS 11312, pp. 181–203, 2018.
https://doi.org/10.1007/978-3-030-04693-4_12

1   Introduction

We consider a fundamental scenario in the analysis of big graphs, in which we would like to implement query access to parts of a solution to a combinatorial problem on some graph G. Consider for example, the maximum matching problem. In order to detect the structure of a solution, we would like to be able to query every vertex for an incident edge (if any) that belongs to an optimal (or approximately optimal) solution. In the classical model of algorithmic analysis, this task can be solved using global computations, by running an algorithm that takes as its input the entire graph, performs some computations, and returns the output. For massive graphs such an approach is not feasible—just reading the entire input may be too costly. In this paper, we consider the scenario where in order to answer the query, we have only local access to the input: we have *probe access to G* and some *additional, strictly limited amount of auxiliary memory* at our disposal. A probe[1] specifies a vertex v and a port number p, and the reply to the probe is the ID of v's p^{th} neighbor, and its degree. To measure the quality of performance of our algorithms, we consider four parameters:

- the number of probes made to the graph per query (which also serves as a proxy for runtime),
- the size of the additional auxiliary memory,
- the preprocessing time, and
- the quality of the solution.

Specifically, the focus of this work is the trade-offs between these four parameters. Our main goal is to study local algorithms that use only very limited additional resources (either sublinear or no additional auxiliary memory) and implement fast replies to queries used in the analysis of big graphs.

2-Coloring Bipartite Graphs. To make our study more precise, we focus on a concrete, representative problem: *2-coloring a bipartite graph*. Given a graph $G = (V, E)$, and a coloring of the vertices $c : V \to \{\text{red}, \text{blue}\}$, we say an edge $(u, v) \in E$ is *monochromatic* if $c(u) = c(v)$.

Given an uncolored graph $G = (V, E)$, $|V| = n, |E| = m$, that is known to be bipartite, the goal is to design an algorithm $\mathcal{A}$ that has probe access to the graph G, takes queries of the form $q \in V$ and outputs a color $c(q) \in \{\text{red}, \text{blue}\}$, such that the following properties hold:

1. the coloring $c = \langle c(v) : v \in V \rangle$ induces at most ϵm monochromatic edges;
2. $\mathcal{A}$ uses at most k probes per query;
3. $\mathcal{A}$ uses at most α words of (auxiliary) memory;
4. $\mathcal{A}$ uses little or no preprocessing.

Naturally, we would like ϵ, k, and α to be as small as possible. If $\epsilon = 0$, then we say the algorithm $\mathcal{A}$ is *exact*.

---

[1] Feige et al. [9], differentiate between strong and weak probes. Our definition of probe is consistent with their definition of weak probe, which is also the definition of probes used in the context of property testing by Goldreich and Ron [14] and others.

Isn't 2-Coloring a Bipartite Graph Trivial? To see the difficulty of this scenario, let us first consider the case that we want $\epsilon = 0$, and G is an n-vertex cycle, with vertex IDs taking the values $\{1, 2, \ldots, n\}$. Assume the reply to the first queried vertex is `red`, and the next queried vertex is "on the other side" of the cycle. We can clearly not hope to give the correct color with $o(n)$ probes per query, regardless of how much memory we have. This example shows the main challenge in designing good algorithms in our setting: *how to coordinate local computations to establish global properties of the input graph, to maintain consistent local information.*

The problem we have posed naturally falls in the framework of *local computation algorithms* (LCAs), as introduced by Rubinfeld et al. [32] (see Sect. 2 for a formal definition). This work differs from previous work on LCAs, see, e.g., [8,17,23,25], in several ways:

1. We (sometimes) allow a preprocessing stage in which the LCA can write to the auxiliary memory. If this stage is short enough (i.e., uses a small enough number of probes) we can incorporate it into the algorithm's first query.
2. We use the auxiliary memory to store additional information about the graph. In previous work the auxiliary memory was used exclusively to store a random seed that is needed for consistency in randomized LCAs, see, e.g., [9,26,31].
3. We allow our algorithms to use new random bits in every query. This could potentially lead to different results for the same query if different random bits are used; we ensure that this can not happen.

We note that the new features of LCAs that we use are not disallowed by previous definitions, they have just not been used previously. We discuss this further in Sect. 2.

1.1 Results

We begin with a simple algorithm that exploits the use of linear-time preprocessing.

Theorem 1. *There exists an exact 2-coloring algorithm that, for any $\alpha > 0$ and any bipartite graph $G = (V, E)$, uses $O(m + n)$ preprocessing probes, $\alpha \log n$ words of memory and performs $O\left(\frac{m}{\alpha}\right)$ probes per query w.h.p.*

The idea behind Theorem 1 is straightforward: if a linear amount of preprocessing is allowed, we use BFS to color each connected component of the graph, and then store a randomly selected set of vertices S along with their color. When given a query q, the algorithm performs a BFS from the queried vertex. We show that the BFS either finds a vertex of S or the entire connected component of q is processed, and it is easy to color q in both cases.

Theorem 1 shows that with sufficiently long (linear-time) preprocessing, there is an efficient exact 2-coloring LCA. If the input is large, however, even linear preprocessing may be infeasible in many scenarios. Our main focus in this paper is therefore on the case when little to no preprocessing is allowed.

Our main technical result is a lower bound that serves to complement Theorem 1.

Theorem 2. *For any $\alpha > 0, k \geq 0$, and $n \geq (k+2)\alpha + 2$, there is no deterministic probe-first LCA for exact 2-coloring of even-length cycles of length n that uses at most k probes per query and less than α bits of memory, even with unbounded preprocessing.*

Theorem 2 holds for a natural family of LCAs: deterministic probe-first LCAs. Probe-first LCAs are the ones whose probes are independent of the memory: the LCA first performs its probes and only then consults the memory. While the memory may provide useful information, it does not provide any encoding of the graph itself. All of the LCAs in this paper can be thought of as probe-first LCAs, and indeed, we are not aware of any deterministic LCA in the literature that is not (or cannot be formulated as one). The randomized LCAs in the literature read the memory to obtain the random seed, and so are not technically probe-first LCAs; we remark upon this further in Sect. 2.

We then consider what can be done *without any auxiliary memory*. We give algorithms for regular graphs with high conductance and for grids (grids have low conductance) that require no preprocessing and no memory. The high level description of both algorithms is similar: given the queried vertex q, we find a path from q to some predetermined vertex (in our case, we choose the vertex with ID 1), whose color is set to be red. If the path between q and 1 is of even length, we color q red, otherwise we color it blue. On regular graphs with high conductance, we perform two random walks, one from q and one from 1, until they intersect. On grids, we show how to efficiently construct a hyperplane containing q and 1-dimensional paths from vertex 1; we use their intersection to find the path. The results on grids immediately apply to tori as well.

For regular graphs that have constant conductance, we give a lower bound of $\Omega(n)$ probes for *deterministic* algorithms to complement the upper bound of $\tilde{O}(\sqrt{n})$ (where n is the number of vertices), showing that randomization is necessary to obtain algorithms with small probe complexity for this problem on graphs with high conductance (we use $\tilde{O}$ to hide logarithmic dependencies). We use complete bipartite graphs to prove our lower bounds; these graphs have the largest conductance of all the bipartite graphs.

Theorem 3. *There exists a randomized exact 2-coloring algorithm of connected regular bipartite graphs with conductance Φ that uses no preprocessing, no memory, and performs $O\left(\frac{\sqrt{n}\log^2 n}{\Phi^2}\right)$ probes per query w.h.p.*

Theorem 4. *There is no deterministic algorithm for exact 2-coloring of complete bipartite graphs that uses no memory and fewer than $\frac{n}{4}$ probes per query.*

Theorem 5. *There exists a deterministic exact 2-coloring algorithm for r-dimensional grids with no preprocessing and no memory that performs $O\left(rn^{\left(\frac{r-1}{r}\right)}\right)$ probes per query.*

Our results for grids and graphs with large conductance implicitly use the fact that these graphs have (relatively) small diameter. What if the diameter is unbounded?

For the special family of trees, we show the following.

Theorem 6. *For any $\alpha > 0$, there exists a 2-coloring algorithm for trees with $O(\alpha)$ preprocessing and $O(\alpha)$ memory that performs $O\left(\frac{n \log n}{\alpha}\right)$ probes per query with probability $1 - \frac{1}{\text{poly}(n)}$ with at most $\alpha - 1$ monochromatic edges.*

We summarize our results in Table 1.

Table 1. Summary of results—n is the number of vertices; m is the number of edges in the input graph. "det." means the bound holds for deterministic algorithms. "det. pf." means the bound holds for deterministic probe-first algorithms.

Graph type	Preproc. time	Memory	Probes	Failure probability[a]	Monoch. edges	
General	$O(m+n)$	$\alpha \log n$	$O(m/\alpha)$	$1/\text{poly}(n)$	0	Theorem 1
Cycle (det. pf.)	any	α (bits)	$\Omega(n/\alpha)$	0	0	Theorem 2
Expander	0	0	$\tilde{O}(\sqrt{n})$	$1/\text{poly}(n)$	0	Theorem 3
Expander (det.)	0	0	$\Omega(n)$	0	0	Theorem 4
r-dim grid	0	0	$O\left(rn^{\left(\frac{r-1}{r}\right)}\right)$	0	0	Theorem 5
Tree	$O(\alpha)$	$O(\alpha)$	$\tilde{O}(n/\alpha)$	$1/\text{poly}(n)$	$\alpha - 1$	Theorem 6

[a]The probability that the LCA performs more than the allotted number of probes.

1.2   Lower Bound Proof Technique

The main technical contribution of this paper is a new technique for proving lower bounds for LCAs. In Sect. 2.2, we give a concise description of (deterministic) LCAs: any LCA can be viewed as a set of functions that map replies to previous probes to a new probe. In order to prove that any LCA for a problem requires x, we explicitly construct an adversarial family of graphs for every LCA $\mathcal{A}$. The graph family is such that $\mathcal{A}$ gives the same reply for each member of this family after x probes, while it should give different replies. We are able to explicitly describe the family of graphs using the description and notation of Sect. 2.2. We use this technique to prove both of the lower bounds in this paper – Theorems 2 and 4.

1.3   Related Work

If preprocessing is allowed, our model is similar in spirit to graph sparsifiers, see, e.g., [5,12,35], in particular graph spanners [4,10,29]: given a connected, edge-weighted graph $G = (V, E)$, a spanner is a sparse subgraph H of G that approximately preserves all pairwise distances. A spanner can be thought of as adding some auxiliary information to the graph: which edges of G are in H. This additional information allows for saving in query reply time. Spanners are useful in routing [30,37], by allowing for small routing tables, and in distance oracles [6,36,38]. For example, Chechik [6] shows that it is possible to augment a graph with $O(kn^{1+1/k})$ bits of memory, such that it is possible to reply to queries of the form "what is the distance between u and v?" in time $O(1)$, where the reply is a $2k - 1$ approximation to the real distance. The two main conceptual differences between this work and the previous work on spanners are (1) we are interested in performing only *sublinear* or even *no preprocessing* (all of the above works use preprocessing that is polynomial in the input size), and (2) we wish to augment the graph with *sublinear memory*.

Rubinfeld et al. [32] formally introduced the LCA model, though several well studied models fit within the framework. For example, *local reconstruction* [1,33]: given access to a function g that is close to having a certain property (e.g., monotonicity) the goal is to reply to a query x with some value $f(x)$ such that f is close to g, using few probes to g. *Locally decodable codes* e.g., [18,39] allow the decoding of part of a code without decrypting it in its entirety. In the past few years, many papers have studied LCAs for maximal independent set (e.g., [3,13]), maximal and approximate maximum matching (e.g., [23,27]) and coloring (e.g., [8,11]). Most of the work has been on bounded degree graphs; recently Feige et al. [9] considered LCAs on graphs of unbounded degree, employing sparsification techniques to obtain LCAs for weak coloring and approximate maximum matching. London et al. [24] showed how to apply LCAs to convex optimization to obtain distributed algorithms for e.g., network utility maximization, that are robust to link failures. Levi et al. [21,22] and Lenzen and Levi [20] describe LCAs that reply to queries of the form "is this edge in a sparse spanner of G?"

A main difference between LCAs and distributed algorithms in the LOCAL model (in which vertices can only communicate with their neighbors; see, e.g., [28]) is that LCAs are allowed to use *remote probes*—probes to vertices that were neither given as the query nor received as a reply to a probe. Göös et al. [16] show that for certain problems, remote probes do not help. We use remote probes in our LCAs for regular graphs and tori; Levi et al. [23] use remote probes to find a good random seed for their approximate matching LCA. We are unaware of other LCAs that use remote probes.

Graph 2-coloring has been used as a benchmark in the area of *property testing*, where a large body of work has been devoted to testing bipartiteness, see, e.g.,[7, 14,15,19]. Property testing differs conceptually from this line of work in that we are *guaranteed* that the input is bipartite and our goal is to provide local evidence of this fact. Testing for bipartiteness typically involves sampling a subgraph and coloring it; it is not clear whether it is possible to adapt these testers to LCAs, while ensuring the consistency of the coloring. It is interesting to extend the results of the paper to graphs that are only guaranteed to be "almost" bipartite.

2   Preliminaries

We denote the set of integers $\{1, 2, \ldots, n\}$ by $[n]$. Logarithms are natural. Our input is a simple undirected bipartite graph $G = (V, E)$, $|V| = n$, $|E| = m$, in which each vertex has a unique ID in $[n]$. The neighborhood of a vertex v, denoted by $N(v)$, is the set of vertices that share an edge with v: $N(v) = \{u : \{u, v\} \in E\}$. The *degree* of a vertex v is denoted by $d_v = |N(v)|$.

We think of each vertex v as having d_v *ports*, $1, 2, \ldots, d_v$, where d_v is v's degree. Each of v's neighbors is connected to v via a single unique port. There are two probe models for LCAs: *strong* and *weak* [9]. A *strong probe* is given a vertex ID v and returns a list of all neighbors of v. A *weak probe* is given a vertex ID v and a port number i, and returns v's i^{th} neighbor u and u's degree. In this paper *we focus on weak probes*. We remark that it is easy to adapt all of our algorithmic results to the strong probe model, and that Theorem 2 holds in the strong probe model as well.

2.1   Local Computation Algorithms

We define LCAs as follows (cf. [32]).

Definition 1 (Local computation algorithm (LCA)). *A $p(n)$-probe $s(n)$-memory local computation algorithm $\mathcal{A}$ for a computational problem is an algorithm that receives an input of size n. Given a query, $\mathcal{A}$ makes at most $p(n)$ probes to the input in order to reply. $\mathcal{A}$ is allocated a memory of $s(n)$ words in addition to the memory required to reply to each query. $\mathcal{A}$ must be* consistent*; that is, the algorithm's replies to all possible queries must combine into a single feasible solution to the problem.*

If the LCA is randomized, it also has a failure probability $\delta(n)$*, which is the probability per a single query that $\mathcal{A}$ uses more than $p(n)$ probes.*

Our definition of LCAs is subtly different from earlier definitions considered in the literature. The main differences are the following:

1. We allow the LCA to write on the auxiliary memory. Although not explicitly disallowed in most previous work, the enduring memory has thus-far (see, e.g., [3,23,26,27,31]) only been used for storing a random seed.
2. The randomness we use does not need to be identical between queries. In previous work, randomized LCAs fix their randomness before the first query; thereafter, they behave deterministically: if the same query is given several times, the LCA performs exactly the same steps each time. In this work, we allow LCAs to perform different actions if the same query is given.

Regarding the second point, we note that we do not allow the LCA to reply "fail", or return an arbitrary color if too many probes are used. The replies to all probes must be consistent with the same solution, and this implies that the replies to the same query must be the same, even though the LCA may reach the same result in a different way. An LCA *fails* if it exceeds the number of probes it is allowed to make (in this case the LCA still performs the number of probes it needs to obtain the required solution). In contrast to previous work, our failure probability is per single query and not over all possible queries. This is because the number of possible queries is bounded by the input size in previous works, whereas here it is unbounded. Furthermore, as we do not need all queries to use the same randomness, we do not need to reason about random seeds; we simply assume that all nodes have access to (sufficiently many) random bits. Note that this is a weaker requirement than access to a shared random seed, and it is easy to extend our results to a shared pseudorandomness, similarly to e.g., [31].

Definition 2 (Probe-first LCAs). Probe-first *LCAs are LCAs whose choice of probes is not a function of the memory content. In other words, the LCA first performs the probes and only then accesses the memory.*

There are several properties of LCAs that are usually considered to be desirable (cf. [8,32]):

1. *Query-obliviousness:* the replies to different queries do not depend on the order in which the queries are presented to the LCA.
2. *Parallelizability.*

All of the LCAs in this paper possess these two properties. In the following Subsect. 2.2, we discuss a method of describing LCAs that will be later found useful for the proofs of Theorems 2 and 4.

We sometimes allow preprocessing. In that case, we measure the complexity of this phase using the classical notion of running time. Note that this is trivially an upper bound to the number of probes performed in this stage.

2.2   A Concise Description of LCAs

For the purposes of the description, assume that we have a deterministic probe-first LCA that uses exactly k probes per query and α words of memory (which we sometimes call a *key*); it is easy to extend the notation to the case when the LCA is allowed to use at most k probes. A probe is of the form "who is the neighbor at port p of vertex v?" and is represented by a pair $(v, p), v \in V, p \in [\Delta]$, where Δ is the maximal degree of the input graph $G = (V, E)$. The reply to a probe is a vertex ID u and u's degree. We assume for simplicity for the rest of this subsection that the graph is d-regular, and the reply to a probe is only a vertex ID. It is straightforward to extend the notation to non-regular graphs. An LCA is also allowed to write to the memory and can have a pre-processing stage. As we do not use these in the proofs, we omit both of these aspects from the notation.

The LCA consists of a set of functions $f_v^1, \ldots, f_v^k : v \in [n]$ that map sets of IDs (a history of replies to previous probes) to an ID and a port number:

$$f_v^i : V^i \to V \times [d].$$

The LCA includes one more set of functions that map the key and the history of replies to all k probes to a color:

$$f_v^{k+1} : L^\alpha \times V^k \to \{\texttt{red}, \texttt{blue}\},$$

where L is the set of all possible words.

We give an example of a (hypothetical) LCA using this notation. Let $\mathcal{A}$ be an LCA that is defined on cycles on the vertex set $\{1, 2, \ldots, 9\}$. $\mathcal{A}$ does some preprocessing on the graph and writes 2 words of memory before the first query is given. When $\mathcal{A}$ is given a query, it performs 3 probes, accesses the memory and decides on a color for the queried vertex. For readability, instead of denoting the port numbers by the set $\{1, 2\}$, we denote them by $\{\texttt{clockwise}, \texttt{anticlockwise}\}$.

We do not give the entire LCA in this form, as this would require at least $9!/5!$ probe descriptions (there are 9 possible queries, the reply to the first query is one of the 8 vertices that were not queried, and so on). Instead, we give a small sample of the description of $\mathcal{A}$.

When $\mathcal{A}$ receives vertex ID 1 as a query, it probes vertex 1's anticlockwise neighbor:

$$f_1^1 = \langle 1, \texttt{anticlockwise} \rangle.$$

If it receives as a reply that 1's anticlockwise neighbor is 2, it probes vertex 1's clockwise neighbor next. If it receives as a reply that 1's anticlockwise neighbor is 3, it probes 6's anticlockwise neighbor next (this is a remote probe, as $\mathcal{A}$ has not encountered vertex 6 yet):

$$f_1^2(2) = \langle 1, \texttt{clockwise} \rangle$$
$$f_1^2(3) = \langle 6, \texttt{anticlockwise} \rangle.$$

For the last probe example:

$$f_1^2(2, 3) = \langle 3, \texttt{clockwise} \rangle$$

means that given that the replies to the previous two queries were 2 and 3, we next probe vertex 3's clockwise neighbor.

Finally,

$$f_1^4((0, 0), (2, 3, 9)) = \texttt{red}$$

means that if the replies to the three probes were $2, 3, 9$ and the key is $(0, 0)$, $\mathcal{A}$ colors vertex 1 $\texttt{red}$. Note that $f_1^4((1, 0), (2, 3, 1)) = \texttt{red}$, while syntactically correct, is impossible for this particular LCA, as 1 cannot be 3's clockwise neighbor because it is 2's clockwise neighbor. We allow "impossible" histories, to simplify the notation and proof.

This interpretation can be extended in the natural way to general (not probe-first) LCAs, by having the key as part of the input to all of the functions.

2.3   Breadth First Search

Several of the LCAs herein use Breadth-First Search (BFS). During its execution, the BFS algorithm maintains a data structure that contains so called "gray" vertices—vertices that have been encountered, but whose neighbors have not yet been probed. At any time, the vertices in the data structure are all at distance i or $i+1$ from the root of the BFS tree, for some i. In the generic BFS description, when it "pops" a new vertex from the data structure, it arbitrarily chooses one of the vertices at distance i. For consistency, we always break ties by ID; that is, our BFS always chooses the vertex at distance i with the lowest ID. The tie breaking with respect to the ports is done similarly: earlier neighbors in the adjacency list are chosen first.

3   Connected General Graphs with Preprocessing

We begin with a simple result that shows that with linear preprocessing, there is an LCA for 2-coloring of arbitrary connected bipartite graphs that uses k probes and α words of memory such that $k\alpha = \tilde{O}(m)$.

Theorem 1. *For any $\alpha > 0$, and any bipartite graph $G = (V, E)$, there exists an exact 2-coloring LCA for G that uses $O(m + n)$ preprocessing probes, $\alpha \log n$ words of memory and performs $O\left(\frac{m}{\alpha}\right)$ probes per query w.h.p.*

Proof. Let $G = (V, E)$ be a bipartite graph. In the preprocessing stage, 2-color G arbitrarily. This can be done using $O(m + n)$ probes using BFS on each connected component. Uniformly at random, choose a set $E' \subset E$ of $\alpha \log n$ edges. Set $S = \{u : (u, v) \in E'\}$; in other words, for each edge in E', arbitrarily choose an endpoint and denote the chosen vertices by the set S. Save the ID and color of the vertices in S to memory.

For any queried vertex q, perform BFS until either (1) some vertex $s \in S$ is encountered, or (2) q's entire connected component is discovered. If (1) occurs, color q according to the parity of a path between q and s. If (2) occurs, color the vertex with the smallest ID in q's connected component red, and color q accordingly. It is easy to see that this is a proper 2-coloring LCA.

To show that at most k probes are used w.h.p., we consider two cases: If q's connected component has at most k edges, then clearly at most k probes are used, whether a vertex $s \in S$ is encountered or not. If q's connected component has more than k edges, the probability that no vertex $s \in S$ is encountered after k probes is at most the probability that no edge of E' is traversed:

$$\prod_{i=0}^{k-1} \left(1 - \frac{\alpha \log n}{m - i}\right) \le \left(1 - \frac{\alpha \log n}{m}\right)^k .$$

Setting $k = O(\frac{m}{\alpha})$ gives that this probability is at most $e^{-\Omega(\log n)}$. ∎

Remark 1. Although the result and proof are for randomized LCAs, the result also holds for deterministic LCAs, albeit with a worse time bound on the time used by the preprocessing stage (naturally the number of preprocessing probes is the same): if $\alpha \log n$ "good" edges can be selected at random, they can clearly also be selected deterministically.

4   Lower Bound for Probe-First LCAs

In this section we consider deterministic probe-first LCA for exact 2-coloring (cf. Definition 1) and prove the following main result:

Theorem 2. *For any $\alpha > 0, k \geq 0$, and $n \geq (k+2)\alpha + 2$, there is no deterministic probe-first LCA for exact 2-coloring of even cycles of length n that uses at most k probes per query and less than α bits of memory, even with unbounded preprocessing time.*

4.1   Summary of Proof

The high level idea of the proof is the following: assume that there exists an LCA $\mathcal{A}$ that uses k probes per query and $\alpha - 1$ bits of memory and can correctly color all vertices of an n vertex graph. We build a family of 2^α graphs for which $\mathcal{A}$ makes the same k probes and receives the same replies when queried on the vertices $1, \ldots, \alpha$. We use these graphs to encode strings in $\{0,1\}^\alpha$. Specifically, let $\{s_1, \ldots, s_{2^\alpha}\}$ denote the set of all strings $\{0,1\}^\alpha$. We construct 2^α graphs $G(s_1), \ldots, G(s_{2^\alpha})$ such that for all $i \in [\alpha]$, the color of vertex i in $G(s_j)$ is blue iff $s_j(i) = 1$. Alice is then given a graph, corresponding to a string s, and sends $\alpha - 1$ bits to Bob. By the assumption that the LCA can color the vertices $1, \ldots, \alpha$ correctly, we get that Bob can recover s. This is true as Bob already knows the replies to all of the probes the LCA would make, and only needs the $\alpha - 1$ bits to decide the colors of $1, \ldots, \alpha$. But if this is true, we will have given a protocol for encoding α bits of information using $\alpha - 1$ bits, which is information-theoretically impossible. Hence no such LCA can exist.

4.2   Proof of Theorem 2

Proof. Assume there is a probe-first LCA $\mathcal{A}$ for bipartite 2-coloring even cycles on $n = (k+2)\alpha + 2$ vertices that uses (at most) k probes per query and $\alpha - 1$ bits of memory. For simplicity, we assume that k is even; the proof can easily be tweaked to accommodate odd k. Note that each vertex on the cycle has two ports: 0 and 1. For a line segment $I = (v_1, v_2, \ldots, v_m)$, we call v_1 the *first* vertex of I and v_m the *last*. Vertices $v_2, \ldots, v_{m-1}$ have both ports allocated and v_1 and v_m only have one port allocated.

Constructing the Segments. We use $\mathcal{A}$ to construct a set of $\alpha + 1$ line segments as follows. Initialize the segments $I_1, \ldots, I_{\alpha+1}$ to consist of vertices $1, \ldots, \alpha + 1$ respectively (i.e., each segment starts off as a single vertex). Set $\Psi = [n] \setminus \{1, 2, \ldots, \alpha+1\}$, the set of vertices that have not yet been assigned. We simulate $\mathcal{A}$ for the queries $1, \ldots, \alpha$. Whenever $\mathcal{A}$ probes $\langle y, p \rangle$, i.e., port p of vertex y, do the following.

1. If $y \in \bigcup_{j=1}^{\alpha+1} I_j$ and port p of y has been assigned, do nothing.

2. If $y \in \bigcup_{j=1}^{\alpha+1} I_j$ and port p of y has not been assigned, choose some $z \in \Psi$, assign z to port p of y, assign y to some port of z, and remove z from Ψ.

3. If $y \notin \bigcup_{j=1}^{\alpha+1} I_j$, choose the last vertex $z \in I_{\alpha+1}$, assign z to port p of y and assign y to an available port of z.

After simulating k probes for queries $1, \ldots, \alpha$, we have a set of $\alpha+1$ line segments. Denote this set by $\mathcal{I}$. Note that given $\mathcal{I}, \mathcal{A}$ can perform all of the probes when queried on the vertices $1, \ldots, \alpha$.

If $|\mathcal{I}|$ is odd, add $\alpha + 1$ integers from Ψ to $\mathcal{I}$, otherwise add α integers from Ψ. We call these vertices *auxiliary* vertices, and we shortly explain precisely how they are added to $\mathcal{I}$. Note that $|\mathcal{I}|$ is now even. Then add to $\mathcal{I}$ the remaining $(k+2)\alpha + 2 - |\mathcal{I}|$ vertices in Ψ, to ensure that $\mathcal{I}$ has exactly $(k+2)\alpha+2$ vertices. We call these vertices *leftover* vertices, and shortly explain how these are added to $\mathcal{I}$ as well.

Lemma 1. *After the addition of the auxiliary vertices, $\mathcal{I}$ contains at most $(k + 2)\alpha + 2$ vertices.*

Proof. After the initialization of the segments, $|\mathcal{I}| = \alpha + 1$. We perform αk probes, each one removes at most 1 vertex from Ψ, for a total of $\alpha + 1 + k\alpha$. After adding at most $\alpha + 1$ auxiliary vertices to $\mathcal{I}$, $\mathcal{I}$ has at most $(k + 2)\alpha + 2$ vertices. $\qquad\square$

Connecting the Segments. Before adding the auxiliary and leftover vertices, there are $\alpha + 1$ segments $I_1, \ldots, I_\alpha, I_{\alpha+1}$, such that I_i contains vertex i. For every $i \in [\alpha]$, we allow I_i and I_{i+1} to be joined in two ways:

1. By connecting the last vertex in I_i directly to the first vertex of I_{i+1}.
2. By connecting the last vertex in I_i and the first vertex of I_{i+1} to an auxiliary vertex.

After these segments are connected, the last vertex of $I_{\alpha+1}$ is connected to the first vertex of I_1, using the remaining auxiliary and all of the leftover vertices as intermediaries. This completes the cycle, and ensures all cycles have the same number of vertices. The α choices of inserting or not inserting an auxiliary vertex between I_i and I_{i+1}, $i \in [\alpha]$ describe the 2^α possible graphs that we can construct in this fashion from the $\alpha + 1$ intervals.

Encoding the String. We show how to construct the graph $G(x)$ that encodes the string $x = x_1 x_2 \ldots x_\alpha$. Note that the graph is a function of $\mathcal{A}$ and x.

Set $x_{\alpha+1} = 0$. For all $i \in [\alpha]$, if connecting the last vertex of I_i and the first vertex of I_{i+1} results in the path between i and $i + 1$ being of *even* length, and $x_i \neq x_{i+1}$, add an auxiliary vertex between I_i and I_{i+1}, otherwise connect them directly. Similarly, if connecting the last vertex of I_i and the first vertex of I_{i+1}

results in the path between i and $i+1$ being of *odd* length, and $x_i = x_{i+1}$, add an auxiliary vertex between I_i and I_{i+1}, otherwise connect them directly.

Finally, connect the last vertex of $I_{\alpha+1}$ to the first vertex of I_1, using the remaining auxiliary and leftover vertices, completing the cycle.

This cycle encodes x, because vertex i is an odd number of vertices away from vertex $\alpha + 1$ if and only if $x_i = 1$, hence a 2-coloring of the graph that assigns vertex $\alpha + 1$ the color red, will assign all vertices i such that $x_i = 0$ the color red and all vertices i such that $x_i = 1$ the color blue.

Set $|\mathcal{I}| = n$. We have shown a construction of 2^α different graphs $\{G(x) : x \in \{0,1\}^\alpha\}$, each of length n, for which $\mathcal{A}$ performs the same probes, and gets the same replies thereto, when queried on vertices $1, \ldots, \alpha$, but there are no two graphs $G(x) \neq G(x')$ for which it colors all the vertices $1, \ldots, \alpha$ identically.

The Protocol. We are ready to give the one-way $(\alpha - 1)$-word protocol for encoding an α-word string x.

Alice is given x (or $G(x)$ - they are isomorphic) and sends an $(\alpha - 1)$-word encoding of $G(x)$ to Bob. Bob knows the replies to the probes that the LCA $\mathcal{A}$ would make when queried on vertices $1, \ldots, \alpha$. The LCA, having the replies to the probes, can use the key to correctly color all the vertices $1, \ldots, \alpha$ by the assumption that $\mathcal{A}$ is a 2-coloring LCA for even-length cycles of length n. But from this coloring, it is easy to decode x. Therefore such an LCA implies an encoding protocol for α bits using $\alpha - 1$ bits, a contradiction.

5   LCAs for Regular Graphs (No Preprocessing, No Memory)

In this section, we will consider exact 2-coloring LCAs for arbitrary (connected) regular graphs that have no preprocessing, use no auxiliary memory (stored between the queries), and so the goal is to optimize the number of probes per query. We show in Subsect. 5.1 that for graphs with conductance Φ, one can design an LCA that requires only $O\left(\frac{\sqrt{n}\log^2 n}{\Phi^2}\right)$ probes w.h.p. We note that Φ is assumed to be an input to the LCA; it is easy to verify that an approximation to Φ suffices. In Subsect. 5.2, we show that even when the conductance is the best possible (i.e., on a complete balanced bipartite graph), any *deterministic exact 2-coloring* that uses no preprocessing and no memory requires $\Omega(n)$ probes. This proves a separation between randomized and deterministic algorithms for regular bipartite graphs with a large (constant) conductance and shows that randomness is indeed necessary for obtaining a sublinear probe complexity if no auxiliary memory is used.

5.1   Upper Bound

We design a *randomized exact 2-coloring* that uses no preprocessing and no memory, and characterizes the number of probes used as a function of the *conductance* of the graph. Let A be the adjacency matrix of G such that $a_{u,v} = 1$

iff $(u, v) \in E$, and let $a(S) = \sum\limits_{u \in S, v \in V} a_{u,v}$. The *conductance* Φ of an undirected graph $G = (V, E)$ is defined as follows (see, e.g., [34]):

$$\Phi(G) = \min_{S \subset V : 0 \leq a(S) \leq a(V)/2} \frac{\sum_{u \in S, v \notin S} a_{u,v}}{a(S)}.$$

Our randomized LCA uses random walk approach, and so we begin with some basic definitions. Consider a discrete-time random walk (Markov chain) $\mathcal{M}$ on an undirected graph $G = (V, E)$ with symmetric transition probability matrix P. The state space of the chain is V. The chain is said to be *irreducible* if it is possible to get from any state to any state using a finite number of transitions; it is *aperiodic* if for any $v, \in \Omega$, $\gcd\{t : P^t(v, v) > 0\} = 1$; it is *lazy* if for all $v \in \Omega, P(v, v) \geq 1/2$. A fundamental theorem of stochastic processes states that an irreducible and aperiodic Markov chain on G converges to a unique *stationary distribution* π over V, i.e., $\lim_{t \to \infty} P^t(u, v) = \pi(v')$ for all $u, v' \in V$. If in addition P is symmetric, then π is uniform over V (cf. [2]).

We are interested in the rate at which a Markov chain converges to its stationary distribution π. We define the *mixing time from a vertex* v to be

$$\tau_v(\delta) = \min\{\bar{t} : d_{TV}(P^t(v, \cdot), \pi) \leq \delta \text{ for all } t \geq \bar{t}\}, \tag{1}$$

where d_{TV} is the total variation distance: for any two distributions μ, ν on V,[2]

$$d_{TV}(\mu, \nu) = \max_{S \subseteq V} |\mu(S) - \nu(S)|. \tag{2}$$

We further define the *mixing time of the Markov chain* to be $\tau(\delta) = \max\limits_{\sigma} \tau_\sigma(\delta)$. We say that a Markov chain is *rapidly mixing* if $\tau(1/2e)$ is logarithmic in n. The constant $1/2e$ is arbitrary, as a bound on $\tau(1/2e)$ implies a bound on $\tau(\delta)$ for any $\delta > 0$ (e.g., [2]) by,

$$\tau(\delta) \leq (1 - \log \delta) \cdot \tau(1/2e).$$

Let $G = (V, E)$ be a d-regular connected bipartite graph. We define the Markov chain $\mathcal{M}$ for a random walk on G (in standard way) as follows: for every vertex v, move to each of v's neighbors with probability $\frac{1}{2d}$, and stay at v with probability $\frac{1}{2}$. It is well known that this chain is aperiodic and irreducible. As its transition matrix is symmetric, its stationary distribution is the uniform distribution, i.e., $\forall v \in V : \pi(v) = \frac{1}{n}$.

We use the following well-known result (see, e.g., [34]).

Lemma 2. *Let P be the transition matrix of a lazy, reversible, irreducible Markov chain that describes a random walk on a graph G. Then*

$$\tau(\epsilon) \leq \frac{2}{\Phi(G)^2} \log\left(\frac{1}{\epsilon \pi_{\min}}\right),$$

where $\pi_{\min} = \min_{v \in \Omega} \pi(v)$.

---

[2] Alternatively, it is known that we can define it as $d_{TV}(\mu, \nu) = \frac{1}{2} \sum_{\sigma \in \Omega} |\mu(\sigma) - \nu(\sigma)|$.

Our main positive result of this section is the following theorem describing a randomized exact 2-coloring LCA:

Theorem 3. *There exists a randomized exact 2-coloring LCA for connected regular bipartite graphs with conductance Φ that uses no preprocessing, no memory, and performs $O\left(\frac{\sqrt{n}\log^2 n}{\Phi^2}\right)$ probes per query with probability at least $1 - \frac{1}{n}$.*

We note that it is straightforward to increase the success probability to $1 - \frac{1}{\text{poly}(n)}$.

Proof. Set $\tau = \frac{6}{\Phi(G)^2}\log n$ and $\epsilon = \frac{1}{n^2}$, and since $\pi_{\min} = \frac{1}{n}$, we have that $\tau(\epsilon) \le \tau$ by Lemma 2.

Our LCA is the following: color vertex 1 **red**. Given a vertex q as a query, perform the random walk $\mathcal{M}$ for $4\tau\sqrt{n}$ steps starting at vertex 1. Color the encountered vertices **blue** if they are at an odd distance from vertex 1 and **red** otherwise. Store IDs and colors of the vertices encountered on this random walk; call this set S. Next, perform a random walk from vertex q until one of the vertices of S is encountered. This creates a path p between q and some vertex $s \in S$. If p is of even length, return the color of s, otherwise return its complement.

The algorithm above clearly returns a valid 2-coloring (since G is bipartite). Lemma 3 below shows that $O(\tau\sqrt{n}\log n)$ probes are used w.h.p. in the second walk.

Lemma 3. *Let S be the set of vertices encountered in the first random walk. The probability that none of these vertices is encountered in the second random walk after $4\tau\sqrt{n}\log n$ steps is at most $\frac{1}{n}$.*

Proof. For the purposes of the proof, let the second random walk be such that it is carried out for $4\tau\sqrt{n}\log n$ steps regardless of whether a vertex in S is encountered. Consider the vertices at intervals of τ steps from one another in the first and second random walks. Denote these sets by $T_1 \subset S$ and T_2 respectively. We bound the probability that $T_1 \cap T_2 = \emptyset$, which is an upper bound to the probability that none of the vertices in the second random walk are in S. We do this by standard methods, but need to be careful with the analysis as the vertices of T_1 and T_2 are not selected uniformly at random from V, but rather from a distribution that is almost uniform. Let X_1 denote the probability that there are at least $2\sqrt{n}$ distinct vertices in T_1; let X_2 denote the event that there are at least $2\sqrt{n}\log n$ distinct vertices in T_2. Let X_3 be the event that $T_1 \cap T_2 \ne \emptyset$.

We bound $\Pr[X_1]$ and $\Pr[X_2]$: There are fewer than x distinct vertices in a set S if all vertices in S are chosen from the same subset $S' \subseteq V$ of size x; there are $\binom{n}{x}$ such subsets and the measure of each such subset is at most $(x+1/n)/n$, by the fact that the total variation distance of the distribution reached by the random walk from uniform is at most $\frac{1}{n^2}$.

$$\Pr[\neg X_1] \leq \binom{n}{2\sqrt{n}} \left(\frac{2\sqrt{n} + 1/n}{n} \right)^{4\sqrt{n}} \leq \left(\frac{ne}{2\sqrt{n}} \right)^{2\sqrt{n}} \left(\frac{2\sqrt{n} + 1/n}{n} \right)^{4\sqrt{n}}$$

$$\leq \left(\frac{e\sqrt{n}}{2} \right)^{2\sqrt{n}} \left(\frac{4}{\sqrt{n}} \right)^{4\sqrt{n}} < \frac{1}{n^2}.$$

The proof that $\Pr[X_2] \geq 1 - \frac{1}{n^2}$ is similar and is omitted.

We bound $\Pr[\neg X3|X_1 \wedge X_2]$, the probability that none of the vertices of the first walk is encountered in the second walk after t steps, given that X_1 and X_2 both hold. This is the probability that every unique vertex in T_2 is not in T_1: after verifying that $i - 1$ vertices from T_2 are not in T_1, the probability that the next chosen vertex is in T_1 would be least $\frac{|T_1|}{n-i}$ if the vertices were all chosen from the uniform distribution. The total variation distance between D_1 and D_2 is at most $\frac{2}{n^2}$, where D_1 and D_2 are the distributions over V from which the vertices are sampled in the first and second random walks respectively. This is because $d_{TV}(D_1, U)$ and $d_{TV}(D_2, U)$ are both at most $\frac{1}{n^2}$. Hence the additional $\frac{2}{n}$ in the numerator of the first inequality.

$$\Pr[\neg X3|X_1 \wedge X_2] \leq \prod_{i=1}^{|T_2|} \left(1 - \frac{2\sqrt{n} - 2/n}{n - i} \right) \leq \prod_{i=1}^{|T_2|} \left(1 - \frac{\sqrt{n}}{n} \right) \leq \left(1 - \frac{1}{\sqrt{n}} \right)^{2\sqrt{n}\log n} \leq \frac{2}{n^2}.$$

Therefore,

$$\Pr[\neg X_3] \leq \Pr[\neg X3|X_1 \cap X_2] + \Pr[\neg X_1] + \Pr[\neg X_2] \leq \frac{1}{n},$$

implying $\Pr[X_3] \geq 1 - \frac{1}{n}$, as required.

5.2   Lower Bound

We prove the lower bound:

Theorem 4. *There is no deterministic LCA for exact 2-coloring of complete bipartite graphs that uses no memory and fewer than $\frac{n}{4}$ probes.*

Proof. Let n be a multiple of 4. Consider any LCA $\mathcal{A}$ on the vertex set $[n]$ that uses at most $k = \frac{n}{4} - 1$ probes per query. Assume w.l.o.g. that $\mathcal{A}$ uses exactly k probes per query.

We construct two complete balanced (i.e., both sides have $\frac{n}{2}$ vertices) bipartite graphs G_a and G_b such that the following hold:

- $\mathcal{A}$ receives the same replies to the k probes it makes on G_a and G_b when it is given vertices 1 or 2 as queries.
- In G_a, vertices 1 and 2 are neighbors, and hence should be colored differently, while in G_b, they are not, and should be colored the same.

In order to construct G_a and G_b, we first construct two subgraphs, G_1 and G_2. Intuitively, G_1 and G_2 will consist of the vertices that $\mathcal{A}$ will encounter when probing vertices it has already seen when queried on vertices 1 and 2 respectively.

In addition, G_1 and G_2 will also be used for remote probes made by $\mathcal{A}$ when queried on vertices 2 and 1 respectively.

We initialize $G_1 = (\{1\}, \emptyset)$, $G_2 = (\{2\}, \emptyset)$. Set $\Psi = V \setminus \{1, 2\}$, the set of "free" vertices – vertices that we have not yet assigned to G_1 or G_2.

We first consider the set of probes when $\mathcal{A}$ is queried on vertex 1. Consider the jth probe, $j \in \{1, \ldots, k\}$ made by $\mathcal{A}$; denote the replies to the previous probes by $v_1, \ldots, v_{j-1}$. Using the notation of Sect. 2.2, the jth probe is represented as $f_1^j(v_1, \ldots, v_{j-1}) = \langle y, p \rangle$, where y is a vertex and p is a port number (in our case, $p \in \{1, \ldots, \frac{n}{2}\}$). In other words, given that the replies to the previous probes were $v_1, \ldots, v_{j-1}$, $\mathcal{A}$ now probes the pth port of vertex y. We (possibly) extend G_1 or G_2 as follows:

1. If $y \in G_1 \cup G_2$, and port p of y has been assigned, do nothing.
2. If $y \in G_1 \cup G_2$, and port p of y has not been assigned, choose some $z \in \Psi$, assign z to port p of y, assign y to some port of z, and remove z from Ψ.
3. If $y \notin G_1 \cup G_2$, choose some vertex $z \in G_2$, assign z to port p of y and assign y to an available port of z. As z initially has $\frac{n}{2}$ free ports, and $2k < \frac{n}{2}$, there is at least one available port for z.

After considering these k probes, we have graphs G_1 and G_2, which comprise together at most $k + 2$ vertices. Next, we consider the set of probes made when $\mathcal{A}$ is queried on vertex 2. We continue in the same fashion, except that if $y \notin G_1 \cup G_2$, choose some vertex $z \in G_1$, assign z to port p of y and assign y to an available port of z. G_1 and G_2 now comprise at most $2k + 2 = \frac{n}{2}$ vertices. Note that G_1 and G_2 are bipartite and disjoint, by their construction.

G_a is the following: let G_1 and G_2 be subgraphs of G_a such that vertices 1 and 2 are on the same side and arbitrarily fill in the rest of the vertices and edges. For G_b, fix G_1 and G_2 so that vertices 1 and 2 are on different sides, again arbitrarily filling in the vertices and edges. It is possible to construct both G_a and G_b in this fashion regardless of $\mathcal{A}$ as the total number of vertices in G_1 and G_2 is at most $\frac{n}{2}$. Executing $\mathcal{A}$ on G_a and G_b with queries 1 and 2 gives the same replies to the probes. As there is no key (additional memory), the color given to vertices 1 and 2 must be the same in G_a and G_b, hence there does not exist a deterministic LCA for exact 2-coloring complete bipartite graphs that uses no memory and fewer than $\frac{n}{4}$ probes.

6   LCAs for Grids and Tori (No Preprocessing, No Memory)

The LCA from Theorem 3 shows that for graphs with high conductance there is an exact 2-coloring LCA with very low number of probes per query, and therefore in this section we focus now our attention to a representative class of graphs with very low conductance: *grids* and *tori*. We describe our results for grids, noting that they hold for tori as well. Designing algorithms for a tori is intuitively at least as hard as for grids, as there is no notion of an end, and hence no absolute position, which could hypothetically be leveraged by algorithms for grids.

6.1   Warm Up : 2-Dimensional Grids

We assume that a $\sqrt{n} \times \sqrt{n}$ grid is arranged so that each interior vertex has four neighbors—to the north, south, east and west. However, the vertices do not know which neighbor is in which direction; locally, each vertex just sees four neighbors. When we say that it is possible to *land on* a vertex we mean to reach it and "know that we are there" (to distinguish from the usual notion of reachability).

The central feature of our algorithms is that we can efficiently follow any single line on a grid (Corollaries 1 and 2), and that we can detect efficiently all vertices on a single hyperplane of a multidimensional grid (Lemma 7). We begin with the following lemma.

Lemma 4. *Assume w.l.o.g. that vertex v's neighbor at port p is u and u is to the north of v. It is possible to land on the vertex north of u using at most 36 weak probes.*

Proof. Vertex u has at most three neighbors in addition to v. All of u's neighbors (that are not v) are at distance 2 from v, but if we do not count paths that pass through u, the north neighbor is at distance 4 (and the other two are still at distance 2).

It takes it most 16 probes to find out whether x, a neighbor of u, is at distance 2 from v—four probes to find all of x's neighbors and four more per neighbor that is not u. Therefore 16 probes per vertex are sufficient to either confirm or rule out any of u's neighbors as the northern neighbor. It suffices to check any two of them (if both are ruled out, the third is the northern neighbor). Adding the 4 probes made by u completes the claim.

It is also possible to land on the vertex south of v using the same reasoning. It is easy to know when we reach the edge of the grid, as the edge vertices have one fewer neighbor; the following is immediate.

Corollary 1. *It is possible to traverse the graph in a straight line from side to side through any vertex using $36\sqrt{n}$ probes.*

It is not possible to know in which direction the graph is being traversed (north-south or east-west), however it is not necessary.

Theorem 7. *There exists a deterministic exact 2-coloring LCA for 2-dimensional grids with no preprocessing and no memory that performs $O(\sqrt{n})$ probes per query.*

Proof. When queried on a vertex q, traverse the graph in both directions (north-south and east-west). Traverse the graph in one direction from the vertex with ID 1 (it will be either north-south or east-west). The line described by the traversal from vertex 1 must intersect exactly one of the lines emanating from v (unless v's ID is 1, in which case it intersects both). Coloring vertex 1 **red** determines v's color.

6.2 General Grids

Each interior vertex of an r-dimensional grid is connected to $2r$ other vertices. The LCA is the following: When a vertex q is queried, the LCA discovers all the vertices of some hyperplane passing through vertex 1; the LCA then traverses the grid along the r straight lines that pass through q. At least one of these lines intersects the hyperplane; coloring vertex 1 red determines q's color. The proof follows the same reasoning as the proof for 2-dimensional grids.

Lemma 5. *Let $G = (V, E)$ be a r-dimensional grid, $r \geq 1$ and let u and v be two interior vertices such that $(u, v) \in E$. Then there exists a single $w \in N(u) \setminus \{v\}$ such that there does not exist a path of length 2 from w to v that does not include u.*

Proof. Denote $W = N(u) \setminus \{v\}$. There exists one vertex $w \in W$ for which u, v, w lie on a line. The shortest path between v and w that does not include u must have length 4. For all other $w' \in W$, u, v, w define a two dimensional plane, and there is a vertex x such that v, u, w, x comprise a square. The path w, x, v has length 2.

For simplicity, we give a loose upper bound in the next lemma instead of an exact bound as in the previous subsection; it is straightforward to compute a tight bound.

Lemma 6. *Given an edge (v, u), assume w.l.o.g. that u is to the north of v. It is possible to land on the vertex north of u using at most $(2r)^3$ weak probes.*

Proof. Probe all $2r$ neighbors of u. By Observation 5, all but one of them has a path of length 2 to v that doesn't pass through u. For each of the neighbors of u, it is possible to enumerate all length 2 paths using $(2r)^2$ probes.

Note that it is also possible to land on the vertex to the south of v using the same reasoning, therefore:

Corollary 2. *It is possible to traverse the r-dimensional grid in a straight line from side to side through any vertex using $O\left(r^3 n^{\frac{1}{r}}\right)$ probes.*

We show how to build on the line construction to construct a hyperplane.

Lemma 7. *Given a r-dimensional grid $G = (V, E)$ and a vertex $v \in V$, it is possible to construct a hyperplane that passes through v using at most $O\left(r^4 n^{\frac{1}{r}}\right) + O\left(r n^{\left(\frac{r-1}{r}\right)}\right)$ weak probes.*

Proof. Construct $r-1$ lines passing through v that traverse the grid, $p_1, \ldots p_{r-1}$. This takes $O\left(r^4 \sqrt[r]{n}\right)$ probes. We can complete these lines to a hyperplane as follows: Set $P = \bigcup_{i=1}^{r-1} p_i$. As long as there exist two vertices $u, w \in P$, such that $N(u) \cap N(w) \neq \emptyset$, add the vertex $x \in N(u) \cap N(w)$ (there is exactly one such neighbor) to P. This takes at most r probes per vertex in the hyperplane.

We need to show that (1) all vertices of the hyperplane are added to P, and (2) no vertices that are not in the hyperplane are added to P. To show (1), assume that some vertices in the hyperplane are not added to P. Consider the distances from v in the metric space defined by the grid. Let x be the closest vertex to v in the hyperplane (there may be more than one) that is not in P. If x has a single neighbor in P then x must be on p_i for some i, hence it is added to the hyperplane. If x is not on p_i for any i, it must hold that there are two neighbors of x in the plane that contains v and x that are closer to v and are in P (as x is the closest vertex not in P). By the construction of the hyperplane, x would have been added too. To show (2), note that any vertex not in the hyperplane has at most one neighbor in the hyperplane at distance 1, and hence will not be added to P.

Theorem 5. *There exists a deterministic exact 2-coloring LCA for r-dimensional grids with no preprocessing and no memory that performs $O\left(rn^{\left(\frac{r-1}{r}\right)}\right)$ probes per query.*

Proof. If the queried vertex q's ID is 1, color q **red**. Otherwise, construct a hyperplane through vertex 1, and r orthogonal lines that traverse the grid that pass through q. The hyperplane must intersect exactly one of these lines. Color 1 **red**, and color q accordingly.

7 LCAs for Trees (with a Small Number of Monochromatic Edges)

In this section, we demonstrate that for trees, significantly better results can be obtained if we allow for a small number of monochromatic edges. We show that, for any α, we can find a coloring (using only two colors) that has at most $\alpha - 1$ monochromatic edges using $O(\alpha)$ preprocessing, $O(\alpha)$ memory and $O(\frac{n \log n}{\alpha})$ probes per query w.h.p.

Our LCA $\mathcal{A}$ is the following. Given a tree $T = (V, E)$ and α, choose a set $S \subset V, |S| = \alpha$, of *focii* uniformly at random, and store their IDs. (Implicitly) color the focii **red**. The preprocessing and memory required are therefore proportional to α. Whenever a vertex q is queried, if $q \notin S$, perform a BFS. Color q consistently with the first encountered focus, which we call q's focus.

Define the *skeleton* of the tree to be the union of the paths between the focii. For any vertex v in the skeleton, its *skeletal subtree* is the set of vertices W not on the skeleton, for which any path connecting $w \in W$ to any vertex on the skeleton passes through v. Note that some vertices on the skeleton may have an empty skeletal subtree.

Lemma 8. *If v is on the skeleton, and v's focus is u, then u is the focus of every vertex in v's skeletal subtree.*

Proof. Let w be some vertex in v's skeletal subtree. Clearly a focus is closest to v if and only if it is closest to w. Furthermore, the BFS tie-breaking rule is global, guaranteeing that the focus that will be found first (out of all the closest focii) in the BFS from v is the same one that will be found by BFS from w.

Lemma 9. *There are at most $\alpha - 1$ monochromatic edges in T, when T is colored according to $\mathcal{A}$.*

Proof. From Claim 8, edge violations can only be on the skeleton; we restrict our attention to the skeleton. Call any vertex of degree at least 3 that is not a focus, a *junction*. To show that at there are at most $\alpha - 1$ monochromatic edges, we use a charging process, in which we charge monochromatic edges to focii. We describe an iterative process by which we contract edges to shrink the skeleton, maintaining the invariant that the skeleton remains a tree, and charge monochromatic edges to focii. At the start of an iteration, consider some focus f that is a leaf of the skeleton. There must be at least one, as the skeleton is a tree. From f, follow the skeleton until we either reach (i) a junction, (ii) another focus or (iii) a monochromatic edge.

If (i): we reach a junction j, we contract the path, and let f replace j (the degree of f is now 1 less than the degree of j, and f is not a leaf any more).

If (ii): we reach another focus f', remove f and the path between f and f'. If the degree of f' was 2 it is now a leaf.

If (iii): we reach a monochromatic edge, we charge the monochromatic edge e to f, and remove f, e and the path between them. We also remove the path from e to the nearest junction, j or focus f', whichever is closer. Clearly, if we reach another focus f', no edge between e and f' will be monochromatic. It remains to show that no edge between e and the junction j can be monochromatic. Once e and f are removed, the path between e and j is part of j's skeletal subtree, and therefore by Claim 8 (with $v = j$), all of the vertices on the path from e to j have the same focus, hence there is no monochromatic edge on this path.

Lemma 10. *The number of probes used to reply to any query is $O\left(\frac{n \log n}{\alpha}\right)$ with probability $1 - \frac{1}{\text{poly}(n)}$.*

The proof is similar to the proof of Theorem 1 and is omitted. The algorithm description together with Lemmas 9 and 10 imply Theorem 6.

Acknowledgements. We thank the anonymous reviewers for their useful comments and suggestions.

References

1. Ailon, N., Chazelle, B., Comandur, S., Liu, D.: Property-preserving data reconstruction. Algorithmica **51**(2), 160–182 (2008)
2. Aldous, D.J.: Random walks on finite groups and rapidly mixing Markov chains. Séminaire de probabilités de Strasbourg **17**, 243–297 (1983). http://eudml.org/doc/113445

3. Alon, N., Rubinfeld, R., Vardi, S., Xie, N.: Space-efficient local computation algorithms. In: Proceedings of 22nd ACM-SIAM Symposium on Discrete Algorithms (SODA), pp. 1132–1139 (2012)
4. Althöfer, I., Das, G., Dobkin, D., Joseph, D., Soares, J.: On sparse spanners of weighted graphs. Discrete Comput. Geom. **9**(1), 81–100 (1993)
5. Benczúr, A.A., Karger, D.R.: Approximating s-t minimum cuts in $\tilde{O}(n2)$ time. In: Proceedings of the Twenty-Eighth Annual ACM Symposium on Theory of Computing, STOC 1996, pp. 47–55 (1996)
6. Chechik, S.: Approximate distance oracles with constant query time. In: Symposium on Theory of Computing, STOC, pp. 654–663 (2014)
7. Czumaj, A., Monemizadeh, M., Onak, K., Sohler, C.: Planar graphs: random walks and bipartiteness testing. In: Proceedings of 52nd Annual IEEE Symposium on Foundations of Computer Science (FOCS), pp. 423–432 (2011)
8. Even, G., Medina, M., Ron, D.: Best of two local models: local centralized and local distributed algorithms. CoRR abs/1402.3796 (2014). http://arxiv.org/abs/1402.3796
9. Feige, U., Patt-Shamir, B., Vardi, S.: On the probe complexity of local computation algorithms. CoRR abs/1703.07734 (2017). http://arxiv.org/abs/1703.07734
10. Filtser, A., Solomon, S.: The greedy spanner is existentially optimal. In: Proceedings of the 2016 ACM Symposium on Principles of Distributed Computing, PODC, pp. 9–17 (2016)
11. Fraigniaud, P., Heinrich, M., Kosowski, A.: Local conflict coloring. In: IEEE 57th Annual Symposium on Foundations of Computer Science, FOCS, pp. 625–634 (2016)
12. Fung, W.S., Hariharan, R., Harvey, N.J.A., Panigrahi, D.: A general framework for graph sparsification. In: Proceedings of the 43rd ACM Symposium on Theory of Computing, STOC 2011, San Jose, CA, USA, 6–8 June 2011, pp. 71–80 (2011)
13. Ghaffari, M.: An improved distributed algorithm for maximal independent set. In: Proceedings of the Twenty-Seventh Annual ACM-SIAM Symposium on Discrete Algorithms, SODA 2016, pp. 270–277 (2016)
14. Goldreich, O., Ron, D.: A sublinear bipartiteness tester for bounded degree graphs. Combinatorica **19**(3), 335–373 (1999)
15. Goldreich, O., Ron, D.: Property testing in bounded degree graphs. Algorithmica **32**(2), 302–343 (2002)
16. Göös, M., Hirvonen, J., Levi, R., Medina, M., Suomela, J.: Non-local probes do not help with many graph problems. In: Gavoille, C., Ilcinkas, D. (eds.) DISC 2016. LNCS, vol. 9888, pp. 201–214. Springer, Heidelberg (2016). https://doi.org/10.1007/978-3-662-53426-7_15
17. Hassidim, A., Mansour, Y., Vardi, S.: Local computation mechanism design. ACM Trans. Econ. Comput. **4**(4), 21:1–21:24 (2016). https://doi.org/10.1145/2956584
18. Katz, J., Trevisan, L.: On the efficiency of local decoding procedures for error-correcting codes. In: Proceedings of the Thirty-Second Annual ACM Symposium on Theory of Computing, pp. 80–86 (2000)
19. Kaufman, T., Krivelevich, M., Ron, D.: Tight bounds for testing bipartiteness in general graphs. SIAM J. Comput. **33**(6), 1441–1483 (2004)
20. Lenzen, C., Levi, R.: A local algorithm for the sparse spanning graph problem. CoRR abs/1703.05418 (2017). http://arxiv.org/abs/1703.05418
21. Levi, R., Ron, D., Rubinfeld, R.: Local algorithms for sparse spanning graphs. In: Approximation, Randomization, and Combinatorial Optimization. Algorithms and Techniques, APPROX/RANDOM 2014, pp. 826–842 (2014)

22. Levi, R., Ron, D., Rubinfeld, R.: A local algorithm for constructing spanners in minor-free graphs. In: Approximation, Randomization, and Combinatorial Optimization. Algorithms and Techniques, APPROX/RANDOM, pp. 38:1–38:15 (2016)

23. Levi, R., Rubinfeld, R., Yodpinyanee, A.: Brief announcement: local computation algorithms for graphs of non-constant degrees. In: Proceedings of the 27th ACM on Symposium on Parallelism in Algorithms and Architectures, SPAA, pp. 59–61 (2015)

24. London, P., Chen, N., Vardi, S., Wierman, A.: Distributed optimization via local computation algorithms (2017). http://users.cms.caltech.edu/~plondon/loco.pdf

25. Mansour, Y., Patt-Shamir, B., Vardi, S.: Constant-time local computation algorithms. In: Sanità, L., Skutella, M. (eds.) WAOA 2015. LNCS, vol. 9499, pp. 110–121. Springer, Cham (2015). https://doi.org/10.1007/978-3-319-28684-6_10

26. Mansour, Y., Rubinstein, A., Vardi, S., Xie, N.: Converting online algorithms to local computation algorithms. In: Czumaj, A., Mehlhorn, K., Pitts, A., Wattenhofer, R. (eds.) ICALP 2012. LNCS, vol. 7391, pp. 653–664. Springer, Heidelberg (2012). https://doi.org/10.1007/978-3-642-31594-7_55

27. Mansour, Y., Vardi, S.: A local computation approximation scheme to maximum matching. In: Raghavendra, P., Raskhodnikova, S., Jansen, K., Rolim, J.D.P. (eds.) APPROX/RANDOM -2013. LNCS, vol. 8096, pp. 260–273. Springer, Heidelberg (2013). https://doi.org/10.1007/978-3-642-40328-6_19

28. Peleg, D.: Distributed Computing: A Locality-Sensitive Approach. Monographs on Discrete Mathematics and Applications. SIAM, Philadelphia (2000)

29. Peleg, D., Schäffer, A.A.: Graph spanners. J. Graph Theory **13**(1), 99–116 (1989)

30. Peleg, D., Upfal, E.: A trade-off between space and efficiency for routing tables. J. ACM **36**(3), 510–530 (1989)

31. Reingold, O., Vardi, S.: New techniques and tighter bounds for local computation algorithms. J. Comput. Syst. Sci. **82**(7), 1180–1200 (2016)

32. Rubinfeld, R., Tamir, G., Vardi, S., Xie, N.: Fast local computation algorithms. In: Proceedings of 2nd Symposium on Innovations in Computer Science (ICS), pp. 223–238 (2011)

33. Saks, M.E., Seshadhri, C.: Local monotonicity reconstruction. SIAM J. Comput. **39**(7), 2897–2926 (2010)

34. Sinclair, A., Jerrum, M.: Approximate counting, uniform generation and rapidly mixing Markov chains. Inf. Comput. **82**(1), 93–133 (1989)

35. Spielman, D.A., Srivastava, N.: Graph sparsification by effective resistances. SIAM J. Comput. **40**(6), 1913–1926 (2011)

36. Thorup, M., Zwick, U.: Approximate distance oracles. In: Proceedings on 33rd Annual ACM Symposium on Theory of Computing, (STOC), pp. 183–192 (2001)

37. Thorup, M., Zwick, U.: Compact routing schemes. In: SPAA, pp. 1–10 (2001)

38. Wulff-Nilsen, C.: Approximate distance oracles with improved preprocessing time. In: Proceedings of the Twenty-Third Annual ACM-SIAM Symposium on Discrete Algorithms, SODA, pp. 202–208 (2012)

39. Yekhanin, S.: Locally decodable codes. Found. Trends Theor. Comput. Sci. **6**(3), 139–255 (2012)

Bin Packing Games with Weight Decision: How to Get a Small Value for the Price of Anarchy

Gyorgy Dosa$^{1(\boxtimes)}$, Hans Kellerer^2, and Zsolt Tuza^{3,4}

^{1} Department of Mathematics, University of Pannonia, Egyetem u. 10,
Veszprém 8200, Hungary
dosagy@almos.vein.hu
^{2} Institut für Statistik und Operations Research,
Universität Graz, Universitätsstraße 15, 8010 Graz, Austria
hans.kellerer@uni-graz.at
^{3} Alfréd Rényi Institute of Mathematics, Hungarian Academy of Sciences,
Reáltanoda u. 13–15, Budapest 1053, Hungary
^{4} Department of Computer Science and Systems Technology,
University of Pannonia, Egyetem u. 10, Veszprém 8200, Hungary
tuza@dcs.uni-pannon.hu

Abstract. A selfish bin packing game is a variant of the classical bin packing problem in a game theoretic setting. In our model the items have not only a size but also a positive weight. The cost of a bin is 1, and this cost is shared among the items being in the bin, proportionally to their weights. A packing is a Nash equilibrium (NE) if no item can decrease its cost by moving to another bin, and OPT means a packing where the items are packed optimally (into minimum number of bins). Without any misunderstanding we denote by NE both the packing and the number of bins in the packing, and the same holds for the OPT packing. We are interested in the Price of Anarchy (PoA), which is the limsup of NE/OPT ratios. Recently there is a growing interest for games where the PoA is low.

We give a setting for the weights where the PoA is between 1.4646 and 1.5. The lower bound is valid also for the special case of the game where the weight of any item is the same as its size, and any item has size at most one half. The previous bound was about 1.46457. Next we give another setting where the PoA is at most $16/11 \approx 1.4545$. This value is better than any previous, that was got for such games.

Keywords: Selfish bin packing · Price of anarchy
Algorithmic game theory

1 Introduction

The bin packing problem [7] is one of the classical combinatorial optimization problems. We are given a set $I = \{1, 2, ..., m\}$ of m items, the size of item i is

© Springer Nature Switzerland AG 2018
L. Epstein and T. Erlebach (Eds.): WAOA 2018, LNCS 11312, pp. 204–217, 2018.
https://doi.org/10.1007/978-3-030-04693-4_13

$s_i \in (0,1]$. The goal is to pack the items into a minimum number of bins, so that any bin can contain only items with total size at most 1. For a survey on approximation algorithms for bin packing problems see [2].

In a selfish bin packing game (in our model) an item i has not only a size but also a positive weight, denoted by g_i. Each item plays the role of a selfish agent, and any agent/item pays some cost for being in a bin. The cost of a bin is 1, and this cost is shared among the items being in the bin, proportionally to their weight g_i; i.e., any item i packed into a bin (with total size at most 1) pays the cost $g_i/g(B)$ where $g(B)$ is the total weight of the items in the bin. Analogously, we denote by $s(B)$ the total size of items being packed into this bin, also called *level* of the bin.

Any item is interested in decreasing its own cost. So, if an item fits into another bin, and moving there the cost of this item will be strictly smaller than before, we call this move an *improving step*.

A *Nash equilibrium* (NE, for short) is a packing where an improving step, as described above, does not exist for any item. Such a NE is also called a *stable packing*. It is easy to see that a NE packing is reached from any initial (valid) packing after finitely many selfish steps.

Observation 1. *The game terminates in a NE after a finite number of selfish moves, starting from any initial packing.*

Proof. Define a potential function ψ for all feasible packings $\mathcal{B}$ as

$$\psi(\mathcal{B}) = \sum_{B \in \mathcal{B}} (g(B))^2$$

where the summation is taken over all used bins B. Under a selfish move the contents of two bins, say B_i and B_j, are modified to B_i' and B_j' such that

$$g(B_i') + g(B_j') = g(B_i) + g(B_j) \qquad \text{and} \qquad |g(B_i') - g(B_j')| > |g(B_i) - g(B_j)|.$$

It follows that the value of ψ increases after every selfish move, and consequently no $\mathcal{B}$ can occur more than once during a game. Since each input admits only finitely many possible packings, the process surely terminates.   □

Note that in case of a NE packing, the number of used bins can be much larger than necessary for packing the items in an optimal solution for classical bin packing. Therefore, it is interesting to compute the price of anarchy (PoA) [8] of the game which is in general given by the maximum ratio between the maximum social cost of any NE, and the minimum social cost of any situation. In our case, it is defined as

$$PoA = \limsup_{N \to \infty} (\sup_I \max_{NE \in \mathcal{NE}(I)} \{\frac{NE(I)}{OPT(I)} \mid OPT(I) = N\}),$$

where I is an arbitrary (finite) set of items, $\mathcal{NE}(I)$ is the set of all Nash equilibria for I, $NE(I)$ is the number of used bins in a Nash equilibrium packing NE of the items, and $OPT(I)$ denotes the number of bins used in an optimal packing of the items. If it is clear from the context, we will write NE instead of $NE(I)$ and OPT instead of $OPT(I)$, respectively.

Next, we provide a short review about selfish bin packing games. The first such game was defined by Bilò [1] for $g_i = s_i$. Consequently, the cost of an item i for being in bin B is $s_i/s(B)$ in this case. Bilò proved that $8/5 \leq PoA \leq 5/3$ holds in this game. Epstein and Kleiman [4] tightened the gap between the lower and upper bound to $1.6416 \leq PoA \leq 1.6428$.

Ma et al. [9] defined another model. If there are k items in a bin, each item in this bin pays $1/k$ cost for being in this bin. This corresponds to the case $g_i = 1$. It is shown that any NE packing is the output of a run of the First Fit (FF) algorithm (which packs the items according to some given list, and the next item is always packed into the first bin where it fits [7]). From this, it directly follows that the PoA of this version of the game is at most 1.7. Later Dosa and Epstein [3] gave a more exact estimation, showing that the PoA is in the interval $[1.6966, 1.6994]$.

There are also many other results which correspond to the models listed above; see, e.g., [4–6,12]. In fact there is also a model in the literature which generalizes all previous selfish bin packing games [11].

Returning to our game, we have seen that the two different models (sharing the cost proportionally to the sizes, or equally, regardless of item sizes) give two different PoA values. For the former this is at most 1.6428, but for the latter it is bigger than 1.6966 [3,4]. Note that both these cost-sharing mechanisms are special cases of our more general model, where the cost is shared according to the weights of the items. It means that the weight matters. Notice that in [3] it is proven that by any choice of the weights, the PoA is at most 1.7, and there exists a choice where this value is exactly 1.7.

In this work we attack the problem from the opposite side, namely we investigate the following question.

Problem 1. How should one define the weights for the items, so that the PoA will be as small as possible?

This problem—in a more general context—sounds as follows: Can one determine the rules of a structure[1] (in our case it is the current game) in such a way that the behavior of participants will be advantageous for the whole community? If the answer is yes, how to do this? In the language of games: Can one construct a game where the PoA is very small?

A recent work in this direction is [13] which defines a cost-sharing mechanism with $PoA \approx 1.467$. Another similar work is [14] where the PoA is between 1.47407 and 1.4748. Another recent work defining a bin packing game with a price of anarchy of $3/2$ is [10]. It means that we see a growing interest in defining

---

[1] An existing example on the large scale is that different types of products include different percentages of VAT in their price, hence orienting the consumption habits of people to some extent.

such games where the PoA is low. In this paper we define a game with an even smaller PoA.

We will consider two settings for the weights in this paper. In the first one, in most cases the weight of an item is equal to its size, except that the items having sizes above one half get larger weight, namely their weight will be set to 1. For this special setting we will see that the PoA is already at most 1.5. Although the PoA in this case is above 1.4646, the principle of the construction provides a frame for later improvements and hence in this aspect it is crucial.

It also turns out that our construction (given for this special setting) also works for the special case of the game were the weight of any item is the same as its size (i.e. for the original version of Bilò [1]), and the size of any item is at most $1/2$. This parametric model (where the size of any item is at most $1/t$ for some integer t) is investigated in [6]. So, as a by-product, for the case of $t = 2$ we get a slightly improved lower bound.

We also define and investigate a more complicated setting, where the PoA is at most $16/11 \approx 1.4545$, and at least 1.4528.

It is important to note that the PoA cannot be very small, namely we shall prove that it can never be smaller than $4/3$.

The paper is structured as follows. In Sect. 2 we present a general lower bound for the price of anarchy. Section 3 contains the first simple setting which has PoA of at most $3/2$. In Sect. 4 we present the more complicated setting with PoA at most $16/11$. We finish in Sect. 5 with short conclusions and open problems.

As the page limit does not allow to present all material of this work, some proof is omitted and will be published in the journal version of this paper.

2   A General Lower Bound

In this section we present a general lower bound for the PoA, which universally is valid no matter how the weights for the items are chosen. The construction is based on the property that no item can move to another bin since it does not fit.

Theorem 2. *The PoA of the game is at least 4/3.*

Proof. We make the following construction. Let n be divisible by four, and let ε be a sufficiently small rational number. We construct a packing into n bins that is a NE. Then we show that the items can be packed into $(3/4)n + 1$ bins proving our claim.

In any NE bin there are two items, a medium item and a small item, denoted by M_i and S_i, $1 \le i \le n$, respectively. In the first $n/2$ bins the medium items have the same size, $M_i = 1/3 + (2n - 1)\varepsilon$, for $1 \le i \le n/2$, these are the biggest items in our construction. In the other bins the medium items have sizes as follows: $M_i = 1/3 + (\frac{5}{2}n - i)\varepsilon$, for $n/2 + 1 \le i \le n$ (then $M_{n/2+1}$ has the same size as the previous medium items, and the sizes of the medium items are slightly decreasing from this point.)

Now we give the sizes of the small items in the bins.

For any $n/2+1 \leq i \leq n$, let $S_i = 1/3 - (\frac{3}{2}n - i)\varepsilon$. Hence, $M_i + S_i = 2/3 + n\varepsilon$. Note that the smallest small item among these items is $S_{n/2+1} = 1/3 - (n-1)\varepsilon$. This is the size of any further small items as well, so $S_i = 1/3 - (n-1)\varepsilon$ for $1 \leq i \leq n/2$. (The biggest small item has size $S_n = 1/3 - (n/2)\varepsilon$.)

Then the levels of the bins are as follows. We have level $1/3 + (2n-1)\varepsilon + 1/3 - (n-1)\varepsilon = 2/3 + n\varepsilon$ in the first $n/2$ bins and also exactly $2/3 + n\varepsilon$ in the last $n/2$ bins. It is trivial that this packing is a NE, as no item fits into another bin (to see this, it is enough to see that the smallest item does not fit into any other bin).

Now we show that the items can be packed into $(3/4)n + 1$ bins.

We create $n/2 - 1$ bins as follows. For any $n/2 + 2 \leq i \leq n$, the bin contains M_i, moreover S_{i-1} and finally $S_{i-n/2}$. Let us realize that $M_i + S_{i-1} = 2/3 + (n-1)\varepsilon$, so after packing these two items in the bin, the remaining space in the bin is just the size of $S_{i-n/2}$. So these three items fit into a common bin.

The largest medium items are packed pairwise, and so we get $n/4$ further bins.

There remain one medium item and two small items unpacked (namely $M_{n/2+1}$, S_n and finally $S_{n/2}$). We pack them into two further bins and get in total $(n/2 - 1) + n/4 + 2 = (3/4)n + 1$ bins.   □

Observation 3. *In the previous construction the level of any NE bin is exactly $2/3 + n\varepsilon$, and the level of any not full optimal bins is exactly $2/3 - (4n-2)\varepsilon$.*

Since the PoA is computed as a limes superior, it does not matter if a certain property does not hold for a bounded number of bins. Bins for which a certain property does not hold, will be called *irregular* in the following. The bins which are not irregular are called *regular* bins.

At the end of this section we provide a simple lemma which will be useful several times.

Lemma 1. *Suppose that the level of any NE bin is at least $l > 0$ with at most C irregular bins. Then $NE \leq (1/l) \cdot OPT + C$.*

Proof. We have $OPT \geq S \geq l \cdot (NE - C)$ where S denotes the total level of the bins. From this we get $NE \leq (1/l) \cdot OPT + C$.   □

3   A Simple Setting: $S1$

In this section we will examine a simple setting, denoted as setting $S1$, and we redefine the classes (i.e. big and small). The weights are defined as follows. Let the items be called *big* and *small* (denoted by B and S) if their sizes are larger than $1/2$, or at most $1/2$, respectively. The big items get weight 1. Each small item gets weight equal to its size. The weights in setting $S1$ are illustrated in Table 1. In the following we give lower and upper bounds for the PoA of setting $S1$.

Table 1. Weights for setting $S1$

Class	Size s_i	Weight g_i
Big	$s_i > \frac{1}{2}$	$g_i = 1$
Small	$s_i \leq \frac{1}{2}$	$g_i = s_i$

3.1   A Simple Lower Bound for $S1$

As a first step, in this subsection we prove the following lower bound for the PoA of $S1$. The method will be developed further afterwards, to obtain a stronger estimate.

Theorem 4. *For setting $S1$ the PoA is at least $\frac{4}{3} + \frac{1}{3} \cdot \sum_{k=1}^{\infty} \frac{1}{3 \cdot 2^k - 1} \approx 1.4589$.*

Proof. We extend our construction from Theorem 2. The original items of that instance, say I, are called *old* items. Let us recall that in the optimum packing of I, the $n/2$ largest medium items are packed pairwise into a set of $n/4$ bins, which we shall denote by $\mathcal{B}$. The bins in $\mathcal{B}$ have some space left for further items. Let us consider any bin of $\mathcal{B}$. The unused space of such a bin is $1/3 - (4n - 2)\varepsilon$ by Observation 3, that is, a bit below $1/3$. We will use that space for packing smaller items.

Let $k \geq 4$ be a very large integer. We create a new instance I' which augments I by the following *new* items: $n/4$ items of size $a_1 = 1/6 + n\varepsilon$, $n/4$ items of size $a_2 = 1/12 + n\varepsilon$, and so on. In general, we get for any $1 \leq i \leq k$, $n/4$ new items of size $a_i = \frac{1}{3 \cdot 2^i} + n\varepsilon$ so that the smallest item a_k has size $a_k = \frac{1}{3 \cdot 2^k} + n\varepsilon$.

We get a packing for the items in I' by adding to each bin of $\mathcal{B}$ items of sizes $a_1, a_2, \ldots, a_k$, exactly one item for each size. The total size of the items in any such bin is

$$1 - \frac{1}{3 \cdot 2^k} + (kn + 4n - 2)\varepsilon \leq 1 - \frac{1}{3 \cdot 2^k} + 2kn\varepsilon.$$

We choose ε such that the items fit together, so let $\varepsilon \leq 1/(6kn \cdot 2^k)$. In this way all bins are "almost" completely full.

The optimal packing for I used $(3/4)/n + 1$ bins. Consequently, the optimal packing for I' does also use not more than $(3/4)/n + 1$ bins.

Now we construct a NE for the instance I'. Without loss of generality, we suppose that n is divisible by $3 \cdot 2^k - 1$ for any $1 \leq i \leq k$. The old items are packed as before. In any new bin, items of the same size a_i $(i = 1, \ldots, k)$ are packed. That means, there are bins containing five a_1 items, also there are bins containing eleven a_2 items, and so on. Generally, there are bins containing $3 \cdot 2^i - 1$ items of type a_i, $1 \leq i \leq k$. It follows that the number of newly created bins is

$$\frac{n}{4} \cdot \sum_{i=1}^{k} \frac{1}{3 \cdot 2^i - 1}.$$

The total number of bins in the packing is

$$n + \frac{n}{4} \cdot \sum_{i=1}^{k} \frac{1}{3 \cdot 2^i - 1},$$

while there is an optimal packing using only $(3/4)n + 1$ bins. Thus the lower bound follows, once we prove that this packing is a NE.

We show that no item can make an improving step. It is trivial that no old item can move as it fits into no other bin. Let us suppose that the bins containing the new items are ordered in nonincreasing order of their levels. Then, we have:

- No new item fits into a bin to the right.
- No new item wants to move to the left.

This completes the proof of the theorem.   □

Remark 1. In [6] a better lower bound is proved. In fact, [6] considers the proportional cost sharing, the general and also the parametric case, where each item has size at most $1/t$ for an integer t. If $t = 2$, this means that each item has size at most $1/2$. For this case their (stronger) lower bound is 1.464571. This lower bound applies also to our case, as for items having sizes at most $1/2$ we give the same weights as in [6], namely the weight of any item is the same as its size. In the previous theorem we gave our very simple construction to give intuition. It provides an easy lower bound and allows a simple proof. In the next subsection we give a stronger lower bound, but it will need a much more careful analysis.

3.2   A More Difficult Lower Bound Construction

Here we give an improved lower bound for the *PoA* as follows.

Theorem 5. *The PoA is at least $\frac{3543193}{2419209} \approx 1.46460806$.*

Proof. The better lower bound is achieved by a more tricky construction, again using the one from Theorem 2. Its *old* items are given in n optimal bins. Among these optimal bins there are $n/4$ bins, each having a room of $1/3 - (4n - 2)\varepsilon$ (see Observation 3). We will use this room for packing *new* items, and we call these bins *recycled* bins.

The new items will have sizes approximately $1/6$, $1/12$, $1/24$ and $1/48$. There are three subtypes from each type, these subtypes are denoted by the symbols "-","+", and "++". The exact sizes are as follows:

$$\begin{array}{lll}
1/6^+ = 1/6 + 19\alpha & 1/6^- = 1/6 - 20\alpha & 1/6^{++} = 1/6 + 999\alpha \\
1/12^+ = 1/12 + 19\beta & 1/12^- = 1/12 - 20\beta & 1/12^{++} = 1/12 + 999\beta \\
1/24^+ = 1/24 + 19\gamma & 1/24^- = 1/24 - 20\gamma & 1/24^{++} = 1/24 + 999\gamma \\
1/48^+ = 1/48 + 19\delta & 1/48^- = 1/48 - 20\delta & 1/48^{++} = 1/48 + 999\delta
\end{array}$$

where the values of α, β, γ, and δ are as follows: $\alpha = (4n - 2)\varepsilon$, $\beta = 1000\alpha$, $\gamma = 1000\beta$ and $\delta = 1000\gamma$. We ensure that δ is still very small, for example $\delta < 10^{-100}$ applies.

We give the types of the recycled bins of the optimal packing. We list only the new items in these bins (any such recycled bin already contains two old medium sized items.) We will need new parameters of a, b, c, d, e.

- There are a bins containing one $1/6^+$ and one $1/6^-$-item.
- There are b bins containing one $1/6^{++}$-item, one $1/12^+$ and one $1/12^-$-item.
- There are c bins containing one $1/6^{++}$ and one $1/12^{++}$-item, moreover one $1/24^+$ and one $1/24^-$-item.
- There are d bins containing one $1/6^{++}$, one $1/12^{++}$ and one $1/24^{++}$-item, and one $1/48^+$ and one $1/48^-$-item.
- There are e bins containing one $1/6^{++}$, one $1/12^{++}$, one $1/24^{++}$ and one $1/48^{++}$-item.

It is easy to see that based on the definition of the sizes of the items, all such bins are valid (i.e. the items fit into the bins); in fact, all these bins are fully packed, except the last subtype where among other items the $1/48^{++}$-items are packed. Now we give the new bins that are created in the NE. For this we need some new parameters, as x, y, z, u, t, s, v, w.

- There are x bins containing one $1/6^-$ and four $1/6^+$-items, and y bins containing three $1/6^-$ and two $1/6^{++}$-items.
- There are z bins containing four $1/12^-$ and seven $1/12^+$-items, and u bins containing nine $1/12^-$ and two $1/12^{++}$-items.
- There are t bins containing ten $1/24^-$ and thirteen $1/24^+$-items, and s bins containing twenty-one $1/24^-$ and two $1/24^{++}$-items.
- There are v bins containing twenty-two $1/48^-$ and twenty-five $1/48^+$ -items, and w bins containing forty-five $1/48^-$ and two $1/48^{++}$-items.

We need to ensure that the number of items of each subtype is the same in the two (i.e. optimal and NE) packings. For this we get the next system of equations:

$$b + c + d + e = 2y \tag{1}$$
$$a = 4x \tag{2}$$
$$a = x + 3y \tag{3}$$
$$c + d + e = 2u \tag{4}$$
$$b = 7z \tag{5}$$
$$b = 4z + 9u \tag{6}$$
$$d + e = 2s \tag{7}$$
$$c = 13t \tag{8}$$
$$c = 10t + 21s \tag{9}$$
$$e = 2w \tag{10}$$
$$d = 25v \tag{11}$$
$$d = 22v + 45w \tag{12}$$

Here (1) stands for the number of $1/6^{++}$-items in the OPT packing and in the NE packing on the left-hand side and the right-hand side, respectively. Similarly, (4), (7), and (10) count the number of $1/12^{++}$-items, $1/24^{++}$-items and $1/48^{++}$-items, respectively.

Moreover, (2) stands for the number of $1/6^{+}$-items, in the OPT packing and in the NE packing on the left-hand side and the right-hand side, respectively. Similarly, (3) stands for the number of $1/6^{-}$-items. Following the list, (5) and (6) count the number of $1/12^{+}$ and $1/12^{-}$-items, (8) and (9) count the number of $1/24^{+}$ and $1/24^{-}$-items, and finally (11) and (12) count the number of $1/48^{+}$ and $1/48^{-}$-items.

We have 12 equations and 13 variables, the degree of freedom is one. Let us choose $w = 8$, then the unique solution for the variables is as follows: $w = 8$, $v = 120$, $s = 1508$, $t = 10556$, $u = 70122$, $z = 210366$, $x = y = 806403$, moreover $a = 3225612$, $b = 1472562$, $c = 137228$, $d = 3000$, $e = 16$. Note that $a + b + c + d + e = 4838418$ while $x + y + z + u + t + s + v + w = 1905486$.

Now let us see what is the new lower bound. The number of optimal bins is $(3/4)n + 1$. Within these bins, there are $n/4$ recycled bins. Let us choose a new integer variable g, such that $n/4 = 4838418g$. Then there are $1905486g$ new bins in the NE packing, thus $OPT = (3/4)n + 1$ while $NE' = n + 1905486g$ (here NE' means the new value of the bins in the equilibrium). It means that by letting $g \to \infty$, our improved lower bound is

$$PoA \geq \frac{NE'}{OPT - 1} = \frac{4 \cdot 4838418 + 1905486}{3 \cdot 4838418} = \frac{3543193}{2419209} \approx 1.46460806.$$

We still need to prove that the new packing is really a stable packing. We will call the new bins in the NE as x-bins, y-bins, ..., w-bins, according to the number of those bins. By easy calculation we get the levels of the different types of bins. This is given in the table below. For the sake of simplicity, we use the notation $l(x)$ for the level of x-bins, $l(y)$ for the level of y-bins, and so on.

$$l(x) = 5/6 + 56\alpha \quad l(y) = 5/6 + 1938\alpha$$
$$l(z) = 11/12 + 53\beta \quad l(u) = 11/12 + 1818\beta$$
$$l(t) = 23/24 + 47\gamma \quad l(s) = 23/24 + 1578\gamma$$
$$l(v) = 47/48 + 35\delta \quad l(w) = 47/48 + 1098\delta$$

In what follows, we show that no item moves in the NE packing. A k-*bin* means a bin with exactly k items, and a k^{+}-*bin* means a bin with at least k items. We illustrate the principle with the 5-bins (i.e. the x-bins and y-bins) and items therein.

A $1/6^{+}$-item or a $1/6^{-}$-item does not want to go to a 2-bin, since there the level of the bin is $2/3 + n\varepsilon < 2/3 + \alpha$ (see Observation 3), and the size of an x-item is at most $1/6 + 19\alpha$, thus the increased level of the target bin would be at most $5/6 + 20\alpha$, which is smaller than $l(x)$. A $1/6^{++}$-item would create a higher level, but still lower than $l(y)$, therefore an item of that size does not want to move either. A $1/6$-type item does not fit into any other 5-bin, neither into any 11^{+}-bin. We conclude that no $1/6$-type item moves.

In a very similar way, which is a matter of routine to check, it can be verified that the other items do not move either. Details are left to the reader.   □

Notes. We do not apply any big item in our construction. The reason is that the weight of a big item itself is 1, and the weight of any bin not containing a big item is at most 1. It follows that if there is at least one bin containing a big item, any small item wants to move there. As a consequence, the bin of any big item will be almost completely full in any NE packing (more exactly, if x is the size of the smallest item being packed into a bin without a big item, then the level of any bin with a big item is bigger than $1 - x$). Hence, it seems that applying big items in a construction is not advantageous if we want to get a large NE/OPT ratio in the considered setting.

On the other hand, since we do not apply any big item in the construction, our improved lower bound (i.e. 1.46460806) applies also for the parametric PoA where each item has size at most $1/2$, for the original version of the bin packing game. For this case [6] provides a lower bound of about 1.464571, hence our lower bound is a little better.

We also note that our construction applies to items of sizes around $1/6$, $1/12,...,1/48$. If we go further and include also items of sizes around $1/96$ (or even smaller ones) we will get certain improvement on the lower bound, but this improvement is really very small. If the smallest items were about $1/24$ (instead of about $1/48$), the implied lower bound would be about 1.464599.

3.3 An Upper Bound of 3/2 for $S1$

We prove the following bound.

Theorem 6. *For setting $S1$ the PoA is at most $3/2$.*

Proof. Let us consider an arbitrary NE packing. Recall that an item is called big if its size is above $1/2$ otherwise it is small. We define the following two bin types:

B1: The bin contains one big item, and possibly several further (small) items.
B2: The bin contains only small items.

Since all big items have weight 1, the weight of any B1-bin is at least 1, while the weight of any B2-bin is at most 1. It follows that any item of a B2-bin has the intention to move into a B1-bin. Since the considered packing is a NE, no item of a B2-bin fits into a B1-bin.

If there is no B2-bin, then every NE bin contains a big item, thus the packing is optimal. If there is only one B2-bin, still $NE \leq OPT + 1$ holds. Thus, let us suppose that there exist at least two B2-bins. Consider the B2-bin with the smallest level—let l_1 denote this level—and let X be the smallest item in this bin. Now consider the B2-bin with the second smallest level—let l_2 denote this level, then $l_2 \geq l_1$—and let Y be the smallest item in this bin.

We distinguish two cases.

Case 1: $l_2 > 2/3$.

It means that the level of any B2-bin is above 2/3, except possibly the bin containing X. If also the level of any B1-bin is above 2/3, we are done by an averaging argument. Thus let us suppose that there is a B1-bin with smaller level. Since no item of a B2-bin fits into a B1-bin, it follows that any small item in the B2-bins has size above 1/3. Let us denote the number of B1 and B2-bins by x and y, respectively. Obviously, $OPT \geq x$.

If $y \leq x/2$, we have

$$NE = x + y \leq \frac{3}{2}x \leq (3/2)OPT.$$

Suppose now that $y > x/2$. Every B2-bin contains exactly two items, both are larger than 1/3, and any B1-bin contains a big item. Two big items cannot be packed into a common bin, and a big item can share a bin with at most one small item which is larger than 1/3. Thus $OPT \geq x + (2y - x)/2 = x/2 + y$. Hence,

$$\frac{NE}{OPT} \leq \frac{x+y}{x/2+y} = 1 + \frac{x/2}{x/2+y} \leq 1 + \frac{x/2}{x/2+x/2} = 3/2.$$

Case 2: $l_1 \leq l_2 \leq 2/3$.

Under this assumption it follows that $X > 1/3$, as X would like to go into the bin of Y, but cannot move, since X does not fit into the bin of Y.

Assume $Y \leq 1/3$. Since Y does not fit into any other bin apart from the bin of X, the level of any bin apart from the bins of X and Y is above 2/3. Thus, $NE \leq (3/2)OPT + 2$ and we are done.

Therefore, also $Y > 1/3$ holds. If there are at least two items in the bin of X, then $X \leq 1/3$ would follow, also if there were at least two items in the bin of Y. We conclude that both X and Y are the only items in their bins. This is a contradiction, since both are small items. They rather share their bins than stay alone. This completes the proof of the theorem.   □

4   A More Complicated Special Setting: S2

We found quite close bounds in the considered simple setting $S1$. In this section we consider another, more complicated setting, denoted as $S2$, where the PoA is a bit smaller. For this, we give larger weights to certain smaller items as well, but then it is necessary to make a more detailed analysis.

We choose the item weights as follows. Let the items be denoted as big, medium, small and tiny if their sizes are larger than 1/2, larger than 1/3 and at most 1/2, larger than 1/4 and at most 1/3, and finally at most 1/4, respectively. The classes and their weights are described in Table 2.

Table 2. Weights for setting $S2$

Class	Size s_i	Weight g_i
Big	$1/2 < s_i$	$g_i = 1$
Medium	$1/3 < s_i \leq 1/2$	$g_i = 1/2$
Small	$1/4 < s_i \leq 1/3$	$g_i = s_i$
Tiny	$s_i \leq 1/4$	$g_i = s_i$

4.1 A Lower Bound for $S2$

Here we give a lower bound for the PoA as

$$PoA \geq \frac{4}{3} + \frac{1}{11} + \frac{1}{3} \cdot \sum_{k=3}^{\infty} \frac{1}{3 \cdot 2^k - 1} \approx 1.4528$$

Theorem 7. *For setting $S2$ the PoA is at least 1.4528.*

Proof. Consider the construction given in Theorem 4 where we gave a packing for instance I' which was a NE for setting $S1$. This packing is *not* a NE for $S2$ since any bin containing a medium item has level close to $2/3$ and weight close to $5/6$. Thus, we would get an improving step by moving an item of size $a_1 = 1/6 + n\varepsilon$ into that bin.

Hence, we construct a new instance I'' where each item of size a_1 is replaced by two items of size $a_2 = 1/12 + n\varepsilon$ and the other items of I' are left unchanged.

Again, there is an optimal packing using only $(3/4)n + 1$ bins. We use the same packing strategy for the items in I'' as for the items in I'.

Now in the NE packing the a_1 items are missing, but there are three times more bins of type B_2 containing eleven a_2 items. The number of newly created bins is therefore

$$\frac{n}{4}\left(\frac{3}{3 \cdot 2^2 - 1} + \sum_{i=3}^{k} \frac{1}{3 \cdot 2^k - 1}\right).$$

The total number of bins in the packing is

$$n + \frac{n}{4} \cdot \left(\frac{3}{3 \cdot 2^2 - 1} + \sum_{i=3}^{k} \frac{1}{3 \cdot 2^k - 1}\right).$$

Thus, the lower bound follows, once we show that this packing is a NE. We show that no item can make an improving step. Now the difference is only regarding the option that a new item perhaps wants (and can) move into a bin of a medium item. It is sufficient to see that an item a_2 does not want to move into a bin of a medium item. In such a target bin the total weight is maximal if the biggest small item is packed together with a medium item. Hence the maximum weight is bounded by $1/2 + S_n = 1/2 + 1/3 - (n/2)\varepsilon = 5/6 - (n/2)\varepsilon$. The weight of a current bin of an a_2 item is $11 \cdot (1/12 + n\varepsilon)$. Should the a_2-item move, the weight in the target bin would become $5/6 - (n/2)\varepsilon + 1/12 + n\varepsilon = 11/12 + (n/2)\varepsilon$, which is smaller than the weight of the current a_2-bin. Hence, this packing gives a NE.   □

Let us remark that in the previous construction there are no big items as their bins would be filled too much. We have medium items, and considerably smaller items which are close to $1/12$, and even smaller items. It means that if we want to define a "better" weighting system which makes the PoA (much) smaller, it does not help if we give larger weight for the items between $1/3$ and $1/12 + \delta$ for some $\delta > 0$.

4.2   An Upper Bound of $16/11 \approx 1.4545$ for $S2$

The main result of Sect. 4 is summarized in the following theorem. The proof is quite long and detailed, so due to page limit we cannot provide it here.

Theorem 8. *The PoA of the game in setting S2 is at most* $16/11$.

We mention only one more thing here. We have seen that in our simpler setting the PoA is at least 1.4646, but for the more difficult setting the PoA is at most 1.4545. This clearly distinguishes the two settings, as the PoA of the simpler setting and the PoA of the more difficult setting cannot be the same, as they are in two, disjoint intervals.

5   Conclusions

We have investigated the price of anarchy in selfish bin packing games under two types of weight assignments. In the setting of weights which is more sensitive to item size, the PoA is proven to be between 1.4528 and $16/11 \approx 1.4545$. The lower bound is definitely not tight, as it can slightly be improved by a more complicated construction. Also, it is possible that the upper bound could be decreased a bit.

We also mention that along the lines of our more difficult lower bound construction for setting $S1$ (see Sect. 3.2.) one can get improved lower bounds for the parametric game of Bilò (i.e. weight = size), not only for $t = 2$ as we got the bound but for any bigger t values as well. As this was not our main purpose in this work, we leave this option (to improve) for others.

Beyond S1 and S2. Another option would be to introduce a refinement of setting S2: We follow the harmonic sequence, define weight 1 for items having sizes above $1/2$, weight $1/2$ for items having sizes at most $1/2$ and bigger than $1/3$, weight $1/3$ for items having sizes above $1/4$ but at most $1/3$, and the weight of any smaller item equals the size of the item. In this setting (called S3) possibly the PoA is smaller, but the investigation seems hard.

Acknowledgements. Gyorgy Dosa acknowledges the financial support of Széchenyi 2020 under the EFOP-3.6.1-16-2016-00015. Research of the first and third authors was supported in part by the National Research, Development and Innovation Office – NKFIH under the grant SNN 116095. All three authors are supported by Stiftung Aktion Österreich-Ungarn 99öu1.

References

1. Bilò, V.: On the packing of selfish items. In: Proceedings of the 20th International Parallel and Distributed Processing Symposium (IPDPS 2006), 9 pages. IEEE (2006)
2. Coffman, E.G., Csirik, J., Galambos, G., Martello, S., Vigo, D.: Bin packing approximation algorithms: survey and classification. In: Pardalos, P.M., Du, D.-Z., Graham, R.L. (eds.) Handbook of Combinatorial Optimization, pp. 455–531. Springer, New York (2013). https://doi.org/10.1007/978-1-4419-7997-1_35
3. Dosa, G., Epstein, L.: Generalized selfish bin packing. arXiv:1202.4080, 43 pages (2012)
4. Epstein, L., Kleiman, E.: Selfish bin packing. Algorithmica **60**(2), 368–394 (2011)
5. Epstein, L., Kleiman, E., Mestre, J.: Parametric packing of selfish items and the subset sum algorithm. In: Leonardi, S. (ed.) WINE 2009. LNCS, vol. 5929, pp. 67–78. Springer, Heidelberg (2009). https://doi.org/10.1007/978-3-642-10841-9_8
6. Epstein, L., Kleiman, E., Mestre, J.: Parametric packing of selfish items and the subset sum algorithm. Algorithmica **74**(1), 177–207 (2016)
7. Johnson, D.S., Demers, A., Ullman, J.D., Garey, M.R., Graham, R.L.: Worst-case performance bounds for simple one-dimensional packing algorithms. SIAM J. Comput. **3**, 299–325 (1974)
8. Koutsoupias, E., Papadimitriou, C.: Worst-case equilibria. In: Meinel, C., Tison, S. (eds.) STACS 1999. LNCS, vol. 1563, pp. 404–413. Springer, Heidelberg (1999). https://doi.org/10.1007/3-540-49116-3_38
9. Ma, R., Dosa, G., Han, X., Ting, H.-F., Ye, D., Zhang, Y.: A note on a selfish bin packing problem. J. Glob. Optim. **56**(4), 1457–1462 (2013)
10. Nong, Q.Q., Sun, T., Cheng, T.C.E., Fang, Q.Z.: Bin packing game with a price of anarchy of $\frac{3}{2}$. J. Comb. Optim. **35**, 632–640 (2018)
11. Wang, Z., Han, X., Dosa, G., Tuza, Z.: Bin packing game with an interest matrix. Algorithmica **80**(5), 1534–1555 (2018)
12. Yu, G., Zhang, G.: Bin packing of selfish items. In: Papadimitriou, C., Zhang, S. (eds.) WINE 2008. LNCS, vol. 5385, pp. 446–453. Springer, Heidelberg (2008). https://doi.org/10.1007/978-3-540-92185-1_50
13. Zhang, C., Zhang, G.: Cost-sharing mechanisms for selfish bin packing. In: Gao, X., Du, H., Han, M. (eds.) COCOA 2017. LNCS, vol. 10627, pp. 355–368. Springer, Cham (2017). https://doi.org/10.1007/978-3-319-71150-8_30
14. Chen, X., Nong, Q., Fang, Q.: An improved mechanism for selfish bin packing. In: Gao, X., Du, H., Han, M. (eds.) COCOA 2017. LNCS, vol. 10628, pp. 241–257. Springer, Cham (2017). https://doi.org/10.1007/978-3-319-71147-8_17

Probabilistic Embeddings of the Fréchet Distance

Anne Driemel[1]([⊠]) and Amer Krivošija[2]([⊠])

[1] Department of Mathematics and Computer Science, TU Eindhoven, Postbus 513,
5600 MB Eindhoven, The Netherlands
a.driemel@tue.nl
[2] Department of Computer Science, TU Dortmund, 44221 Dortmund, Germany
amer.krivosija@tu-dortmund.de

Abstract. The Fréchet distance is a popular distance measure for curves
which naturally lends itself to fundamental computational tasks, such as
clustering, nearest-neighbor searching, and spherical range searching in
the corresponding metric space. However, its inherent complexity poses
considerable computational challenges in practice. To address this problem we study distortion of the probabilistic embedding that results from
projecting the curves to a randomly chosen line. Such an embedding
could be used in combination with, e.g. locality-sensitive hashing. We
show that in the worst case and under reasonable assumptions, the discrete Fréchet distance between two polygonal curves of complexity t in
$\mathbb{R}^d$, where $d \in \{2, 3, 4, 5\}$, degrades by a factor linear in t with constant
probability. We show upper and lower bounds on the distortion. We also
evaluate our findings empirically on a benchmark data set. The preliminary experimental results stand in stark contrast with our lower bounds.
They indicate that highly distorted projections happen very rarely in
practice, and only for strongly conditioned input curves.

Keywords: Fréchet distance · Metric embeddings
Random projections

1 Introduction

The Fréchet distance is a distance measure for curves which naturally lends itself
to fundamental computational tasks, such as clustering, nearest-neighbor searching, and spherical range searching in the corresponding metric space. However,
their inherent complexity poses considerable computational challenges in practice. Indeed, spherical range searching under the Fréchet distance was recently
the topic of the yearly ACM SIGSPATIAL GISCUP 2017 competition [34], high-
lighting the relevance and the difficulty of designing efficient data structures for

A. Driemel was funded by NWO Veni project "Clustering time series and trajectories
(10019853)". A. Krivošija has been partly supported by DFG within the Collaborative
Research Center SFB 876 "Providing Information by Resource-Constrained Analysis",
project A2. We thank Kevin Buchin for useful discussions on the topic of this paper.

© Springer Nature Switzerland AG 2018
L. Epstein and T. Erlebach (Eds.): WAOA 2018, LNCS 11312, pp. 218–237, 2018.
https://doi.org/10.1007/978-3-030-04693-4_14

this problem. At the same time, Afshani and Driemel showed lower bounds on the space-query-tradeoff in the pointer model [2] that demonstrate that this problem is even harder than simplex-range searching.

The computational complexity of computing a single Fréchet distance between two given curves is a well-studied topic [3, 10–13, 18, 21]. It is believed that it takes time that is quadratic in the length of the curves and this running time can be achieved by applying dynamic programming. In this body of literature, the case of 1-dimensional curves under the continuous Fréchet distance stands out. In particular, no lower bounds are known on computing the continuous Fréchet distance between 1-dimensional curves. It has been observed that the problem has a special structure in this case [14]. Clustering under the Fréchet distance can be done efficiently for 1-dimensional curves [19], but seems to be harder for curves in the plane or higher dimensions. Bringmann and Künnemann used projections to lines to speed up their approximation algorithm for the Fréchet distance [12]. They showed that the distance computation can be done in linear time if the convex hulls of the two curves are disjoint. It is tempting to believe that the curves being restricted to 1-dimensional space makes the problem significantly easier. However, in the general case, there are no algorithms known which are faster for 1-dimensional curves than for curves in higher dimensions. In practice, it is very common to separate the coordinates of trajectories to simplify computational tasks. It seems that in practice the inherent character of a trajectory is often largely preserved when restricted to one of the coordinates of the ambient space. Mathematically, this amounts to projecting the trajectory to a line.

This motivates our study of probabilistic embeddings of the Fréchet distance into the space of 1-dimensional curves. Concretely, we study distortion of the probabilistic embedding that results from projecting the curves to a randomly chosen line. Such a random projection could be used in combination with probabilistic data structures, e.g. locality-sensitive hashing [20], but also with the multi-level data structures for Fréchet range searching given by Afshani and Driemel [2]. See below for a more in-depth discussion of these data structures.

We show that in the worst case and under certain assumptions, the discrete Fréchet distance between two polygonal curves of complexity t in $\mathbb{R}^d$, where $d = \{2, 3, 4, 5\}$, degrades by a factor linear in t with constant probability. In particular, we show upper and lower bounds on the change in distance for the class of c-packed curves. The notion of the c-packed curves was introduced by Driemel, Har-Peled and Wenk in [18] and has proved useful as a realistic input assumption [4, 10, 17]. A curve is called c-packed for a value $c > 0$ if the length of the intersection of the curve with any ball of any radius r is a most cr. While our study is mostly restricted to the discrete Fréchet distance, we expect that our techniques can be extended to the case of the continuous Fréchet distance.

A closely related distance measure, which is popular in the field of datamining, is dynamic time warping (DTW) [16, 31, 33]. The computational complexity of DTW has also been extensively studied, both empirically and in theory [1, 4, 24, 29]. Some of our lower bounds extend to DTW.

1.1   Related Work on Data Structures with Fréchet Distance

The complexity of classic data structuring problems for the Fréchet distance is still not very well-understood, despite several papers on the topic. We review what is known for nearest-neighbor searching and range searching. Indyk [27] gave a deterministic and approximate near-neighbor data structure for the discrete Fréchet distance. A c-approximate nearest-neighbor data structure returns for a given query point q a data point $p \in S$, such that the distance $d(p, q)$ is at most $c \cdot d(p^*, q)$, where $p^* \in S$ is the true nearest neighbor to q. Indyk's data structure for data set S, containing n curves which have at most t vertices, achieves approximation factor $c \in \mathcal{O}(\log t + \log \log n)$ and has query time $\mathcal{O}(\text{poly}(t) \log n)$. This data structure requires large space, as it precomputes all queries with curves with $\sqrt{t}$ vertices. For short curves (with $t \in \mathcal{O}(\log n)$) Driemel and Silvestri [20] described an approximate near-neighbor structure based on locality-sensitive hashing with approximation factor $\mathcal{O}(t)$, query time $\mathcal{O}(t \log n)$, using space $\mathcal{O}(n \log n + tn)$. See [15] for an experimental evaluation of this data structure with improvements by Ceccarello et al. LSH is a technique that uses families of hash functions with the property that near points are more likely to be hashed to the same index than far points. Driemel and Silvestri were the first to define locality-sensitive hash functions for the discrete Fréchet distance. Emiris and Psarros [22] improved their result and also showed how to obtain $(1 + \varepsilon)$-approximation with query time $\tilde{\mathcal{O}}\left(d \cdot 2^{2t} \cdot \log n\right)$ and using space $\tilde{\mathcal{O}}(n) \cdot (2 + d/\log t)^{\mathcal{O}(t \cdot d \cdot \log(1/\varepsilon))}$. No such hash functions are known for the continuous case. It is conceivable that the concept of signatures which was introduced by Driemel, Krivošija and Sohler [19] in the context of clustering of 1-dimensional curves could be used to define an LSH for the continuous case and that this technique could be used in combination with projections to random lines.

De Berg et al. [9] studied range counting data structures for spherical range search queries under the continuous Fréchet distance assuming that the centers of query ranges are line segments. This data structure stores compressed subcurves using a partition tree, using space $\mathcal{O}(s\,\text{polylog}(n))$ and query time $\mathcal{O}((n/\sqrt{s})\,\text{polylog}(n))$ to obtain a constant approximation factor, where $n \leq s \leq n^2$ is a parameter to the data structure which is fixed at preprocessing time.

Afshani and Driemel recently showed how to leverage semi-algebraic range searching for this problem [2]. Their data structure also supports polygonal curves of low complexity and answers queries exactly. In particular, for the discrete Fréchet distance they described a data structure which achieves query time $\mathcal{O}\left(n^{1-1/d} \cdot \log^{\mathcal{O}(t_s)} n \cdot t_q^{\mathcal{O}(d)}\right)$ and uses space $\mathcal{O}\left(n \left(\log \log n\right)^{t_s - 1}\right)$, where t_s denotes the complexity of an input curve, and it is assumed that the complexity of the query curves t_q is upper-bounded by a polynomial of $\log n$. For the continuous Fréchet distance they described a data structure for polygonal curves in the plane which achieves query time $\mathcal{O}\left(\sqrt{n} \log^{\mathcal{O}(t_s^2)} n\right)$ and uses space $\mathcal{O}\left(n(\log \log n)^{\mathcal{O}(t_s^2)}\right)$. For the case where the curves lie in dimension higher than

2 and the distance measure is the continuous Fréchet distance, no data structures for range searching or range counting are known.

1.2 Related Work on Metric Embeddings

Given metric spaces (X, d_X) and (Y, d_Y), we call a metric embedding an injective mapping $f : X \to Y$. We call c, $c \geq 1$, the distortion of the embedding f [28] if there is an $r \in (0, \infty)$ such that for all $x, y \in X$ it is $r \cdot d_X(x, y) \leq d_Y(f(x), f(y)) \leq c \cdot r \cdot d_X(x, y)$.

The work that is perhaps closest to ours is a recent result by Backurs and Sidiropoulos [5]. They gave an embedding of the Hausdorff distance into constant-dimensional ℓ_∞ space with constant distortion. More precisely, for any $s, d \geq 1$, they obtained an embedding for the Hausdorff distance over point sets of size s in d-dimensional space, into $\ell_\infty^{s^{O(s+d)}}$ with distortion $s^{O(s+d)}$. No such metric embeddings are known for the discrete or continuous Fréchet distance. It has been shown that the doubling dimension of the Fréchet distance is unbounded, even in the case when the metric spaces is restricted to curves of constant complexity [19]. A result of Bartal $et\ al.$ [8] for doubling spaces implies that a metric embedding of the Fréchet distance into an ℓ_p space would have at least super-constant distortion, but it is not known how to find such an embedding.

We discuss what is known on two variations of the metric embedding problem that are most studied. The first is to find the smallest distortion for any metric from the given class. Matoušek [30] showed that any metric on a point set of size s can be embedded into d-dimensional Euclidean space with multiplicative distortion $\mathcal{O}\left(\min\{s^{2/d} \log^{3/2} s, s\}\right)$, but not better than $\Omega\left(s^{1/\lfloor (d+1)/2 \rfloor}\right)$. For $d = 1$ this implies that the distortion is linear in the worst case.

The second problem is to find the smallest approximation factor to a minimal distortion for a given metric over a point set of size s. We call a spread Δ a maximum/minimum ratio of the distances of the input point set X. Badoiu $et\ al.$ [6] gave an $\mathcal{O}\left(\Delta^{3/4} c^{11/4}\right)$-approximation to the embedding to a line, where c is the distortion of embedding of the input set onto the line. They also showed that it is hard to approximate this problem up to a factor $\Omega\left(n^{1/12}\right)$, even for a weighted tree metrics with polynomial spread. Assuming a constant distortion c and a polynomial spread Δ, Nayyeri and Raichel [32] gave a $\mathcal{O}(1)$-approximation algorithm to the minimal distortion of the embedding to a line, in time polygonal in s and Δ. See the work of Badoiu $et\ al.$ [7], Fellows $et\ al.$ [23], Håstad $et\ al.$ [25], and Indyk [26] for further reading.

1.3 Our Results

Given two polygonal curves P and Q with t vertices each from $\mathbb{R}^d$, where $d \in \{2, 3, 4, 5\}$. Consider sampling a unit vector $\mathbf{u}$ in respective $\mathbb{R}^d$ uniformly at random, and let P' and Q' be the projections of the two curves to the line supporting $\mathbf{u}$. We observe that Fréchet distance always decreases when the curves are projected to a line (Lemma 3). We show that if the curves P and Q are c-packed for constant c, then, with constant probability, the discrete Fréchet distance between the curves P and Q, denoted by $d_F(P, Q)$, degrades by at most a linear factor in t. This is stated by Theorem 1 for $d \in \{2, 3\}$, and by Theorem 2 for $d \in \{4, 5\}$.

Theorem 1. *Given $c \geq 2$, for any two polygonal c-packed curves P and Q from $\mathbb{R}^2$ or $\mathbb{R}^3$, and for any $\gamma \in (0, 1)$ it holds that*

$$\Pr\left[\frac{d_F(P, Q)}{d_F(P', Q')} \leq \frac{12c + 16}{\gamma} \cdot t\right] \geq 1 - \gamma.$$

Theorem 2. *Given $c \geq 2$, for any two polygonal c-packed curves P and Q from $\mathbb{R}^4$ or $\mathbb{R}^5$, and for any $\gamma \in (0, 1)$ it holds that*

$$\Pr\left[\frac{d_F(P, Q)}{d_F(P', Q')} \leq \left(1 + \frac{2}{\pi}\right) \cdot \frac{12c + 16}{\gamma} \cdot t\right] \geq 1 - \gamma.$$

We also present a lower bound on the ratio of the two distances. The construction of the lower bound uses c-packed curves with $c < 3$.

Theorem 3. *Given $c \geq 2$, there exist polygonal c-packed curves P and Q, such that for any $\gamma \in (0, 1/\pi)$*

$$\Pr\left[\frac{d_F(P, Q)}{d_F(P', Q')} \geq \frac{5\pi\gamma}{6} \cdot t\right] \geq 1 - \gamma.$$

Theorem 3 holds for the continuous Fréchet distance and for dynamic time warping distance as well.

We also show that there exist polygonal curves P and Q that are not c-packed for sublinear c and their (continuous or discrete) Fréchet distance degrades by a linear factor for any projection line (i.e. with probability 1). Theorem 4 presents this result.

Theorem 4. *There exist the curves $P = \{p_1, \ldots, p_t\}$ and $Q = \{q_1, \ldots, q_t\}$, such that if P' and Q' respectively are their projections to the one-dimensional space that supports the unit vector chosen uniformly at random on the unit hypersphere, then it holds that*

$$\frac{d_F(P, Q)}{d_F(P', Q')} \geq f(t),$$

where $f(t) \in \Omega(t)$.

Please refer to the full version of the paper for the omitted proofs.

2 Preliminaries

Throughout the paper we use the following notational conventions. Consider two polygonal curves $P = \{p_1, p_2, \ldots, p_t\}$ and $Q = \{q_1, q_2, \ldots, q_t\}$ in $\mathbb{R}^d$ given by their sequences of vertices. We choose a unit vector $\mathbf{u}$ in $\mathbb{R}^d$ by choosing a point on the $(d-1)$-dimensional unit hypersphere uniformly at random. We denote with L the line through the origin that supports the vector $\mathbf{u}$. Let $P' = \{p'_1, p'_2, \ldots, p'_t\}$ and $Q' = \{q'_1, q'_2, \ldots, q'_t\}$ be the projections of P and Q to L, defined by $p'_i = \langle p_i, \mathbf{u} \rangle$ and $q'_j = \langle q_j, \mathbf{u} \rangle$, for all $1 \leq i \leq t$ and $1 \leq j \leq t$. We denote $\delta_{i,j} = \|p_i - q_j\|$ and $\delta'_{i,j} = \|p'_i - q'_j\|$, for all $1 \leq i \leq t$ and $1 \leq j \leq t$, i.e. $\delta_{i,j}$ and $\delta'_{i,j}$ are the pairwise distances of the vertices for the input curves P and Q and for their respective projections P' and Q'.

We define the discrete Fréchet distance of P and Q as follows: we call a *traversal* T of P and Q a sequence of pairs of indices (i,j) of vertices $(p_i, q_j) \in P \times Q$ such that

(i) the traversal T starts with $(1,1)$ and ends with (t,t), and
(ii) the pair (i,j) of T can be followed only by one of $(i+1,j)$, $(i,j+1)$ or $(i+1,j+1)$.

We notice that every traversal is monotone. If $\mathcal{T}$ is the set of all traversals T of P and Q, then the discrete *Fréchet distance* between P and Q is defined as

$$d_F(P,Q) = \min_{T \in \mathcal{T}} \max_{(i,j) \in T} \|p_i - q_j\|. \tag{1}$$

Furthermore, we define a directed, vertex-weighted graph $G = (V, E)$ on the node set $V = \{(i,j) : 1 \leq i, j \leq t\}$. A node (i,j) corresponds to a pair of vertices p_i of P and q_j of Q and we assign it the weight $\delta_{i,j}$. The set of edges is defined as $E = \{((i,j),(i',j')) : i' \in \{i, i+1\}, j' = \{j, j+1\}, 1 \leq i, i', j, j' \leq t\}$. The set of paths in the graph G between $(1,1)$ and (t,t) corresponds to the set of traversals $\mathcal{T}$. We call a path in G which does not start in $(1,1)$ or end in (t,t) a *partial traversal* of P and Q.

It is useful to picture the nodes of the graph G as a matrix, where rows correspond to the vertices of P and columns correspond to the vertices of Q. For any fixed value $\Delta > 0$, we define the free-space matrix[1] $F_\Delta = (\phi_{i,j})_{1 \leq i,j \leq t}$ with

$$\phi_{i,j} = \begin{cases} 1 & \text{if } \|p_i - q_j\| < \Delta \\ 0 & \text{if } \|p_i - q_j\| \geq \Delta. \end{cases}$$

Overlaying the graph with the free-space matrix for $\Delta > d_F(P,Q)$, we can observe that there exists a path in the graph from $(1,1)$ to (t,t) that visits only the matrix entries with value 1. Moreover, the existence of such a path in the free-space matrix for some value of Δ implies that $\Delta > d_F(P,Q)$.

We define c-packedness of curves as follows.

---

[1] Note that the conventional definition of the free-space matrix for parameter Δ is slightly different, since usually there is an 1-entry iff $\|p_i - q_j\| \leq \Delta$. We are using this definition since it better suits our needs.

Definition 1 (c-packed curve). *Given $c > 0$, a curve $P \in \mathbb{R}^d$ is c-packed if for any point $p \in \mathbb{R}^d$ and any radius $r > 0$, the total length of the curve P inside the hypersphere $ball(p, r)$ is at most $c \cdot r$.*

The proof of the following three lemmas can be found in the full version of the paper. For a general problem in much higher dimension d, the probability stated by Lemmas 1 and 2 cannot be bounded by a linear function in φ, due to the measure concentration around $\pi/2$.

Lemma 1. *If two points p and q are projected to the straight line L, which supports the unit vector chosen uniformly at random on the unit hypersphere in $\mathbb{R}^2$ or $\mathbb{R}^3$, the probability that the distance of their projections will be reduced from the original distance by a factor greater than φ is at most φ.*

Lemma 2. *If two points p and q are projected to the straight line L, which supports the unit vector chosen uniformly at random on the unit hypersphere in $\mathbb{R}^4$ or $\mathbb{R}^5$, the probability that the distance of their projections will be reduced from the original distance by a factor greater than φ is at most $(1 + 2/\pi) \cdot \varphi$.*

Lemma 3. *Given two curves $P = \{p_1, \ldots, p_t\}$ and $Q = \{q_1, \ldots, q_t\}$ in $\mathbb{R}^d$, and let $P' = \{p'_1, \ldots, p'_t\}$ and $Q' = \{q'_1, \ldots, q'_t\}$ respectively be their projections to the straight line L which supports the vector $\boldsymbol{u}$ chosen uniformly at random on the unit hypersphere in $\mathbb{R}^d$. It holds that $d_F(P, Q) \geq d_F(P', Q')$.*

3   Upper Bound

3.1   Guarding Sets

The discrete Fréchet distance between curves P and Q is realized by some pair (p_i, q_j) of vertices $p_i \in P$ and $q_j \in Q$, being at the distance $\|p_i - q_j\| = \delta$. We would like to apply Lemma 1 to this pair of vertices to show that the distance is preserved up to some constant factor. However, it is possible that the pairwise distances in the projection are such that a cheaper traversal is possible that avoids the pair (p_i, q_j) altogether. Therefore, we will apply the lemma to a subset of pairs of vertices of P and Q whose distance is large (e.g. larger than $\Delta = \delta/\theta$ for some small value of $\theta \geq 1$) and such that the chosen set forms a hitting set for the set of traversals $\mathcal{T}$. To this end we introduce the notion of the *guarding set* by the following definition.

Definition 2 (Guarding set). *For any two polygonal curves $P = \{p_1, \ldots, p_t\}$ and $Q = \{q_1, \ldots, q_t\}$ and a given parameter $\theta \geq 1$, a θ-guarding set $B \subseteq V$ for P and Q is a subset of the set of vertices of G that satisfies the following conditions:*

(a) (distance property) for all $(i, j) \in B$, it holds that $\delta_{i,j} \geq d_F(P, Q)/\theta$, and
(b) (guarding property) for any traversal T of P and Q, it is $T \cap B \neq \emptyset$.

Note that the set B "guards" every traversal of P and Q in the sense that any path in G from $(1,1)$ to (t,t) has non-empty intersection with B. In other words, B is a hitting set for the set of traversals $\mathcal{T}$.

For a guarding set B we define the subset of vertices $S_B \subseteq V$ that can be reached by a path in G starting from $(1,1)$ without visiting a vertex of B. We also define the subset of vertices $H_B = V \setminus (B \cup S_B)$. A guarding set B thus defines a vertex partition of the graph G into three subsets $V = S_B \cup B \cup H_B$.

We show the following simple lemma for $d \in \{2,3\}$, and its counterpart for $d \in \{4,5\}$, given by Lemma 5, which is proven analogously.

Lemma 4. *Given parameter $\theta \geq 1$, if B is a θ-guarding set for the given curves $P = \{p_1, \ldots, p_t\}$ and $Q = \{q_1, \ldots, q_t\}$ from $\mathbb{R}^2$ or $\mathbb{R}^3$, and if P' and Q' are their projections to the straight line L, whose support unit vector $\mathbf{u}$ is chosen uniformly at random on the unit hypersphere, then for any $\beta > 1$ it holds that*

$$\frac{d_F(P', Q')}{d_F(P, Q)} \geq \frac{1}{\beta \cdot \theta \cdot |B|}$$

with positive constant probability at least $1 - 1/\beta$.

Proof. Let $\mathbf{u}$ be the unit vector which is chosen uniformly at random on the unit hypersphere in $\mathbb{R}^d$ with $d \in \{2,3\}$, and let $\mathbf{u}$ be supported by the projection line L. Let $\alpha_{i,j}$ be the angle between $\mathbf{u}$ and the vector $q_j - p_i$, for $i, j \in \{1, \ldots, t\}$. If we consider the distances of the pairs of the points $(p_i, q_j) \in P \times Q$, represented by the elements $(i,j) \in B$, then the probability of the event that some of these distances of the points of P and Q is reduced by a factor greater than $\beta \cdot |B|$ (the "bad" event) when projected to L can be bounded by the union bound inequality and by Lemma 1 for $\varphi = \frac{1}{\beta|B|}$ as:

$$\Pr\left[(\exists(i,j) \in B) : \frac{\delta'_{i,j}}{\delta_{i,j}} < \frac{1}{\beta|B|}\right] \leq \sum_{(i,j) \in B} \Pr\left[\frac{\delta'_{i,j}}{\delta_{i,j}} < \frac{1}{\beta|B|}\right] \leq \sum_{(i,j) \in B} \frac{1}{\beta|B|} = \frac{1}{\beta}$$

$$(2)$$

for any $\beta > 1$.

Since by Definition 2 any traversal T of P and Q has nonempty intersection with B, the Fréchet distance of P and Q has to be at least as big as the distance of some pair $(i,j) \in T \cap B$. These pairs of vertices have distance at least $d_F(P,Q) / \theta$, and they are going to be reduced at most by the factor $\beta \cdot |B|$ (with positive constant probability). The traversal T' of P' and Q' that realizes $d_F(P', Q')$ has to contain at least one of the pairs of B by Definition 2, since the pairs of the traversal T' are simultaneously the pairs of the traversal T of P and Q (that contains the pairs of the vertices of P and Q in the same order as the pairs of their projections in P' and Q'). Thus $d_F(P', Q') \geq d_F(P,Q) / (\beta \cdot \theta \cdot |B|)$, which proves the lemma. $\qquad\square$

Lemma 5. *Given parameter $\theta \geq 1$, if B is a θ-guarding set for the given curves $P = \{p_1, \ldots, p_t\}$ and $Q = \{q_1, \ldots, q_t\}$ from $\mathbb{R}^4$ or $\mathbb{R}^5$, and if P' and Q' are* *their projections to the straight line L, whose support unit vector $\mathbf{u}$ is chosen uniformly at random on the unit hypersphere, then for any $\beta > 1$ it holds that*

$$\frac{d_F(P', Q')}{d_F(P, Q)} \geq \frac{1}{(1 + 2/\pi) \cdot \beta \cdot \theta \cdot |B|}$$

with positive constant probability at least $1 - 1/\beta$.

Intuitively we think of $\delta'_{i,j}$ as an approximation to $\delta_{i,j}$. Lemma 4 yields a naive $(\beta \cdot t^2)$-approximation for any $\beta > 1$ and $\theta = 1$. Let B be the set of all pairs $(i, j) \in \{1, \ldots, t\} \times \{1, \ldots, t\}$ such that $\|p_i - q_j\| = \delta_{i,j} \geq d_F(P, Q)$. In the worst case B could contain all t^2 pairs. Set B is a 1-guarding set. The correctness of the condition (a) of Definition 2 is provided directly by the definition of B. The condition (b) follows by contradiction. If there would exist some traversal T such that $T \cap B = \emptyset$, then for all pairs $(i, j) \in T$ it would have to hold that $\|p_i - q_j\| < d_F(P, Q)$. But then the traversal T would witness that $d_F(P, Q) \leq \max_{(i,j) \in T} \|p_i - q_j\| < d_F(P, Q)$, a contradiction.

One could obtain better constant β by more technical argument, which we omit here. Clearly, the approximation factor of Lemma 4 can be improved by the better choice of the set B. This question we explore in the following section.

3.2 Improved Analysis for C-packed Curves

In order to ensure that the number of the pairs of the indices that take part in the sum in the union bound inequality in (2) is not quadratic but at most a linear one in terms of the input size, we have to carefully select a small subset that satisfies the guarding set properties.

Building of the Initial Guarding Set. We first give the simple construction of a θ-guarding set for any $\theta \geq 1$ by Algorithm 1.

Lemma 6. *The set B obtained by Algorithm 1 is a θ-guarding set, for any $\theta \geq 1$.*

Proof. We have to show that the resulting set B satisfies the conditions of Definition 2. In the case that the distance $\delta_{1,1} \geq \delta/\theta$, it suffices to assign $B = \{(1, 1)\}$, since any traversal of the curves P and Q has to include the pair $(1, 1)$. For the rest of the proof let $\delta_{1,1} < \delta/\theta$.

Algorithm 1 selects into B only the pairs (i', j') with $\delta_{i',j'} \geq \delta/\theta$ in the line 12, and that are reached by an edge from a pair (i, j) with $\delta_{i,j} \leq \delta/\theta$. Thus the condition (a) of Definition 2 is satisfied by the yielded set. For the condition (b) we show by induction the following invariant: in each point of time during the BFS, any traversal T contains either a vertex of B or a vertex in the queue $\mathcal{Q}$. The BFS starts with $(1, 1) \in \mathcal{Q}$ with $\delta_{1,1} < \delta/\theta$. While processing the pair in $(i, j) \in \mathcal{Q}$ with $\delta_{i,j} < \delta/\theta$ during the BFS (lines 7 and 8) the traversal T may use one of the pairs $(i+1, j)$, $(i, j+1)$ or $(i+1, j+1)$ (connected by the edges in E).

Algorithm 1. Computing the θ-guarding set, $\theta \geq 1$

Data: $\delta = d_F(P, Q)$, vertex-weighted graph $G = (V, E)$
Result: set B

1 $B \leftarrow \emptyset$
2 **if** $\delta_{1,1} \geq \delta/\theta$ **then**
3 $\quad \lfloor \ B \leftarrow \{(1,1)\}$
4 **else**
5 $\quad$ FIFO-Queue $\mathcal{Q} \leftarrow \{(1,1)\}$ /* Breadth-First-Search on $G = (V, E)$ */
6 $\quad$ **while** $\mathcal{Q} \neq \emptyset$ **do**
7 $\quad\quad (i,j) \leftarrow \text{pop}(\mathcal{Q})$
8 $\quad\quad$ **foreach** $((i,j),(i'j')) \in E$ **do**
9 $\quad\quad\quad$ **if** $\delta_{i,j} < \delta/\theta$ and $\delta_{i',j'} < \delta/\theta$ **then**
10 $\quad\quad\quad\quad \lfloor \ \text{push}(\mathcal{Q},(i',j'))$
11 $\quad\quad\quad$ **else if** $\delta_{i,j} < \delta/\theta$ and $\delta_{i',j'} \geq \delta/\theta$ **then**
12 $\quad\quad\quad\quad \lfloor \ B \leftarrow B \cup \{(i',j')\}$

13 $\quad$ **return** B

The next pair in the traversal T is either added into $\mathcal{Q}$ (line 10), or added into B (line 12). In both cases the invariant remains valid. Since the queue is empty at the end, this means that any traversal contains a vertex in B, as claimed. $\square$

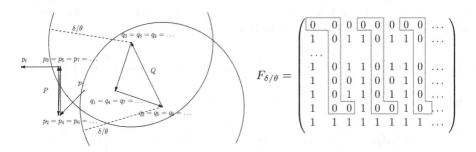

Fig. 1. The curves P and Q (left) that yield a "fork-like" free-space matrix $F_{\delta/\theta}$ for some $\theta \geq 1$ (right). The pairs selected into B by Algorithm 1 are marked with the red bound. (Color figure online)

Unfortunately, the set B built by Algorithm 1 can have a quadratic number of elements in terms of the input size, like the one in Fig. 1 (marked with the red bound). If the free-space matrix $F_{\delta/\theta}$ would have the "fork-like" structure for some $\theta \geq 1$, such that for every column j with $j \mod 3 = 1$ it holds for all pairs $\delta_{i,j} < \delta/\theta$ and thus $\phi_{i,j} = 1$ (except for $\delta_{t,j} \geq \delta/\theta$), and for every column j with $j \mod 3 = 2$ there are all pairs with $\delta_{i,j} \geq \delta/\theta$ and thus $\phi_{i,j} = 0$ (except for $\delta_{1,j} < \delta/\theta$). For the columns with $j \mod 3 = 0$ let $\phi_{1,j} = 1$, $\phi_{2,j} = 0$ and $\phi_{t,j} = 0$ (the rest may be filled arbitrarily). Then the set B built by Algorithm 1 would contain $(t-1) \cdot t/3 = \mathcal{O}(t^2)$ entries. We note that this cannot happen if the curves P and Q are c-packed for some constant c, $c \geq 2$, as it will be discussed in the further text.

On the Structure of the Distance Matrix. Lemma 7 states one property of the c-packed curves, which we apply in Lemma 8.

Lemma 7. *Given point p and a c-packed curve $Q = \{q_1, \ldots, q_t\}$ from $\mathbb{R}^d$, then for any value $b > 0$ there exists a value $r \in [b/2, b]$, such that the hypersphere centered at p with radius r intersects or is tangent to at most $2c$ edges of Q.*

Proof. Assume for the sake of contradiction that there exists $c' > 2c$, such that for any $r \in [b/2, b]$ there are at least c' edges of Q that intersect or are tangent the surface of the hypersphere $\texttt{ball}(p, r)$. Let the *event points* be the points in $\texttt{ball}(p, b) \setminus \texttt{ball}(p, b/2)$, such that they are either

(i) vertices q_i of Q or
(ii) the points $q' \in \overline{q_i q_{i+1}}$, such that $\overline{pq'} \perp \overline{q_i q_{i+1}}$.

Let the set of events be $\mathcal{R} = \{R_1, \ldots, R_\ell\}$, and let $r_i = \|p - R_i\|$ for all $1 \leq i \leq \ell$. We may assume that the events R_i are sorted ascending by r_i. Let $r_0 = b/2$ and $r_{\ell+1} = b$, thus $r_0 \leq r_1 \leq \ldots \leq r_{\ell+1}$.

The number of the edges of Q that intersect or are tangent to $\texttt{ball}(p, r)$ is equal for all $r' \in [r_i, r_{i+1})$ and for all $0 \leq i \leq \ell$, since the number of such edges changes only in event points. After assumption there are at least c' edges of Q that intersect $\texttt{ball}(p, r')$, for any $r' \in [r_i, r_{i+1})$ and for any $0 \leq i \leq \ell$. The length of the curve Q within $\texttt{ball}(p, b) \setminus \texttt{ball}(p, b/2)$ is

$$\sum_{i=0}^{\ell} \|Q \cap (\texttt{ball}(p, r_{i+1}) \setminus \texttt{ball}(p, r_i))\| = \left\|Q \cap \left(\texttt{ball}(p, b) \setminus \texttt{ball}\left(p, \frac{b}{2}\right)\right)\right\| \leq c \cdot b$$

since Q is c-packed. But on the other side it is

$$\sum_{i=0}^{\ell} \|Q \cap (\texttt{ball}(p, r_{i+1}) \setminus \texttt{ball}(p, r_i))\| \geq \sum_{i=0}^{\ell} c' \cdot |r_{i+1} - r_i| = c' \cdot \left(b - \frac{b}{2}\right) > c \cdot b,$$

a contradiction. $\qquad\square$

Lemma 8. *Given point p and a c-packed curve $Q = \{q_1, \ldots, q_t\}$ from $\mathbb{R}^d$, and given a value $b > 0$, then for any pairwise disjoint set of intervals*

$$I \subseteq \{[i_1, i_2] \mid i_1 \leq i_2 \in \mathbb{N}, 1 \leq i_1 \leq i_2 \leq t\}$$

with $d(p, q_i) \geq b$ for all $i \in [i_1, i_2] \in I$, there exists a value of $r \in [b/2, b]$ and a pairwise disjoint set of intervals

$$J \subseteq \{[j_1, j_2] \mid j_1 \leq j_2 \in \mathbb{N}, 1 \leq j_1 \leq j_2 \leq t\}$$

with the following properties:

(i) $|J| \le c+1$
(ii) $\forall \, [j_1, j_2] \in J \, \exists \, i_1 \le i_2 < i_3 \le i_4 \; : \; [i_1, i_2], [i_3, i_4] \in I \wedge j_1 = i_1 \wedge j_2 = i_4$
(iii) $\forall \, i \in [j_1, j_2] \in J \; : \; d(p, q_i) \ge r$

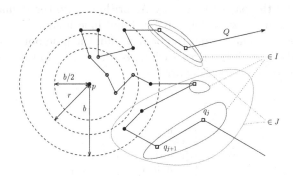

Fig. 2. The process of Lemma 8 for the vertex p and the curve Q

Proof. We set r to be the value of the same variable as in Lemma 7. Now we construct the set J by merging intervals of I as follows. Initially J is empty. We iterate over the intervals of I in the order of their starting points. Consider the first interval $[i_1, i_2]$ and the next interval in the order $[i_3, i_4]$, we merge them into one interval $[i_1, i_4]$ if there exists no point q_j with $i_2 < j < i_3$ such that $d(p, q_j) < r$. We continue merging this interval with the intervals in I until we found a point q_j such that $d(p, q_j) < r$. Then, we add the current merged interval to J and take the next interval from I and merge it with the proceeding intervals in the same manner. When there are no intervals left in I, we also add the current interval to J. Each time we add an interval to J (except possibly for the last one), we encountered two edges of Q that intersect the sphere of radius r centered at p. By Lemma 7 we have added at most $c+1$ intervals to J (including the last interval). The other properties stated in the lemma follow by construction of J. Figure 2 illustrates the merging process. □

Avoidable Pairs

Definition 3 (Avoidable pair). *Let B be the θ-guarding set produced by Algorithm 1, and let $V = S_B \cup B \cup H_B$ be the partition of V implied by B. The pair $(i, j) \in B$ is called avoidable if there exist a pair $(i', j') \in B$ and two partial traversals T_1 and T_2 of P and Q from $(1, 1)$ to (i', j'), such that:*

(i) *$\forall (i'', j'') \in (T_1 \cup T_2) \setminus \{(i', j')\}$ it holds that $(i'', j'') \in S_B$,*
(ii) *there exist pairs $(i, y_1) \in T_1$ and $(i, y_2) \in T_2$, with $y_1 < j < y_2$,*
(iii) *there exist pairs $(x_1, j) \in T_2$ and $(x_2, j) \in T_1$, with $x_1 < i < x_2$.*

We notice that for the pair to be avoidable, it suffices to have the conditions (i) and (ii), or (i) and (iii), since the remaining condition is implied by the monotonicity of the traversals. The definition of the avoidable pair (i, j) implies that any partial traversal of P and Q from (i, j) to (t, t) has to have a nonempty intersection with $T_1 \cup T_2$.

Figure 3 shows the pairs selected by Algorithm 1 into the θ-guarding set B, for some $\theta \geq 1$, marked with polygonal red and blue bounds. The pairs within the red bound are avoidable, and the pairs within the blue bound are not. Two partial traversals T_1 and T_2 in S_B that make the red bounded pairs avoidable (as in Definition 3) are marked by arrows.

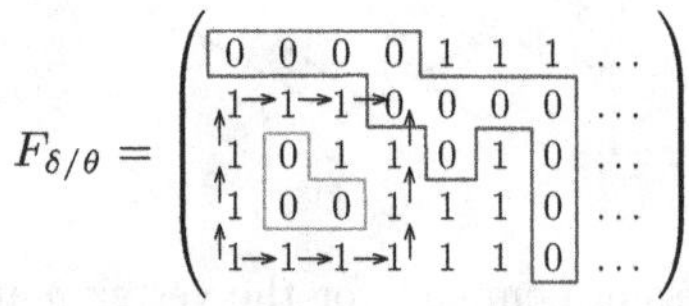

$$
F_{\delta/\theta} =
\begin{pmatrix}
\boxed{0 \;\; 0 \;\; 0 \;\; 0} \; 1 \;\; 1 \;\; 1 & \cdots \\
1 \to 1 \to 1 \to 0 \;\; 0 \;\; 0 \;\; 0 & \cdots \\
1 \; \boxed{0 \;\; 1 \;\; 1} \; 0 \; \boxed{1} \; 0 & \cdots \\
1 \; \boxed{0 \;\; 0 \;\; 1} \; 1 \;\; 1 \;\; 0 & \cdots \\
1 \to 1 \to 1 \to 1 \; 1 \;\; 1 \; \boxed{0} & \cdots
\end{pmatrix}
$$

Fig. 3. Avoidable pairs from the θ-guarding set B (for some $\theta \geq 1$) are marked with red bound. Not avoidable pairs are marked with blue bound. (Color figure online)

Lemma 9. *Given parameter $\theta \geq 1$ and the θ-guarding set B. Let $B' \subseteq B$ be the set of the avoidable pairs. Then $B \setminus B'$ is a θ-guarding set.*

Trimming the Reachable Area of a Guarding Set. Let B be a 1-guarding set for two curves P and Q. We now want to modify B to shrink the number of pairs while maintaining the guarding property. It turns out that we can do this if we relax the approximation quality of the guarding set (which we denoted with θ). We perform this trimming in three phases:

(1) Remove all avoidable pairs from B.
(2) Trim the reachable area of B row by row.
(3) Trim the reachable area of B column by column.

In the following, we describe the trimming operation on a single row. Consider a vertex p_i of the curve P and consider the intersection of B with the row of the distance matrix associated with p_i. Let I_i denote the set of intervals of the column indices that represent this intersection. We now apply Lemma 8 with parameter $b = d_F(P, Q)$ to obtain a set of intervals J_i that can be used to trim the reachable area of B with respect to the ith row. Each interval in J_i covers a set of intervals of I_i. Let A_i be the subset of pairs of the ith row of which the column index is contained in an interval of J_i, but not contained in any interval of I_i. We call A_i the *filling pairs* of the row. We now want to trim the reachable area S_B defined by B along the vertices of the reachability graph which correspond to pairs of A_i. For this we will remove all vertices of B_i that are reachable from A_i and add the pairs of A_i to B. See Algorithm 2 for the pseudocode of this trimming operation. Figure 4 illustrates the process with an example. The trimming operation for a single column is analogous, except that we use $b = d_F(P, Q)/2$ as a parameter to Lemma 8.

Algorithm 2. Trimming the reachable area for one row

Data: guarding set B, row index i, value of $b > 0$
Result: modified guarding set B

1   $I_i := \{[j,j] \mid (i,j) \in B\}$ /* pairs of B in the ith row */
2   Let J_i be the set of intervals obtained from Lemma 8 using I_i and $b = d_F(P,Q)$
3   $A_i := S_B \cap \left\{(i,j) \mid j \in \left(\bigcup_{[j_1,j_2] \in J_i}[j_1,j_2] \setminus \bigcup_{[i_1,i_2] \in I_i}[i_1,i_2]\right)\right\}$ /* Compute filling pairs */
4   FIFO-Queue $\mathcal{Q} \leftarrow A_i$ /* find guarding pairs reachable from A_i via BFS */
5   **while** $\mathcal{Q} \neq \emptyset$ **do**
6      $(i,j) \leftarrow \text{pop}(\mathcal{Q})$
7   **foreach** $(i',j') \in \{(i+1,j),(i+1,j+1)\}$ **do**
8   **if** $(i',j') \in B \setminus \mathcal{Q}$ **then**
9             $B \leftarrow B \setminus \{(i',j')\}$ /* remove them from B */
10   **else**
11             $\text{push}(\mathcal{Q}, (i',j'))$

12 $B \leftarrow B \cup A_i$ /* add pairs of A_i to B */

$$F_b^{\text{before}} = \begin{pmatrix} \cdots & & & & \\ 0 & 0 & \boxed{0} & \boxed{0} & \boxed{0} & \cdots \\ 0 & \boxed{0} & 1 & 1 & \boxed{0} & \cdots \\ 0 & 1 & 1 & \boxed{0} & 0 & \cdots \\ \boxed{0} & 1 & 1 & \boxed{0} & \boxed{0} & \cdots \\ 1 & 1 & 1 & 1 & 1 & \cdots \end{pmatrix} \qquad F_{b/2}^{\text{after}} = \begin{pmatrix} \cdots & & & & \\ 0 & 0 & \textcircled{0} & \textcircled{0} & \textcircled{0} & \cdots \\ 0 & \textcircled{0} & 1 & 1 & \textcircled{0} & \cdots \\ 0 & 1 & 1 & \textcircled{0} & 0 & \cdots \\ \boxed{0} & \boxed{0} & \boxed{0} & \boxed{0} & \boxed{0} & \cdots \\ 1 & 1 & 1 & 1 & 1 & \cdots \end{pmatrix}$$

Fig. 4. The elements of a guarding set (marked with boxes) before (left) and after (right) applying of Algorithm 2 to the second row. The removed pairs are marked by circles

Lemma 10. Let B be a 1-guarding set.

(i) After the first phase of the algorithm, which removes all avoidable pairs, the modified set B is a 1-guarding set.

(ii) *After the second phase of the algorithm, which applies the trimming operation to each row with $b = d_F(P, Q)$, the modified set B is a 2-guarding set.*

(iii) *After the third phase of the algorithm, which applies the trimming operation to each column with $b = d_F(P, Q)/2$, the modified set B is a 4-guarding set.*

Proof. The first part of the lemma follows directly from Lemma 9. We now prove the second part of the lemma statement. Condition (iii) of Lemma 8 ensures that any pair of a set A_i added to B corresponds to a pair of vertices $p \in P$ and $q \in Q$ with $d(p, q) \geq b/2 = d_F(P, Q)/2$. Indeed, the column indices of the pairs of A_i are contained in intervals of J_i. Therefore, after the second phase, the modified set B satisfies property (a) in the definition of guarding sets if we set $\theta = 2$. Secondly, we argue that property (b) is not invalidated after the trimming operation was applied to a row. Let B denote the guarding set before the trimming operation applied to the ith row and let B' denote the modifed guarding set after trimming. Clearly, the trimming operation does not add any avoidable pairs to B. Therefore we can assume that throughout the second phase no avoidable pairs are present.

Assume for the sake of contradiction that there exists a traversal T that contains a pair of B, but does *not* contain a pair of B'. Let (i', j') be the first pair along T that was removed from B during the trimming operation and let (i, j_2) be a pair of A_i that has a BFS-path to (i', j'). T must contain a pair (i, j_1) in the ith row and this pair cannot be contained in an interval of J_i (otherwise T would contain a pair of B'). Let T_1 be the partial traversal (path in G) of T that starts in $(1, 1)$ goes via (i, j_1) and ends in (i', j'). Since (i', j') was the first vertex along T in B, it follows that T_1 only visits vertices that are in S_B. Note that $i' > i$ since the BFS only visits row indices strictly greater i. Since $A_i \subseteq S_B$, there must be a path T_2 in G from $(1, 1)$ via (i, j_2) to (i', j') that only contains vertices of S_B. Now, condition (ii) of Lemma 8 implies that there must be a vertex (i, j'') in B, such that either $j_1 < j'' < j_2$ or $j_2 < j'' < j_1$. This implies that (i, j'') must be avoidable with respect to B. However, this contradicts the fact that B does not contain any avoidable pairs. This proves (ii). The third part of the lemma follows by a symmetric argument applied to the columns.   □

3.3   Bounding the Complexity of the Modified Guarding Set

Given set B after the algorithm of Lemma 10. For every row of B (presented as matrix) let the pairwise disjoint set of intervals $R_i \subseteq \{[j_1, j_2] \,|\, j_1 \leq j_2 \in \mathbb{N}, 1 \leq j_1 \leq j_2 \leq t\}$ be a set of intervals on $\{1, \ldots, t\}$ of minimal size, such that for any $1 \leq j' \leq t$ there exist j_1 and j_2 with $j' \in [j_1, j_2] \in R_i$ if and only if $(i, j') \in B$. We can analogously define such pairwise disjoint sets C_j over the columns of B.

Lemma 8 implies that for every row i there is a set of pairwise disjoint intervals J_i constructed by line 2 of Algorithm 2, with $|J| \leq c + 1$. Algorithm 2 takes into B only the pairs that belong to the subsets of the intervals of J_i that were in

```
         ...
    ...  s  s  b  b  h  h  h  h  h  ...
    ...  s  s  s  b  h  h  h  h  h  ...
    ...  s  b  h  h  h  h  h  h  h  ...
    ...  s  b  h  h  h  b  b  b  b  h  ...
    ...  s  b  h  h  s  s  s  b  h  ...
    ...  s  b  b  b  s  s  s  b  h  ...
    ...  s  s  s  s  s  s  s  b  h  ...
```

Fig. 5. The pairs of the guarding set B (red) and its extended group (blue) within one column. The pairs denoted with s, b, and h are from S_B, B and H_B respectively (Color figure online)

S_B too. But since the pairs $(i,j) \in H_B$ such that $j \in [j_1, j_2] \in J_i$ have the property that any traversal using these pairs has to contain a pair in B prior to (i,j), we could have added such pairs too into B and then it would be $J_i = R_i$. Since we took only its subsets, it holds that for every $[j_1, j_2] \in R_i$ there is $[j_3, j_4] \in J_i$ with $j_3 \le j_1 \le j_2 \le j_4$. By counting all intervals of R_i that are subset of one interval from J_i as one, we say that all such intervals R_i build one *extended group* of consecutive pairs within ith row. It follows that there are at most $c+1$ extended groups within i-th row. This process gets repeated over columns as well. See Fig. 5 for an illustration.

We have to note that the filling pairs added into B also imply the removal of a pair in B that lies in the same row but with higher column index, except possibly for the last pair in the row. This can happen at most once per row, adding one pair (and one extended group) to the row. We obtain the following lemma.

Lemma 11. *In the guarding set produced by Algorithm 1 and modifed by the algorithm of Lemma 10, there are at most $c+1$ extended groups within a column, and $c+2$ extended groups within a row.*

To finally bound the complexity of our guarding set by Lemma 13, we show first Lemma 12.

Lemma 12. *For the guarding set produced by Algorithm 1 and after every phase of algorithm of Lemma 10 the following invariant holds: for every pair $(i,j) \in B$ there exists a pair $(i',j') \in S_B$ such that $((i',j'),(i,j)) \in E$.*

Proof. We call the pair (i',j') the predecessor pair. The construction of the guarding set B Algorithm 1 guarantees that a pair (i,j) is added into B if it is visited over an edge $((i',j'),(i,j)) \in E$, where $(i',j') \notin B$. Thus $((i',j') \in S_B$ as claimed.

The first phase of the algorithm of Lemma 10 removes the avoidable pairs from B, thus for the pairs that remain in B the invariant holds. The second phase runs Algorithm 2 upon a row and adds into B only pairs which were already in S_B, thus have also a predecessor in S_B. For every pair (i',j') which was in S_B before and is in H_B after Algorithm 2 it holds that the BFS passes it and then visits and subsequently removes the pairs from B. Therefore the invariant remains valid for the pairs that remain in B, as for the pairs that were already in B their predecessors remain in S_B, so their status is not changed. The third phase is equivalent to the second one, and the invariant remains valid. □

Lemma 13. *The set B obtained by the algorithm of Lemma 10 is a 4-guarding set, containing at most $(3c + 4) \cdot t$ pairs.*

Proof. For every pair $(i, j) \in B$ one of the following holds true:

   (i) the index j is the smallest index of an extended group over the ith row;
  (ii) the index i is the smallest index of an extended group over the jth column;
 (iii) none of the above.

We argue that if neither (i) nor (ii) holds true, then it must be that $i - 1$ is the smallest index of an extended group over the jth column. Indeed, note that if neither (i) nor (ii) holds true, then $(i - 1, j)$ and $(i, j - 1)$ are part of an extended group and such groups can only contain pairs of B or H_B. Therefore, the pair $(i - 1, j - 1)$ must be in S_B because Lemma 12 implies that (i, j) must have an ingoing edge from a pair in S_B. Now, since pairs of S_B and H_B cannot be directly connected by an edge of G, it must be that $(i - 1, j)$ and $(i, j - 1)$ are both in B. Thus, $i - 1$ is the smallest index of an extended group over the jth column.

We charge elements of B of type i) and of type ii) to their respective extended intervals. We charge elements of type iii) it to their extended interval over the column. Thus, extended intervals in the column are charged at most twice. By Lemma 11 we have at most $(c + 1)$ extended intervals per column and at most $(c + 2)$ extended intervals per row. This implies that altogether $|B| \leq (3c + 4) \cdot t$, as claimed. □

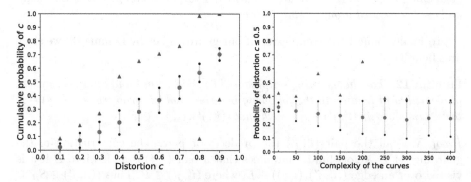

Fig. 6. The cumulative probability distribution of the distortion (left). Given $c \leq 0.5$, the cumulative probability of distortion is shown as a function of the complexity t of the curves, for $t \in \{10, 50, 100, 150, 200, 250, 300, 350, 400\}$ (right). The means μ of the values denoted by red circles. The intervals $[\mu - \sigma, \mu + \sigma]$ denoted by black dots, where σ is the standard deviation. The minima and maxima denoted by blue triangles. (Color figure online)

Lemmas 4 and 13 imply the correctness of Theorem 1. The proof of Theorem 2 is analogous to the proof of Theorem 1, while Lemmas 1 and 4 are replaced by Lemmas 2 and 5, respectively. The rest of the proof can be taken verbatim.

4   Conclusions

We studied the behavior of the discrete Fréchet distance between two polygonal curves under projections to a random line. Our results show that in the worst case and under reasonable assumptions, the discrete Fréchet distance between two polygonal curves of complexity t in $\mathbb{R}^d$, where $d \in \{2, 3, 4, 5\}$, degrades by a factor linear in t with constant probability. One can see this as a negative result, since we hoped that the Fréchet distance would be more robust under such projections. We also performed some preliminary experiments on the dataset of the 6th ACM SIGSPATIAL GISCUP 2017 competition [34] (please refer to the full version of the paper). The cumulative probability distribution of the distortion[2] $c = d_F(P', Q')/d_F(P, Q)$ (Fig. 6, left) suggests that for realistic input curves we can expect that $\Pr[c \leq \gamma] \leq \gamma$. This holds independently of the complexity t of the input curves, as illustrated by Fig. 6 (right) for the given threshold $\gamma = 0.5$. This implies that with probability of at least 0.5 we expect that the discrete Fréchet distance will be reduced at most by a factor 2 when projected to a line chosen uniformly at random, independently of the input complexity. These results stand in stark contrast with our lower bounds. They indicate that highly distorted projections happen very rarely in practice, and only for strongly conditioned input curves.

References

1. Abboud, A., Backurs, A., Williams, V.V.: Quadratic-time hardness of LCS and other sequence similarity measures. CoRR abs/1501.07053 (2015)
2. Afshani, P., Driemel, A.: On the complexity of range searching among curves. In: Proceedings of the 29th ACM-SIAM Symposium on Discrete Algorithms, SODA, pp. 898–917 (2018). https://doi.org/10.1137/1.9781611975031.58
3. Agarwal, P.K., Ben Avraham, R., Kaplan, H., Sharir, M.: Computing the discrete Fréchet distance in subquadratic time. SIAM J. Comput. 43(2), 429–449 (2014). https://doi.org/10.1137/130920526
4. Agarwal, P.K., Fox, K., Pan, J., Ying, R.: Approximating dynamic time warping and edit distance for a pair of point sequences. In: Fekete, S., Lubiw, A. (eds.) 32nd International Symposium on Computational Geometry, SoCG. Leibniz International Proceedings in Informatics (LIPIcs), vol. 51, pp. 6:1–6:16. Schloss Dagstuhl-Leibniz-Zentrum für Informatik, Dagstuhl, Germany (2016). https://doi.org/10.4230/LIPIcs.SoCG.2016.6

---

[2] Technically speaking, this is the inverse of the distortion as defined in the introduction. We choose this definition to simplify the presentation, since this definition ensures that $c \in [0, 1]$.

5. Backurs, A., Sidiropoulos, A.: Constant-distortion embeddings of Hausdorff metrics into constant-dimensional l_p spaces. In: Approximation, Randomization, and Combinatorial Optimization. Algorithms and Techniques, APPROX/RANDOM, pp. 1:1–1:15 (2016). https://doi.org/10.4230/LIPIcs.APPROX-RANDOM.2016.1

6. Badoiu, M., Chuzhoy, J., Indyk, P., Sidiropoulos, A.: Low-distortion embeddings of general metrics into the line. In: Proceedings of the 37th Annual ACM Symposium on Theory of Computing, STOC, pp. 225–233 (2005). https://doi.org/10.1145/1060590.1060624

7. Badoiu, M., et al.: Approximation algorithms for low-distortion embeddings into low-dimensional spaces. In: Proceedings of the 16th Annual ACM-SIAM Symposium on Discrete Algorithms, SODA, pp. 119–128 (2005)

8. Bartal, Y., Gottlieb, L., Neiman, O.: On the impossibility of dimension reduction for doubling subsets of lp. In: ACM Symposium on Computational Geometry, SoCG, pp. 60–66 (2014). https://doi.org/10.1145/2582112.2582170

9. de Berg, M., Cook, A.F., Gudmundsson, J.: Fast Fréchet queries. Comput. Geom. **46**(6), 747–755 (2013). https://doi.org/10.1016/j.comgeo.2012.11.006

10. Bringmann, K.: Why walking the dog takes time: Fréchet distance has no strongly subquadratic algorithms unless SETH fails. In: Proceedings of the 55th Annual IEEE Symposium on Foundations of Computer Science, FOCS, pp. 661–670 (2014). https://doi.org/10.1109/FOCS.2014.76

11. Bringmann, K., Künnemann, M.: Quadratic conditional lower bounds for string problems and dynamic time warping. In: IEEE 56th Annual Symposium on Foundations of Computer Science, FOCS, pp. 79–97 (2015). https://doi.org/10.1109/FOCS.2015.15

12. Bringmann, K., Künnemann, M.: Improved approximation for Fréchet distance on c-packed curves matching conditional lower bounds. Int. J. Comput. Geom. Appl. **27**(1–2), 85–120 (2017). https://doi.org/10.1142/S0218195917600056

13. Buchin, K., Buchin, M., Meulemans, W., Mulzer, W.: Four Soviets walk the dog-with an application to Alt's conjecture. In: Proceedings of the 25th Annual ACM-SIAM Symposium on Discrete Algorithms, pp. 1399–1413 (2014). https://doi.org/10.1137/1.9781611973402.103

14. Buchin, K., et al.: Folding free-space diagrams: computing the Fréchet distance between 1-dimensional curves (multimedia contribution). In: 33rd International Symposium on Computational Geometry, SoCG, pp. 64:1–64:5 (2017). https://doi.org/10.4230/LIPIcs.SoCG.2017.64

15. Ceccarello, M., Driemel, A., Silvestri, F.: FRESH: Fréchet similarity with hashing. CoRR abs/1809.02350 (2018)

16. Ding, H., Trajcevski, G., Scheuermann, P., Wang, X., Keogh, E.: Querying and mining of time series data: experimental comparison of representations and distance measures. Proc. VLDB Endow. **1**(2), 1542–1552 (2008). https://doi.org/10.14778/1454159.1454226

17. Driemel, A., Har-Peled, S.: Jaywalking your dog - computing the Fréchet distance with shortcuts. SIAM J. Comput. **42**(5), 1830–1866 (2013). https://doi.org/10.1137/120865112

18. Driemel, A., Har-Peled, S., Wenk, C.: Approximating the Fréchet distance for realistic curves in near-linear time. Discrete Comput. Geom. **48**(1), 94–127 (2012). https://doi.org/10.1007/s00454-012-9402-z

19. Driemel, A., Krivošija, A., Sohler, C.: Clustering time series under the Fréchet distance. In: Proceedings of the 27th Annual ACM-SIAM Symposium on Discrete Algorithms, SODA, pp. 766–785 (2016). https://doi.org/10.1137/1.9781611974331.ch55

20. Driemel, A., Silvestri, F.: Locally-sensitive hashing of curves. In: 33st International Symposium on Computational Geometry, SoCG, pp. 37:1–37:16 (2017). https://doi.org/10.4230/LIPIcs.SoCG.2017.37

21. Eiter, T., Mannila, H.: Computing discrete Fréchet distance. Technical report, CD-TR 94/64, Christian Doppler Laboratory (1994)

22. Emiris, I.Z., Psarros, I.: Products of Euclidean metrics and applications to proximity questions among curves. In: Speckmann, B., Tóth, C.D. (eds.) 34th International Symposium on Computational Geometry, SoCG, pp. 37:1–37:13 (2018). https://doi.org/10.4230/LIPIcs.SoCG.2018.37

23. Fellows, M.R., Fomin, F.V., Lokshtanov, D., Losievskaja, E., Rosamond, F.A., Saurabh, S.: Distortion is fixed parameter tractable. TOCT **5**(4), 16:1–16:20 (2013). https://doi.org/10.1145/2489789, https://doi.org/10.1145/2489789

24. Gold, O., Sharir, M.: Dynamic time warping and geometric edit distance: Breaking the quadratic barrier. In: 44th International Colloquium on Automata, Languages, and Programming, ICALP, pp. 25:1–25:14 (2017). https://doi.org/10.4230/LIPIcs.ICALP.2017.25

25. Håstad, J., Ivansson, L., Lagergren, J.: Fitting points on the real line and its application to RH mapping. J. Algorithms **49**(1), 42–62 (2003). https://doi.org/10.1016/S0196-6774(03)00083-X

26. Indyk, P.: Algorithmic applications of low-distortion geometric embeddings. In: 42nd Annual Symposium on Foundations of Computer Science, FOCS, pp. 10–33 (2001). https://doi.org/10.1109/SFCS.2001.959878

27. Indyk, P.: Approximate nearest neighbor algorithms for Fréchet distance via product metrics. In: Symposium on Computational Geometry, SoCG, pp. 102–106 (2002). https://doi.org/10.1145/513400.513414

28. Indyk, P., Matoušek, J.: Low-distortion embeddings of finite metric spaces. In: Goodman, J.E., O'Rourke, J. (eds.) Handbook of Discrete and Computational Geometry, pp. 177–196. CRC Press (2004). https://doi.org/10.1201/9781420035315.ch8

29. Keogh, E., Ratanamahatana, C.A.: Exact indexing of dynamic time warping. Knowl. Inf. Syst. **7**(3), 358–386 (2005)

30. Matoušek, J.: On the distortion required for embedding finite metric spaces into normed spaces. Isr. J. Math. **93**(1), 333–344 (1996)

31. Müller, M.: Dynamic time warping. In: Müller, M. (ed.) Information Retrieval for Music and Motion, pp. 69–84. Springer, Heidelberg (2007). https://doi.org/10.1007/978-3-540-74048-3_4

32. Nayyeri, A., Raichel, B.: Reality distortion: exact and approximate algorithms for embedding into the line. In: Guruswami, V. (ed.) IEEE 56th Annual Symposium on Foundations of Computer Science, FOCS, pp. 729–747 (2015). https://doi.org/10.1109/FOCS.2015.50

33. Rakthanmanon, T., et al.: Searching and mining trillions of time series subsequences under dynamic time warping. In: The 18th ACM SIGKDD International Conference on Knowledge Discovery and Data Mining, pp. 262–270 (2012). https://doi.org/10.1145/2339530.2339576

34. Werner, M., Oliver, D.: ACM SIGSPATIAL GIS Cup 2017: range queries under Fréchet distance. SIGSPATIAL Spec. **10**(1), 24–27 (2018). https://doi.org/10.1145/3231541.3231549, http://sigspatial2017.sigspatial.org/giscup2017/home

Algorithms for Dynamic NFV Workload

Yaron Fairstein$^{(\boxtimes)}$, Seffi (Joseph) Naor$^{(\boxtimes)}$, and Danny Raz$^{(\boxtimes)}$

Computer Science Department, Technion, 32000 Haifa, Israel
yyfairstein@gmail.com, {naor,danny}@cs.technion.ac.il

Abstract. The dynamic NFV placement problem captures one of the main challenges facing the telecom industry following the emergence of the Network Function Virtualization (NFV) networking paradigm, that is, deciding the placement of network functions while taking into consideration the dynamic nature of networks and workloads. We model the problem as a generalization of the classic Uncapacitated Facility Location (UFL) problem, where we consider both multiple types of commodities and dynamic clients whose location changes over time.

We show that under reasonable assumptions we are able to develop a 7-approximation algorithm for the Dynamic Facility Location (DFL) problem, improving the logarithmic approximation of Eisenstat et al. [6]. We build upon this result to develop the first virtualized services placement algorithm that accounts for dynamic changes. Our tri-criteria approximation algorithms provide constant approximation factors with respect to the overall performance and size constraints, and logarithmic approximation factors with respect to capacity constraints.

Keywords: Approximation algorithms · Facility location · NFV

1   Introduction

Network Function virtualization (NFV) is an emerging networking paradigm [7] where network functions and networking related applications are implemented in software over Commercially-Off-The-Shelf (COTS) servers located at small data centers that are distributed in the network. This new paradigm is gaining popularity in the telecommunication industry and almost all major operators are reporting initial commercial or proof of concept (POC)[1] deployments. Two of the main reasons for this shift toward NFV and SDN (Software Defined Networks) are the race for the ability to rapidly introduce new services, sometimes referred to as network agility, and the need (of operators) to reduce cost. For this reason the orchestration of network services and the efficient use of resources is gaining

---

[1] Two of the most notable examples are Vodafone's Ocean virtual VPN service (see http://www.mobileeurope.co.uk/news-analysis/ocean-s-40-vodafone-looks-to-minimise-hold-ups-as-it-battles-with-group-wide-nfv-sdn-project) and AT&T's vCPE (see http://www.netmanias.com/en/post/blog/10363/kt-sdn-nfv/sdn-nfv-based-vcpe-services-by-at-t-verizon-and-kt).

© Springer Nature Switzerland AG 2018
L. Epstein and T. Erlebach (Eds.): WAOA 2018, LNCS 11312, pp. 238–258, 2018.
https://doi.org/10.1007/978-3-030-04693-4_15

more and more attention, both in industry and academia, where new forums and open projects are forming to address it [1–3].

Many of the initial NFV deployments concentrated on somewhat static applications like virtual customer premises equipment (vCPE), where demand is either generated in private customers' homes or in business customers' offices, tending to have fixed locations. However, there are new emerging applications and network functions that are much more dynamic in nature, both due to mobility and rapid change in demand. One such an example is vRAN, where the low layer wireless technology is implemented in NFV, and demand depends on the current load of the base stations. Another example is the area of self driving cars and drone controls; here, the mobility of the platform, together with fluctuations in demand, creates a highly dynamic service. A third class of fast growing applications having a very dynamic workload are enhanced video and virtual reality applications. In these applications there is a need for real time latency-bounded services, while the workload can rapidly change over time. This calls for an adjustment of the orchestration modules in order to allow for dynamic allocation of resources in smaller time frames and live migration of services, as needed.

While these new orchestration capabilities are addressed both by the industry and by standardization organizations [1,7], the algorithmic engine behind the actual allocation of resources in real time needs also to adapt to the ever-changing environment, so as to handle dynamic workloads. Current resource optimization algorithms for NFV like [5,8] are static in nature, assuming a fixed demand for network functions which is obtained from static flows, as part of the input. Indeed, one can run these resource allocation algorithms each time demand changes, but that requires reallocation of resources, leading to non-negligible management cost.

We address this important aspect by considering a dynamic model, based on the facility location problem, that captures frequent changes in workload. We develop new algorithms having "built in" capabilities for handling the dynamic nature of networks.

Facility location is a well known family of problems that deals with selecting locations for facilities providing service to a set of clients. It is commonly assumed that each facility has an opening cost and there is a given metric space defining distances between clients and possible locations of facilities (see e.g., [10]). The goal is to open facilities, as well as assigning clients to facilities, minimizing the total cost of opening facilities and serving the clients. Of special relevance to us is the multi-commodity facility location (MCFL) problem, introduced by Ravi et al. [9]. In this problem there is a set of commodities that can be installed (opened) at the facilities, incurring an installation cost for each commodity. Each client requires a subset of the commodities, and the goal is to satisfy the requirements of all the clients by connecting them to facilities that provide the commodities they need, while minimizing the total service and installation costs[2].

---

[2] The specific model considered in [9] captures the set cover problem, and therefore only a logarithmic factor approximation could be obtained for the problems studied therein.

Cohen et al. [5], following the paradigm of MCFL problems, defined a model for the NFV placement problem that takes into account special properties of the NFV setting. The input is a set of clients, representing network flows, each requiring service from a subset of the network functions, or commodities. The network functions are installed on servers at various parts of the network. Locating the functions is performed in a practical environment in which servers have limited space for allocating network functions. The overall cost is defined as the sum of installation costs, reflecting the cost of having VMs that execute a function, and the cost of diverting traffic to these servers. The goal is to locate network functions in a way that minimizes overall network cost, while adhering to limited space (size) for installing functions on servers, and the constraint that each function can serve only a limited number (capacity) of clients.

An additional extension of facility location, dynamic facility location (DFL), was recently presented by Eisenstat et al. [6]. DFL is defined over a time horizon in which a new distance metric is given at each time step, resulting in variable service cost. Clients can be reassigned to new facilities between time steps, however, reassignment of a client incurs a change cost. The goal is to minimize the total opening, service, and change costs. An $O(\log nT)$-approximation algorithm for the problem was given by [6], where n is the number of clients and T is the number of time steps.

1.1   Our Results

Our main contributions are the introduction of a new temporal model for dynamic NFV placement together with approximation algorithms with proven performance guarantees. Our new model contains both NFV placement [5] and dynamic facility location [6], capturing the most important aspects of dynamic NFV placement. We do not make any assumptions on how our network evolves over time, however, we do assume that we are given (ahead of time) the changes in the network between time steps. We note that in the NFV setting it is reasonable to assume that servers remain at fixed locations over time, and only clients are dynamic and can change their location[3]. This allows us to consider the NFV placement problem in a dynamic setting that takes into account the different stages a network goes through, thus developing algorithmic solutions that better fit the NFV setting.

The models we consider generalize facility location which is known to be NP-hard. Therefore, we turn to approximate solutions that can be computed efficiently. We first revisit the dynamic facility location problem studied by [6]. By assuming facilities are static, we obtain an elegant 7-approximation algorithm for this problem, improving over the $O(\log nT)$ approximation algorithm of Eisenstsat et al. [6]. We note that this improvement is essential for constant factor approximations when extending the NFV setting to the dynamic setting.

---

[3] In [6] it is assumed, in contrast, that the full metric (i.e., facilities and clients) changes in each time step.

We distinguish between two versions of the dynamic NFV placement problem: dynamic *uncapacitated* NFV (dynamic UNFV) and dynamic *capacitated* NFV (dynamic CNFV) which extends the former. In dynamic CNFV there is a capacity constraint, a limit on the number of clients that can receive service from a network function. Our results for these problems are summarized in the following theorems.

Theorem 1. *There exists an efficient algorithm for the Dynamic UNFV problem with an $(O(1), O(1))$ bi-criteria approximation algorithm, where we approximate the overall network cost by a constant factor, while exceeding the size constraints of the servers by at most a constant factor.*

The next theorem contains our main contribution, which is also the most technically challenging part of our paper.

Theorem 2. *There exists an efficient algorithm for the Dynamic CNFV problem with an $(O(1), O(1), \log(\min\{n, T\}))$ tri-criteria approximation algorithm, where we approximate the overall network cost by a constant factor, while exceeding the size constraints of the servers by at most a constant factor, and the capacity constraints by at most a factor of $O(\log \min\{n, T\})$. Here, n is the number of clients and T is the number of time steps.*

We also show that the problem of selecting the final assignment of clients to functions in the dynamic CNFV problem reduces to the interval graph list coloring problem. In this problem we are given an interval graph and a palette of colors. For each interval a subset of the color set is given. The goal is to find a coloring of the intervals such that: (i) each interval is colored by a color from its allowed subset of colors; (ii) intersecting intervals are colored differently. As this problem is NP-hard, we turn to find an approximate solution with a bounded size set of intersecting intervals that receive the same color. We prove the following:

Theorem 3. *There exists an efficient algorithm for the Interval Graph List Coloring problem that finds a coloring in which at most $O(\log \min\{k, T\})$ intersecting intervals receive the same color, where k is the size of the largest clique in the interval graph, and T is the number of distinct cliques (time steps) in the interval graph.*

We accompany this results by a proven lower bound for the IGLC problem, showing our algorithm is almost tight.

In order to show that our new algorithms indeed improve network utilization under dynamic workloads, we conduct a real scenario simulation based performance evaluation. In this evaluation we compare the performance of the dynamic algorithm to two variants of the static algorithm (presented by [5]), and to the optimal fractional solution to the dynamic problem computed using an LP-solver. The results indicate that the dynamic algorithm preforms at least 40% better than the static algorithm (in some cases up to 2–3 times better), and the comparison to the fractional solution indicates that in the considered practical scenarios, our algorithm is at most twice the optimal (fractional) solution.

Techniques. Our approximation algorithms are based on linear program rounding, that is, given a fractional solution to the linear relaxation of the problem, our goal is to round it, while bounding the resulting increase in cost. There are several techniques that we use for achieving this goal. For each network function we apply a cover-growing procedure which defines covers of bounded radius around clients. The covers constructed have the property that they are disjoint so that clients do not share opening costs. The covers are also advantageous as they allow us in a sense to "ignore" distances in the network. This turns out to be crucial for computing the final network function placement via a rounding algorithm for the general assignment problem (GAP) [10]. We also use the interval selection technique of Eisenstat et al. [6] to partition the time horizon in the dynamic case into static intervals for each client.

In the dynamic CNFV problem we encounter further difficulties, as capacity constraints do not allow us to assign clients freely to the network functions. Reducing the final assignment of clients to functions to the interval graph list coloring problem, and the algorithm promised by Theorem 3 translates into the violation factor of the capacity in our solution.

2   Preliminaries

Here we formalize our model, as well as present several known procedures we use throughout.

2.1   The Model

The dynamic NFV placement problem is defined over a *time horizon*. We are given a set of demands, each composed of a set of flows, one for each time step, and a set of functions from which it requires service. The functions are to be installed on network servers. Each server has a size constraint; each function has both a size and cost for installing it at each server. In addition, each function has a capacity that bounds the number of flows it can serve. We are also given a change cost paid to change the assignment of a demand to different servers between two consecutive time steps. The objective is to find an assignment that minimizes the total cost of installing the functions at the servers, the sum of the distances of the flows from the servers and the sum of change costs. In our model, demands are represented by clients, servers by facilities, and network functions by network commodities. We now elaborate on our model.

We are given a time horizon T, and an undirected graph (or network) $G = (V, E)$, equipped with a distance function $d^t(\cdot, \cdot)$ between any pair of nodes at each time step $t \in T$. The distance functions induce a metric space over the graph. We are given a set $F \subseteq V$ of m facilities and a set $C \subseteq V$ of n clients. For a facility $i \in F$ and client $j \in C$, i and j indicate both facility and client, respectively, as well as where they reside at each time step. We assume that facilities remain at fixed locations in the network, while the evolution of the metric reflects the movement of clients over time, and thus change their distance to the facilities. There is a set S of k network commodities; for each client $j \in C$, $\delta(j)$ denotes the subset of commodities that it requires. We are also given a change cost g, paid for each change in assignment of a client to a facility.

Each facility $i \in F$ has a total size w_i, and each commodity s occupies size w_{is} on facility i. The installation cost of a commodity s at facility i is denoted by f_{is}. In the capacitated NFV placement problem, each commodity s is associated with a capacity μ_s, a bound on the number of clients it can serve. To accommodate more clients, several copies of commodity s can be installed at facility i, however, each copy occupies size w_{is} and pays cost f_{is}.

In a feasible solution to the dynamic NFV placement problem we find an allocation of commodities to facilities, and an assignment of clients to facilities such that at each time step, each client $j \in C$ is assigned to a subset of facilities that can serve all the commodities in $\delta(j)$. To comply with the constraints, a solution must fulfill the requirement that the sum of the sizes of the commodities installed at a facility does not exceed its size. In the capacitated NFV it is also required that at each time step, the number of clients served by a (copy of a) commodity does not exceed its capacity. The goal is to find a feasible solution minimizing the overall cost, comprising of the sum of installation costs, the sum of distances between the clients and the facilities to which they are assigned (paid for each commodity separately), which we also call connection costs, and the sum of change costs.

We formulate the dynamic NFV placement problem as a linear program (DNFV-LP). We denote by y_{is} the variable indicating the number of copies of commodity s installed in facility i, and by x^t_{ijs} the variable indicating whether facility i serves commodity s to client j at time step t. Variable z^t_{ijs} indicates a change in the assignment.

$$\min \quad \sum_{i \in F} \sum_{s \in S} f_{is} y_{is} + \sum_{t \in [T]} \sum_{j \in C} \sum_{s \in \delta(j)} d^t_{ij} x^t_{ijs} + \sum_{t \in [T]} \sum_{j \in C} \sum_{s \in \delta(j)} z^t_{ijs} \cdot g \quad \text{subject to:}$$

$$x^t_{ijs} \leq y_{is} \qquad \forall i \in F, j \in C, t \in [T], s \in \delta(j) \tag{1}$$

$$\sum_{i \in F} x^t_{ijs} = 1 \qquad \forall j \in C, t \in [T], s \in \delta(j) \tag{2}$$

$$\sum_{s \in S} y_{is} w_{is} \leq w_i \qquad \forall i \in F \tag{3}$$

$$x^t_{ijs} - x^{t+1}_{ijs} \leq z^t_{ijs} \qquad \forall i \in F, j \in C, t \in [T), s \in \delta(j) \tag{4}$$

$$\sum_{j \in C} x^t_{ijs} \leq y_{is} \mu_s \qquad \forall i \in F, t \in [T], s \in S \tag{5}$$

$$y_{is}, x^t_{ijs}, z^t_{ijs} \geq 0 \qquad \forall i \in F, j \in C, t \in [T], s \in S \tag{6}$$

Constraint (1) states that a facility cannot provide service of a commodity unless it is installed in it. Constraint (2) guarantees that each client is served all the commodities it requires. Constraint (3) bounds the total size of the commodities installed in each facility. Constraint (4) charges for connection changes between consecutive time steps. Constraint (5) limits the number of clients each commodity can serve.

2.2 Interval Graph List Coloring

We are given an interval graph $G = (V, E)$, i.e., an intersection graph of intervals, where each vertex $v \in V$ corresponds to an interval I_v on the real line, and there is an edge $e = (u, v) \in E$ if and only if I_u and I_v intersect. In addition, there is a set of colors C, and each vertex v is associated with a subset $C_v \subseteq C$ of colors by which it can be legally colored. The goal is to find a legal coloring of the vertices such that neighboring vertices receive different colors. This is called the interval graph list coloring problem (IGLC).

Let T be the set of points on the real line containing all start and end points of the intervals. Obviously, the set T defines the set of maximal *cliques* (w.r.t. containment) in G. We denote the clique at point $t \in T$ by $I(t)$, i.e., the intervals that contain point t. Let x_{vc} indicate whether vertex v is colored by color c. We define an integer *feasibility* program for (IGLC) with three constraints:

(i) for $v \in V$, $\sum_{c \in C_v} x_{vc} \geq 1$, guaranteeing that vertex v is assigned a color from C_v.
(ii) for $t \in T, c \in C$, $\sum_{I_v \in I(t)} x_{vc} \leq 1$, guaranteeing that at most a single interval in each clique $I(t)$ is assigned a particular color.
(iii) for $v \in V, c \in C$, $x_{vc} \in \{0, 1\}$: integrality constraints for the variables.

Clearly, a feasible solution implies a legal coloring for (IGLC).

In [4] it was shown that the IGLC problem is NP-hard, even though coloring interval graphs can be done efficiently. In Sect. 6 we present approximation algorithms that minimize the size of the largest clique with the same color, i.e., minimizing the violation factor of constraint (ii) in the feasibility program. This can also be viewed as bounding the number of copies of each color. We complement this result by showing that the violation achieved almost matches the gap between the feasibility of the integral and linear programs. We consider this gap as the *integrality gap*.

2.3 Useful Procedures

Our model generalizes several known problems. For example, [5] showed that the generalized assignment problem (below) and the uncapacitated facility location problem (see Introduction) are special cases of the NFV problem. Throughout the paper we use several known procedures briefly defined below. A full description is in Appendix A.

Cover-Growing Algorithm. An LP rounding procedure for the uncapacitated facility location (UFL) problem (see Introduction). Given a fractional solution, the output is a set of disjoint covers, where each cover is defined by a set of fractionally opened facilities, and an assignment of clients to the covers.

Generalized Assignment Problem. In the generalized assignment problem (GAP) we are given m machines and n jobs that need to be assigned to the machines. We use a known rounding procedure for the problem [10]. We define for each cover (created by the cover-growing procedure) a job, and for each facility a machine. The rounding returns an assignment of functions to the facilities.

Interval Selection. In the linear relaxation (presented in Sect. 3) of the DFL problem we pay for fractional changes in the assignment of a client to a facility between consecutive time steps. Eisenstat et al. [6] gave a procedure that outputs a set of intervals, breaking the time horizon, separately for each client, such that in each interval the fractional connection is static. In addition, the fractional change in each interval is bounded. The cost of the new solution is at most twice the optimal fractional solution.

3   Dynamic Facility Location

We consider here the dynamic facility location problem (i.e., single commodity) under the assumption that facilities/servers are static (as in the NFV setting). We obtain a 7-approximation algorithm for this problem, improving over the $O(\log nT)$ approximation algorithm of Eisenstat et al. [6]. This improvement forms the basis for obtaining constant factor approximations when extending the NFV setting to the dynamic setting. Our algorithm uses the interval selection procedure of Eisenstat et al. [6], however, our approach significantly departs from [6], thus enabling us to improve on their results. The linear program is[4]:

$$\min \quad \sum_{i \in F} f_i y_i + \sum_{t \in [T]} \sum_{j \in C} d_t(i,j) x_{ij}^t + \sum_{t \in [T]} \sum_{j \in C} z_{ij}^t \cdot g \qquad \text{subject to:}$$

$$x_{ij}^t \leq y_i \qquad \qquad \forall i \in F, j \in C, t \in [T] \qquad (7)$$

$$\sum_{i \in F} x_{ij}^t = 1 \qquad \qquad \forall j \in C, t \in [T] \qquad (8)$$

$$x_{ij}^t - x_{ij}^{t+1} \leq z_{ij}^t \qquad \forall i \in F, j \in C, t \in [T) \qquad (9)$$

$$y_i, x_{ij}^t, z_{ij}^t \geq 0 \qquad \forall i \in F, j \in C, t \in [T] \qquad (10)$$

We first solve the LP for the problem. Given a fractional solution, we run the interval selection procedure and then apply the cover-growing algorithm. The details are as follows.

---

[4] The variables are the same as in the linear program for the dynamic NFV problem in Sect. 2.1.

DFL-Algorithm

1. Solve the LP and construct the intervals for each client (see Sect. 2.3).
2. For each client and interval, find the average distance from the client to each facility (over the time steps of the interval), and define a cover for the client containing only facilities that, on average, are within distance of at most twice the fractional connection cost.
3. Run the cover-growing algorithm over the covers from the previous step as in Sect. 2.3.

Lemma 1. *DFL-Algorithm provides a 7-approximation factor for the DFL problem.*

Proof. As mentioned in Sect. 2.3, the interval selection procedure can at most double the cost of the solution, yet it allows us to change the assignment (only) between consecutive intervals, while paying at most the fractional change cost. Thus, the total change cost is twice the fractional change cost. The next claim is rather easy to prove:

Claim. For every interval of a client, if two facilities are in the cover of the interval, i.e., the average distance to each of them is smaller than twice the fractional connection cost, $2r$, then the distance between them is at most $4r$.

In Step (3) we run the cover-growing algorithm over the constructed covers. Each client in each interval is either assigned to a facility in its own cover, or in its representative cover. Recall that all facilities in its own cover are at distance of at most twice its fractional connection cost, thus summing up over the interval, the total connection cost is at most twice the fractional connection cost times the number of time steps in the interval. If a client is assigned to a facility in its representative cover, we pay an additional connection cost, from the intersection of the covers to the opened facility. As seen in the above claim, the distance between each pair of facilities inside a cover is at most four times the fractional connection cost over the interval it is defined for, so this additional cost is at most four times the fractional connection cost which defines the cover. Since the representative cover has a smaller fractional connection cost, the overall distance from the client to the opened facility in its representative cover is at most six times its own fractional connection cost. In addition, since we open at most a single facility in each cover, from Markov inequality the total installation cost is at most twice the fractional installation cost. Concluding, after accounting for the $\frac{1}{\theta}$ factor of the interval selection procedure (see Appendix A for the role of θ), we get an approximation ratio of $\frac{6}{\theta}$ for the connection costs, $\frac{2}{\theta}$ for the installation costs, and $\frac{1}{1-\theta}$ for the change costs. By choosing $\theta = \frac{6}{7}$ we get an approximation ratio of 7.

4   Dynamic Uncapacitated NFV

This problem combines the uncapacitated NFV placement problem with the dynamic facility location problem. As mentioned (see Sect. 2.1), for each time step we need to solve an instance of the uncapacitated NFV problem, while paying for changes in the assignment of clients to facilities. We solve this problem by extending the DFL-Algorithm. The algorithm first constructs covers for each commodity separately, but then it uses a rounding algorithm for GAP (instead of the cover-growing algorithm) to decide on the final location of the commodities. The next algorithm together with Lemma 2 prove Theorem 1.

Dynamic-UNFV-Algorithm

1. Solve DNFV-LP (with infinite capacities) and construct the intervals for each client (see Sect. 2.3).
2. For each client and interval, calculate the average distance from the client to each facility (over the time steps of the interval), and define a cover for the client containing only facilities that, on average, are within distance of at most twice the fractional connection cost. The radius of a cover is defined as twice its client's fractional connection cost.
3. Separately, for each commodity, pick a non-intersecting set of representative covers as follows: select a cover with minimum radius and delete all covers that intersect with it. Continue until there are no more covers. Each client whose cover was deleted receives service from the representative cover that "caused" the deletion.
4. Apply the GAP rounding algorithm to the fractional solution defined by the representative covers (see Sect. 2.3): each cover is considered a separate job and each facility a machine.
5. Assign each client at each interval to the facility in its representative cover in which the commodities were installed according to the GAP rounding.

Lemma 2. *Dynamic-UNFV-Algorithm provides $(O(1), O(1))$ bi-criteria approximation factor for the Dynamic UNFV problem.*

Proof. We first notice that the analysis of the connection and change costs remains the same as for the DFL-Algorithm (Sect. 3), assuming the same choice of $\theta = \frac{6}{7}$. This means that the connection costs of the rounded solution are at most 7 times the fractional connection costs, and the change costs are at most 7 times the fractional change costs. For each commodity s, the representative covers do not intersect. This allows us to install a copy of s in each representative cover. It follows that we can treat each cover, a fractional allocation of a commodity to a facility, as a fractional assignment of a job to a set of machines. Each job has a cost equal to its installation cost and a size equal to the commodity's size, and each machine has a size bound equal to the facilities size bound. Thus, the fractional allocation of representative covers to facilities define an instance of GAP. The rounding algorithm for GAP returns an integral solution whose cost is at most the fractional cost and the size constraints are violated by at most a factor of 2. Thus, since we multiplied the installation fractions by $\frac{2}{\theta} = \frac{7}{3}$ (in the interval selection and in the covers construction), the integral solution returned from it has an installation cost of at most $\frac{7}{3}$ times the fractional installation costs, and the size constraints are violated by at most a factor of $\frac{14}{3}$. To conclude, we get a bi-criteria approximation. The approximation ratio of the cost is 7 for the connection costs, $\frac{7}{3}$ for the installation costs and 7 for the change costs, and the size constraints are violated by at most a factor of $\frac{14}{3}$.

5   Dynamic Capacitated NFV

This problem combines the dynamic facility location problem and the capacitated NFV problem. In this problem (see Sect. 2.1) we need to solve an instance of the capacitated NFV placement problem at each time step, while taking into account the cost of changing the assignment of clients to facilities between time steps. Just like the previous algorithm (for dynamic UNFV) we select the intervals and find their covers for each client. The difficulty we encounter here is that we may connect too many covers to the same representative cover, resulting in a violation of the commodities' capacities. We choose the representative covers similarly to Cohen et al. [5], and then reduce the problem to an instance of the interval graph list coloring problem. The next algorithm together with Lemma 3 prove Theorem 2.

Dynamic-CNFV Algorithm

1. Solve DNFV-LP and select intervals.
2. For each client and interval, calculate the average distance from the client to each facility (over the time steps of the interval), and define a cover for the client containing only facilities that, on average, are within distance of at most twice the fractional connection cost. The radius of a cover is defined as twice its client's fractional connection cost.
3. Select a cover with smallest radius and assign all intersecting covers to it. Remove covers that were already assigned at least $\frac{1}{4}$ of their service. Continue till all covers are removed.
4. For each cover, normalize to 1 the total service it gets from the representative covers.
5. Apply the interval graph list coloring rounding algorithm (see Sect. 6) to the problem defined by the covers and representative cover: each cover is considered an interval and each representative cover defines a set of colors (of size equal to its capacity).
6. Apply the GAP rounding algorithm to the fractional solution defined by the representative covers (see Sect. 2.3): each cover is considered a separate job and each facility a machine.
7. Assign each client, at each interval, to the facility in its representative cover in which the commodities were installed according to the GAP rounding.

Lemma 3. *Dynamic-CNFV-Algorithm provides $(O(1), O(1), O(\log \min\{n, T\}))$ tri-criteria approximation factor for the Dynamic UNFV problem.*

Proof. Like previous algorithms, we first select the intervals, paying a factor of $\frac{1}{1-\theta}$ over the change cost, and $\frac{1}{\theta}$ over all other terms. Next we find a cover for each interval and pay another factor of 2 for doubling its radius (for all terms, but the change cost). Afterwards, we assign each cover to a set of representative covers of smaller radius, supplying at least $\frac{1}{4}$ of its service. By normalizing the total service of each cover we multiply all installation fractions by at most 4, increasing the installation cost and size constraint violation.

We are left with two tasks. First, assigning each cover to a single representative cover, and second, choosing in which facility to install each cover, which, like the previous (uncapacitated) algorithm, we solve by running a GAP rounding algorithm. We solve the first task by reducing the problem it defines to the IGLC problem. For each representative cover with a total capacity of μ, we create μ different colors. Each cover is defined for a time interval $[t_1, t_2]$ and is represented by a set of representative covers. Thus, for each cover we create an interval $I = [t_1, t_2]$, associated with the subset of colors created for its representative covers. The fractional service assignment of covers to representative covers can be seen as a fractional coloring of intervals, in which a single copy of each color suffices. This defines an instance of IGLC. From the algorithm for IGLC we get a coloring of the intervals. If the interval defined for cover j is colored by color c, and c is the color created for representative cover i, we set cover i as the final representative of cover j.

What are the properties of the fractional coloring? From the construction of the (IGLC) instance, together with the fact that each interval is fully colored, we can infer that there is a feasible solution for the linear program for (IGLC). This means that every subset of ℓ intersecting intervals has at least ℓ colors available for it, which is exactly the local condition our rounding algorithms requires. Thus, the rounding algorithms for (IGLC) return a solution in which the number of copies of each color is at most $\log(\min\{n, T\})$, leading to a violation of the capacity constraints of the commodities by the same factor. After taking into account the additional violation from the GAP rounding we get that the size constraints are violated by a factor of $\frac{16}{\theta}$. By choosing $\theta = \frac{16}{17}$ we get that the total cost is at most 17 times the fractional cost, and that the size constraints are violated by a factor of 17. Overall we get an approximation of $(17, 17, O(\log(\min\{n, T\})))$: 17-approximation for the costs and size constraints, and $O(\log(\min\{n, T\}))$-approximation for capacities.

6   Interval Graph List Coloring

We provide two approximation algorithms for the (IGLC) problem. Given a legal fractional solution for the feasibility program (see Sect. 2.2), the first algorithm (see Sect. 6.1) finds a solution in which the maximum number of intersecting intervals that get the same color is of size $O(\log k)$, where k is the size of the largest clique. The second algorithm (see Appendix B) finds a solution in which the maximum number of intersecting intervals that get the same color is of size $O(\log T)$. We emphasize that T and k are independent parameters. Combining the two algorithms proves Theorem 3.

We note that the fractional feasibility of the (IGLC) instance guarantees the following: the size of the union of the allowed color sets of any set S of intersecting intervals is at least as large as $|S|$. Thus, the conditions of Hall's theorem are satisfied in the bipartite graph of intersecting intervals vs. colors. This property turns out to be very useful in our algorithms.

We complement the above results and show that the integrality gap of the linear feasibility program is $\Omega(\frac{\log k}{\log \log k})$. This means that given a feasible fractional solution, any integral solution might have to color a clique of size at least $\Omega(\frac{\log k}{\log \log k})$ with the same color.

6.1 Clique Size Dependent Approximation

We can achieve an $O(\log k)$ approximation if we find a coloring of at least half the intervals in each clique (with a constant number of copies of each color). This is achieved by utilizing Hall's theorem. Then, by repeating iteratively, we end up with a full coloring of the intervals.

Claim. There exists a subset of intervals S that can be colored with color c, such that in each clique $I(t)$ having an interval that can be colored by color c, $1 \leq |I(t) \cap S| \leq 2$.

Proof. Let us assume the claim is false and there is no such subset S. Define $S = \{I_v | c \in C_v\}$. Obviously, each clique with an interval that can be colored by color c has an interval in this subset S. There is a point t with at least three intervals (otherwise, we are done). Let $I_1 = [a_1, b_1], I_2[a_2, b_2]$ and $I_3 = [a_3, b_3]$ be three intervals in S that intersect at t. First we notice that all starting points a_i are smaller than t, and all ending points b_i are bigger than t. Without loss of generality let us assume a_1 is the smallest starting point. Now there are two options, either b_1 is bigger than b_2 and b_3, which in this case, obviously, I_1 contains both of I_2 and I_3. The other option is that either b_2 or b_3 has the biggest ending point. W.l.o.g let us assume that $a_2 \geq a_3$. So the range $[a_1, b_2]$ contains I_3. In either case we notice that we can remove at least one of the intervals without uncovering points on the line. We can continue with this process until there are no points with more than two intervals in S.

Using this claim we can iterate through the colors, and color intervals. In each iteration we can guarantee that in each clique $I(t)$ containing an interval that can be colored by color c (the color of the current iteration), some interval $I \in I(t)$ is colored.

Clique-Coloring Algorithm

1. For each color c:
   (a) Create a subset of intervals that can be colored c.
   (b) Until there are no cliques of size three or more: select three intervals in such a clique and remove the interval contained in the other two intervals from the subset.
   (c) Color all intervals in the subset with color c.
2. If there are uncolored intervals, go back to (1).

Lemma 4. *Clique-Coloring Algorithm provides $O(\log k)$ approximation for IGLC.*

Proof. Each clique can be seen as a matching of intervals to colors. We are guaranteed by the fractional solution (to IGLC in Sect. 2.2) that the number of colors available for any set of ℓ intersecting intervals is at least ℓ. Therefore, by Hall's theorem, there is a perfect matching in each clique from intervals to colors.

Whenever we color a subset of intervals by color c, we color at least one interval in every clique with an interval that can be colored with c. Consider a clique $I(t)$ of size s with an interval that received color c. As already mentioned, $I(t)$ has a perfect matching before coloring the intervals. After we color a subset of intervals by color c, at least one of the intervals in $I(t)$ was colored by c, and no other interval in $I(t)$ will receive color c in the same iteration. It may be the case that coloring the interval by color c does not agree with the coloring in the perfect matching of $I(t)$. So, by coloring it we may have lost a potential match, and are left with only $s - 2$ possible matches. In the worst case, for every interval colored, each clique has two less color matches available, one for the colored interval, and one for the interval that was matched to color c in the perfect matching. In addition, any clique without an interval that can be colored by c can still be matched perfectly since the bipartite graph of intervals and colors does not change.

Since we use a single color each time, if an interval can be matched to color c in one clique, it can be matched to it in all cliques (that contain it). Overall we lose a single match in each clique for an interval we color, so we are left without colors only after half of the intervals in each clique were colored. Since this applies to all cliques, after $\log k$ iterations, where k is the size of the largest clique, all cliques are fully colored. Each iteration uses only two copies of each color, thus the result coloring uses at most $O(\log k)$ copies of each color.

6.2 Integrality Gap

We define an instance of (IGLC) where we are given 4 colors denoted by $1, 2, 3, 4$. There are six intervals, $I_1, \ldots, I_6$, as seen in Fig. 1. The intervals' subsets of allowed colors are: $C_1 = \{1, 2\}$, $C_2 = \{3, 4\}$, $C_3 = \{1, 3\}$, $C_4 = \{2, 3\}$, $C_5 = \{1, 4\}$, and $C_6 = \{2, 4\}$. We define a feasible fractional solution in which each interval is colored half-half by its allowed colors. For any integral coloring of I_1 and I_2 there is no legal coloring for one of the other intervals (intervals I_3, I_4, I_5 and I_6, cover all coloring combination of intervals I_1 and I_2). This means we will need at least two copies of some color, and the integrality bgap is 2.

This example can be expanded along the same lines. First, we define a set of x intervals, B_1, each colored equally with its own distinct set of x colors. Next we define B_2, a set of x^x disjoint intervals. Each interval in B_2 requires a different color set, one from each of the intervals in B_1, and is colored equally by them. As seen in the example above, for every coloring of the intervals in B_1, one of

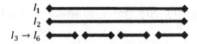

Fig. 1. An IGLC instance with an integrality gap of 2

the intervals in B_2 will be left without a color it can be legally colored with. This results with a clique of two intervals that receive the same color.

Next, we create a copy of the intervals in B_1, each with its own set of distinct colors, and add them to B_1. For each interval $I \in B_2$ we create x^x disjoint intervals in its range. Their colors are chosen in the same way as the colors of the intervals in B_2, but with the colors of the new intervals of B_1. The new intervals are added to B_2 as well. As a result of this process, in any integral coloring, a point t exists such that $I(t)$ contains at least two different pairs of intervals that received the same color.

Applying this process x times results with an instance for which:

(i) in any integral solution, there exists a clique with x different pairs of intervals that received the same color.
(ii) in every clique we used at most $\frac{2}{x}$ of every color, $\frac{1}{x}$ for every level.

An interval that requires the x colors described in (i), that are already used by x pairs of intervals, forces us to use a third copy of some color. We can create the third level B_3 appropriately to guarantee that some interval receives the third color. We can create at most x levels in this manner, as each level requires at most $\frac{1}{x}$ of every color. If the size of the largest clique is $k = O(x^x)$, the integrality gap is $\Omega(\frac{\log k}{\log \log k})$.

A   Useful Procedures

Our model generalizes several known problems. For example, [5] showed that the generalized assignment problem (see Sect. 2.3) and the uncapacitated facility location problem (see Introduction) are special cases of the NFV problem.

Here we describe several known procedures for theses problems which we use throughout the paper.

A.1   Cover-Growing Algorithm

Here we describe a cover-growing algorithm for the uncapacitated facility location (UFL) problem. We present it together with its analysis, since we use it later and take advantage of its local properties. We assume that our input is a fractional solution to the UFL problem. The output is an integral solution, i.e., a set of open facilities such that each client is assigned to an open facility.

We can view a fractional solution to the facility location problem as inducing a probability distribution over the facilities from which a client gets service.

Thus, the fractional connection cost of a client is an *expectation*, since it is a sum of weighted distances (where the service fractions serve as weights). A cover (or ball) around a client, having radius twice the expected distance, contains at least half of the client's fractional service. Thus, by doubling the fractions inside the cover, the client gets all of its service from it.

Rounding Algorithm

1. Define a cover around each client with radius twice the expected distance.
2. Until all clients are satisfied:
   (a) among all unconnected clients, find client i with minimum radius cover.
   (b) open facility f that minimizes the installation cost in the cover.
   (c) for every client i' whose cover intersects the cover of client i (there exists a facility that serves both): connect it to facility f.

It follows from the rounding algorithm that every client is either connected to a facility in its own cover or connected to a facility in an intersecting cover (which does not have a larger radius). In this case we say that a client j is connected to a facility in its *representative cover*. Thus, the connection costs are at most 6 times the sum of the expected distances, and the installation costs are at most twice the fractional installation costs. In total, the approximation factor achieved is 6 for the uncapacitated facility location.

A.2   Generalized Assignment Problem

In the generalized assignment problem (GAP) we are given m machines and n jobs that need to be assigned to the machines. Job j has cost c_{ij} and size w_{ij} on machine i; machine i has total size w_i. Our goal is to assign each job to a machine, without violating machine size constraints, while minimizing the total assignment cost. Assume we are given a feasible fractional solution to GAP. In our algorithms we apply a rounding procedure to the given fractional solution due to Shmoys and Tardos (see [10]). The output of the rounding procedure is an integral solution whose cost is at most the cost of the fractional solution, and the size of every machine is violated by at most the maximum size of a job assigned fractionally to the machine, i.e., by at most a factor of two. GAP is a special case of the NFV placement problem in which all distances in the metric are set to be zero. In this case at most one copy of each function (i.e., a job) is installed, yielding a GAP instance.

A.3   Interval Selection

In the linear relaxation (presented in Sect. 3) of the DFL problem we pay for fractional changes in the assignment of a client to a facility between consecutive time steps. Eisenstat et al. [6] gave a procedure that breaks the time horizon into intervals, separately for each client, such that in each interval the fractional connection is static. The fractional change in each interval is bounded. The idea behind the procedure is to iteratively construct the intervals for each client. An interval terminates at the latest time step t in which the fractional changes that were accumulated through until t are bounded. The procedure for client j is as follows:

1. Set $t_0^j = 1$ and $\ell = 1$.
2. Next interval starts at the maximal t, $t \in (t_\ell^j, T + 1]$, such that $\sum_{i \in F} (\min_{\{t_\ell^j \le u < t\}} x_{ij}^u) \ge \theta$ (where $\theta \in [0, 1]$).
3. If $t = T + 1$, all intervals are selected; otherwise, set $\ell \leftarrow \ell + 1$ and select next interval.

For each interval we set the new static fractions $\hat{x}$ as follows. For each t in the ℓth interval,

$$\hat{x}_{ij}^t = \frac{\min_{\{t_\ell^j \le u < t_{\ell+1}^j\}} x_{ij}^u}{\sum_{i' \in F} \min_{\{t_\ell^j \le u < t_{\ell+1}^j\}} x_{i'j}^u}.$$

It is straightforward to verify that the fractions in the solution are at most multiplied by $\frac{1}{\theta}$, since the numerator is smaller than all fractions in its interval and the denominator is at least $\frac{1}{\theta}$. If we multiply the installation fractions by $\frac{1}{\theta}$, then the solution is feasible. Next, we want to show that fractional changes in each interval are at least $1 - \theta$. If so, since we only change the assignment at the end of each interval, we pay at most $\frac{1}{1-\theta}$ times the change cost. If we consider the ℓth interval, the total fractional change in assignment in the interval is

$$\sum_{i \in F} \sum_{t_\ell^j \le u \le t_{\ell+1}^j} z_{ij}^u \ge \sum_{i \in F} (x_{ij}^{t_\ell^j} - \min_{\{t_\ell^j \le u \le t_{\ell+1}^j\}} x_{ij}^u)$$

$$= 1 - \sum_{i \in F} \min_{\{t_\ell^j \le u \le t_{\ell+1}^j\}} x_{ij}^u \ge 1 - \theta.$$

The first inequality follows since the change is at least the first fractional connection minus the min fractional connection.

B   Horizon Length Dependent Approximation

Assuming there is a feasible fractional solution to the (IGLC) program, we split the time horizon into two independent parts which we color separately, achieving an $O(\log T)$ approximation. Similarly to the previous algorithm, we exploit the fact that each clique can be legally colored.

Time-Split Algorithm

1. Let t be the middle point in T.
2. Color the intervals in clique $I(t)$.
3. Continue recursively on both halves of T. At the ith iteration use the ith copy of the set of colors.

Lemma 5. *Time-Split Algorithm provides $O(\log T)$ approximation for the IGLC problem.*

Proof. Each clique can be seen as a matching of intervals to colors. We are guaranteed by the fractional solution (to IGLC in Sect. 2.2) that the number of colors available for any set of ℓ intersecting intervals is at least ℓ. Therefore, by Hall's theorem, there is a perfect matching in each clique from intervals to colors. Next, at each step we color the middle point with a new set of colors. After removing all colored intervals we are left with two ranges that do not share any intervals thus, we can use the next copy of the set of colors at both of them. Each time the we color a clique we create two ranges that are at most half the size of the original range, so it is obvious we will not need more than $O(\log T)$ copies of the colors to color all intervals.

C   Experiments

We devote this section to test the uncapacitated dynamic NFV algorithm. To this end we consider a subnet of the physical network of Cogent, a tier 1 ISP, which offers us a realistic facilities' deployment (using its publicly available data center locations). We choose ten data centers placed in Europe and defined facilities at their location. Next, we added a hundred clients in random positions, and for each client defined a random walk, taking a step at a random direction of random size. In addition, each client was assigned a random commodity vector describing which commodities, out of a list of five different commodities, it requires. Finally, a random size was assigned to each facility, together with a size and cost for each commodity (at each facility).

In order to evaluate the algorithm, we compare it with previous solutions for the problem, the uncapacitated (static) NFV from [5]. Since the static algorithm does not optimize over the time horizon of the dynamic problem, we define two different variants for using it. The first one uses the clients' position over the time horizon to find the average position of each client. Using this average position we obtain a single time step instance. The second one runs the static algorithm at each time step separately, paying change costs accordingly. These two options define the two extreme options of a fully static solution, in which we do not allow any changes, and a fully dynamic one, which does not integrate between solutions to avoid overpaying for change costs and opening costs.

In Fig. 2 shows the cost percentage of the two versions of the static algorithm compared to the cost of the dynamic algorithm. It can be seen that the intuition for the performance of the algorithm is correct, that is, for small change costs (relative to the connection cost), ignoring it results in higher costs, and for high change costs, it does not necessarily come with a cost. Still, giving consideration to the dynamic nature of the problem does give advantage to the dynamic algorithm, and for high change costs, the performance of the static algorithm and the dynamic algorithm, converge to one another.

In Sect. 2.3 we discussed the interval selection procedure. For each client, we found the time steps so that in between substantial fractional changes in the

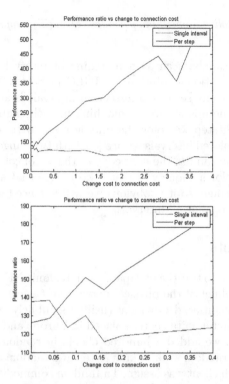

Fig. 2. Performance ratio of the dynamic UNFV algorithm compared to the static UNFV with respect to the ratio between the average connection cost and the change cost. Top presents the full graph while the bottom zooms in on the smaller values.

assignment were accumulated, and we split the time horizon into intervals at these time steps. The question of at which point the "right amount" of fractional change has accumulated arises. We ran the experiments with a small change cost (0.04 of the average connection cost) and a large change cost (which equals the average connection cost). As seen in Fig. 3, this value, denoted by θ, may have significant impact on the performance of the algorithm. If we choose a value too big, we may induce too many assignment changes as we break the time horizon into too many intervals. And for small values of θ, we may end up with a static solution which may lose the advantage of the dynamic algorithm. Another evidence for this can be found in the comparison with the fractional solution. For values of θ in the range $[0.4, 0.6]$, the ratio between the fractional solution and the algorithm's solution is bigger. Usually the optimal solution and the fractional solution are not close. This leads us to assume that in practice, the ratio between the algorithm's solution and the optimal solution is better than 2.

Lastly, in Fig. 4 we can see the performance ratio of the dynamic algorithm as a function of the expected number of commodities that can be installed in each facility. The performance of the algorithm peaks as the size constraint loosens. The lower performance ratio may be the result of tight size constraint which

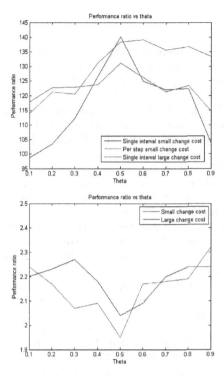

Fig. 3. Performance ratio of the dynamic UNFV algorithm compared to the static UNFV and the fractional solution with respect to theta.

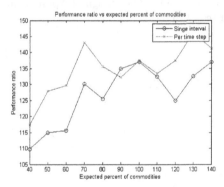

Fig. 4. Performance ratio of the dynamic UNFV algorithm compared to the static UNFV with respect to the facility size.

creates a hard problem without room for much improvement. On the other hand, when the sizes of the facilities are very large, we may install each commodity at several facilities to allow more assignment changes.

References

1. ONAP - Open Network Automation Platform. https://www.onap.org/
2. Gember-Jacobson, A., et al.: OpenNF: enabling innovation in network function control. In: Proceedings of the 2014 ACM Conference on SIGCOMM, SIGCOMM 2014, pp. 163–174. ACM, New York (2014)
3. Bremler-Barr, A., Harchol, Y., Hay, D.: OpenBox: a software-defined framework for developing, deploying, and managing network functions. In: Proceedings of the ACM SIGCOMM 2016 Conference on SIGCOMM (2016)
4. Biro, M., Hujter, M., Tuza, Z.: Precoloring extension. I. Interval graphs. Discrete Math. **100**(1–3), 267–279 (1992)
5. Cohen, R., Lewin-Eytan, L., Naor, J.S., Raz, D.: Near optimal placement of virtual network functions. In: 2015 IEEE Conference on Computer Communications, INFOCOM, pp. 1346–1354. IEEE (2015)
6. Eisenstat, D., Mathieu, C., Schabanel, N.: Facility location in evolving metrics. In: Esparza, J., Fraigniaud, P., Husfeldt, T., Koutsoupias, E. (eds.) ICALP 2014. LNCS, vol. 8573, pp. 459–470. Springer, Heidelberg (2014). https://doi.org/10.1007/978-3-662-43951-7_39
7. ETSI Industry Specification Group (ISG) Network Functions Virtualisation (NFV): Network functions virtualisation. http://www.etsi.org/technologies-clusters/technologies/nfv
8. Feng, H., Llorca, J., Tulino, A., Raz, D., Molisch, A.: Approximation algorithms for the NFV service distribution problem. In: 2017 IEEE Conference on Computer Communications, INFOCOM 2017 (2017)
9. Ravi, R., Sinha, A.: Multicommodity facility location. In: Proceedings of the Fifteenth Annual ACM-SIAM Symposium on Discrete Algorithms, pp. 342–349. Society for Industrial and Applied Mathematics (2004)
10. Williamson, D.P., Shmoys, D.B.: The Design of Approximation Algorithms. Cambridge University Press, Cambridge (2011)

Longest Increasing Subsequence
Under Persistent Comparison Errors

Barbara Geissmann$^{(\boxtimes)}$

Department of Computer Science, ETH Zurich, Zurich, Switzerland
`barbara.geissmann@inf.ethz.ch`

Abstract. We study the problem of computing a *longest increasing subsequence* in a sequence S of n distinct elements in the presence of *persistent* comparison errors. In this model, (Braverman and Mossel, *Noisy sorting without resampling*, SODA, 2008) every comparison between two elements can return the wrong result with some fixed (small) probability p, and comparisons cannot be repeated. Computing the longest increasing subsequence exactly is impossible in this model, therefore, the objective is to identify a subsequence that (i) is indeed increasing and (ii) has a length that approximates the length of the longest increasing subsequence.

We present asymptotically tight upper and lower bounds on both the approximation factor and the running time. In particular, we present an algorithm that computes an $O(\log n)$-approximation in time $O(n \log n)$, with high probability. This approximation relies on the fact that we can approximately sort (Geissmann, Leucci, Liu, and Penna, *Optimal Sorting with Persistent Comparison Errors*, ArXiv e-prints 1804.07575, 2018) n elements in $O(n \log n)$ time such that the maximum dislocation of an element is at most $O(\log n)$. For the lower bounds, we prove that (i) there is a set of sequences, such that on a sequence picked randomly from this set every algorithm must return an $\Omega(\log n)$-approximation with high probability, and (ii) any $O(\log n)$-approximation algorithm for longest increasing subsequence requires at least $\Omega(n \log n)$ comparisons, even in the absence of errors.

Keywords: Longest increasing subsequence
Probabilistic persistent comparison errors · Approximation algorithm
Lower bounds

1 Introduction

When dealing with complex systems and large volumes of information, it is often the case that at least part of the involved data will be inconsistent. These inconsistencies can be *intrinsic*, i.e., they might shed from the fact that the data

---

Research supported by SNF (project number 200021_165524).

L. Epstein and T. Erlebach (Eds.): WAOA 2018, LNCS 11312, pp. 259–276, 2018.
https://doi.org/10.1007/978-3-030-04693-4_16

is obtained from an inherently *noisy* source (this is typically the case in human-produced data), or they might be the result of corruptions caused by random errors (think, for instance, of random memory faults or communication errors). It is therefore important to understand how the classical techniques used to solve basic algorithmic problems can cope with such errors.

In this paper, we consider the problem of computing a *longest increasing subsequence LIS(S)* in a given sequence S of distinct elements –a fundamental task that appears naturally in many areas, such as in probability theory and combinatorics [2,5], scheduling [4,24], and computational biology [11,26]– in presence of *random persistent comparison errors*.

In this model, every comparison between two elements is wrong with some small fixed probability p, and correct with probability $1 - p$. The comparison results are independent over all pairs of elements, and comparisons cannot be repeated. Note that this is equivalent to say that repeating the same comparison multiple times yields each time the same result. Hence, comparison results are persistent: always wrong or always correct. Furthermore, we assume that we cannot inspect the values of the elements, but only use such element comparisons. Because of these comparison errors, it is impossible to compute $LIS(S)$ correctly, instead, we seek to return a sequence that (i) is indeed increasing and that (ii) has some guaranteed minimum length depending on the length of the longest increasing sequence $l := |LIS(S)|$. In particular, we are interested in algorithms that return an increasing sequence of length at least $\frac{1}{r} \cdot l$, where r is the *approximation factor*.

This error model has been first employed by Braverman and Mossel [7], who studied the problem of sorting. Other work on sorting followed (see [15,16,22]) and the model has been studied also for finding the minimum, searching, and linear programming in two dimensions [22]. In this paper, we will present an algorithm that returns an $O(\log n)$-approximation on the longest increasing subsequence in $O(n \log n)$ time, with high probability. Moreover, we will prove that this approximation factor is the best possible as $\Omega(\log n)$ is also a lower bound, regardless of the running time, and that any $(\log n)$-approximation algorithm requires $\Omega(n \log n)$ comparisons, even in the absence of comparison errors.

1.1   Related Work

There are several algorithms to compute a longest increasing subsequence of a sequence S, if no comparison errors happen. Typically, they are based on a common underlying algorithmic idea: They process the elements one by one and maintain for each length found so far the increasing subsequence of this length that ends with the smallest possible element seen so far. We shall call this algorithmic idea the *Core-Algorithm* to compute a longest increasing subsequence. The running time of the Core-Algorithm is $O(n \log n)$ in the decision-tree model (see for instance [6,8,13]). This time complexity is tight, as shown in [13]. In the RAM model, where one can also inspect the values, the algorithm can be implemented to run in $O(n \log \log n)$ time [9,25]. All the results can be parameterized to $O(n \log l)$ or $O(n \log \log l)$, respectively, where l is the length of the longest increasing subsequence.

The longest increasing subsequence of S is also the *longest common subsequence* between S and the sorted sequence of the elements in S. This implies an $O(n^2)$ time (or $O(n^2/\log n)$ time if optimized) algorithm to find the longest increasing subsequence when using the standard dynamic programming technique that is used to find longest common subsequences [13,23].

The model with random persistent comparison errors has been extensively studied for finding the smallest element, for searching, and for sorting (see for instance [7,15,16,22]). A common way to measure the quality of an output sequence in terms of sortedness, is to consider the *dislocation* of the elements. The dislocation of an element is the absolute difference between its position in the output sequence and its position in the correctly sorted sequence (its rank). Typically, one considers the *maximum dislocation* of any element in the output sequence and the *total dislocation* (the sum of the dislocations of all elements). It has been shown for instance in [17], that there is an algorithm with running time $O(n \log n)$ which achieves simultaneously maximum dislocation $O(\log n)$ and total dislocation $O(n)$ with high probability, and that this is indeed the best one can hope for (i.e., there exist matching lower bounds that show that no possibly randomized algorithm can sort such that, with high probability, the maximum dislocation is $o(\log n)$ or the total dislocation is $o(n)$). A maximum dislocation of $O(\log n)$ implies the following: on the positive side, it is possible to derive the correct relative order of two elements whose ranks differ by at least $\Omega(\log n)$; on the negative side, this is not possible for two elements whose ranks differ by less than $O(\log n)$. The results on the maximum dislocation of sorting are of interest for the problem of finding the longest increasing subsequence, because an increasing subsequence is also a sorted subsequence.

An easier variant is a model with *non-persistent* comparison errors, where repeating a comparison can yield different results. In this model, one can sort in $O(n \log(n/q))$ time, where $1 - q$ is the success probability of the algorithm (see for instance [12]). The impact of such errors on classical sorting algorithms such as Insertionsort, Quicksort, and Mergesort have been analyzed in [3,19–21]. Other models restrict the comparisons in which errors can happen. For instance, [1] gives a sorting algorithm when errors occur only between elements whose difference is at most some fixed threshold, and [10] provides an algorithm when the total number of errors is known in advance.

1.2 Our Contribution

We prove asymptotically tight upper and lower bounds on both the approximation factor and the running time for longest increasing subsequence under persistent comparison errors. For the upper bounds, we define an *Approximation-Algorithm* that computes an $O(\log n)$-approximation to the longest increasing subsequence of S. In fact, it even finds the longest possible increasing subsequence under the implication that we cannot sort better than obtaining an order with maximum dislocation $O(\log n)$. Formally, we prove the following result:

Theorem 1 (Upper Bounds). *For any sequence S that contains n distinct elements, our Approximation-Algorithm computes an $O(\log n)$-approximation to the longest increasing sequence of S, in $O(n \log n)$ time, with probability at least $1 - \frac{1}{n}$.*

This result on the upper bound can be generalized to other error models. In fact, if we are given or able to obtain an approximately sorted sequence with maximum dislocation d, then our Approximation-Algorithm will return a $2d$-approximation to the longest increasing subsequence. We discuss this point in the Conclusion (Sect. 6).

To prove our lower bound on the approximation factor of any algorithm solving $LIS(S)$ under persistent comparison errors with high probability, we will identify a small collection of sequences that contain a longest increasing sequence of size $\Theta(\log n)$ and that are likely to look the same in our error model. Then, we show for any algorithm that if it *succeeds* on one sequence of this collection by returning a constant number of elements of this increasing sequence it must fail on another sequence. In particular, we will prove the following theorem:

Theorem 2 (Lower Bound – Approximation Factor). *There exists a collection of sequences $\mathcal{S}$ (permutations of length n) and a probability distribution on $\mathcal{S}$, such that no algorithm can return an $O(\log n)$-approximation (for s suitable hidden constant that depends on p) of the longest increasing subsequence with probability $1 - \frac{1}{n}$.*

We prove a lower bound of $\Omega(n \log n)$ on the number of comparisons (which is also a lower bound on the running time) needed to compute an $O(\log n)$-approximation by considering the easier case in which all comparisons are correct, and by adapting the techniques used in [13] for proving a similar lower bound for exact (i.e., 1-approximate) algorithms:

Theorem 3 (Lower Bound – Running Time). *Any $(\log n)$-approximation algorithm for longest increasing subsequence requires $\Omega(n \log n)$ comparisons, even if no errors occur.*

2   Preliminaries

Since we assume that all elements in the input sequence $S = \langle s_1, s_2, \ldots, s_n \rangle$ are distinct, we can also assume, for easier analysis and readability, that S is a permutation of the numbers (elements) $\{1, \ldots, n\}$. By our error model, the elements in S posses a true linear order, i.e., $S^{sort} := \langle 1, \ldots, n \rangle$, however, this order can only be observed through erroneous comparisons.

For two distinct elements x and y, we will write $x < y$ to denote that x is smaller than y according to the true linear order (resp. $x > y$ to denote that x is larger than y according to the true linear order), and we will write $x \prec y$ (resp. $x \succ y$) to mean that x is observed to be smaller (resp. larger) than y in the comparison result. For a given sequence S and an element $x \in S$, we define $\mathrm{rank}(x, S) = 1 + |\{y \in S : y < x\}|$ to be the *true rank* of element x

in S (note that ranks start from 1), and we define $\text{pos}(x, S) \in [1, |S|]$ to be the *position* of x in S (positions also start from 1). The dislocation of x in S is then $\text{disl}(x, S) = |\text{pos}(x, S) - \text{rank}(x, S)|$, and the *maximum dislocation* of S is $\text{disl}(S) = \max_{x \in S} \text{disl}(x, S)$. For a given sequence S, we let $C \in \{\prec, \succ\}^{\binom{n}{2}}$ denote the comparison outcomes that we can observe. For $C = \langle c_1, \ldots, c_{\binom{n}{2}} \rangle$, this means that if $c_k = c_{(i-1)n+j} = \text{``}\prec\text{''}$ with $1 \leq i < n$ and $i < j \leq n$, then $s_i \prec s_j$ (resp. $s_i \succ s_j$ if $c_k = c_{(i-1)n+j} = \text{``}\succ\text{''}$). Finally, for $z \in \mathbb{R}$, we write $\log z$ for the binary logarithm of z.

We continue the preliminaries with some results on sorting that we will use to prove our upper bound on the approximation factor.

Theorem 4 (Theorem 3 in [17]). *There is an algorithm that approximately sorts, in $O(n \log n)$ worst-case time, n elements subject to random persistent comparison errors so that the maximum dislocation of the resulting sequence is $O(\log n)$, with probability $1 - \frac{1}{n}$.*

Lemma 5. *Let $S^{apx} = \langle apx_1, apx_2, \ldots, apx_n \rangle$. If $\text{disl}(S^{apx}) \leq d$, then for $1 \leq i < n - 2d$, apx_i and apx_{i+2d} are in correct relative order: $\text{pos}(apx_i, S^{sort}) < \text{pos}(apx_{i+2d}, S^{sort})$.*

Proof. Since the maximum dislocation in S^{apx} is at most d, $\text{pos}(apx_i, S^{sort}) \in \{i - d, \ldots, i + d\}$ and $\text{pos}(apx_{i+2d}, S^{sort}) \in \{i + d, \ldots, i + 3d\}$. These intervals intersect in at most one position, and the claim follows since no two elements can appear in the same position. □

3   Upper Bound and Approximation-Algorithm

We will modify the so-called Core-Algorithm (as named in Sect. 1.1, Related work) that computes a longest increasing subsequence in the absence of comparison errors, such that it computes an $O(\log n)$-approximation with high probability in our error model. Before we do so, we first show that it is possible to identify a $2d$-approximation by looking at S and a sequence S^{apx} with maximum dislocation d. Since we can sort such that the maximum dislocation is $O(\log n)$ (see Theorem 4), this implies an $O(\log n)$-approximation on $LIS(S)$.

3.1   Upper Bound

The proof of the upper bound is based on the following fact and observation:

- Without any comparison errors, the problem of finding $LIS(S)$ is equivalent to the problem of finding a *longest common subsequence* between S and S^{sort}, where S^{sort} is the correctly sorted order of the elements in S.
- This leads to the following observation. Let S^{apx} be the sequence obtained from approximately sorting S with comparison errors and consider now S^{apx} as the total order over all elements, i.e., for each pair of elements, their comparison result is *redefined* as their relative order in S^{apx}. Furthermore, let A

be any algorithm that solves $LIS(S)$ in the absence of errors. If A uses the redefined comparison results, it computes the *longest common subsequence* $LCS(S, S^{apx})$ between S and S^{apx}.

The immediate idea of computing $LCS(S, S^{apx})$ comprises some difficulties, since this subsequence is not necessarily increasing and, on top of that, its length might be smaller than $|LIS(S)|$. However, we can still get a first approximation. Assume that S^{apx} has maximum dislocation at most d. Lemma 5 implies that we obtain an increasing subsequence when taking every $2d$-th element of $LCS(S, S^{apx})$. And the maximum dislocation implies that the elements in the subset containing every $2d$-th element of $LIS(S)$ appear in the same relative order in S^{apx}, thus $|LCS(S, S^{apx})| \geq \frac{1}{2d}|LIS(S)|$. When put together, we get a $4d^2$-approximation.

This approximation factor can be improved, and it turns out that considering common subsequences whose elements lie (at least) $2d$ positions apart in S^{apx} is actually a good start: By Lemma 5, a common subsequence between S and S^{apx} is *increasing* if for every pair of adjacent elements in this subsequence their positions in S^{apx} differ by at least $2d$. Therefore, we say that a sequence $S' = (s'_1, s'_2, \ldots, s'_m)$ is $2d$-*distant* in S^{apx} if

$$\text{pos}(s'_i, S^{apx}) + 2d \leq \text{pos}(s'_{i+1}, S^{apx}) \qquad \text{for } 1 \leq i < m. \tag{1}$$

Notice that any (increasing) subsequence of S that is $2d$-distant in S^{apx} is automatically also a common (increasing) subsequence of S and S^{apx}. This observation suggests the following easy recipe to obtain a $2d$-approximation on longest increasing subsequence:

– First, partition the elements into $2d$ subsets, such that every $2d$-th element in S^{apx} gets into the same subset, and obtain $2d$ input subsequences based on this partition.
– Then, on every input subsequence, run any algorithm that computes a longest increasing subsequence if no comparison errors happen, and return the longest result.

By pigeon hole principle and since every input subsequence is now $2d$-distant in S^{apx}, the longest result must be a $2d$-approximation on $|LIS(S)|$. This recipe however is not optimal in the sense that in many cases, we could do better and find a longer subsequence in S that is still $2d$-distant in S^{apx}. In fact, we lose up to a factor $2d$ in the case where $LIS(S)$ is already $2d$-distant in S^{apx}, but these elements are equally distributed among all input subsequences. For this reason, we will define an approximation algorithm that finds the longest increasing subsequence in S that is $2d$-distant in S^{apx}. We conclude this section with the obvious lemma.

Lemma 6. *The longest subsequence S^* of S that is $2d$-distant in S^{apx} has length at least*

$$|S^*| \geq \frac{1}{2d}|LIS(S)|.$$

Algorithm 1. $Core\text{-}Algorithm(S = \langle s_1, \ldots, s_n \rangle)$

---

1 $L[1] \longleftarrow s_1$;
2 $k \longleftarrow 1$;
3 **foreach** $i = 2, \ldots, n$ **do**
4 $\quad x \longleftarrow s_i$;
5 $\quad$ **if** $x < L[1]$ **then** $L[1] \longleftarrow x$;
6 $\quad$ **else**
7 $\qquad j \longleftarrow \max\{j \le k : L[j] < x\}$;
8 $\qquad$ **if** $j = k$ **then** $k \longleftarrow k + 1$;
9 $\qquad L[j + 1] \longleftarrow x$;
10 $\qquad prec[x] \longleftarrow L[j]$;

11 $lis[1] \longleftarrow L[k]$;
12 **foreach** $i = 2, \ldots, k$ **do**
13 $\quad lis[i] \longleftarrow prec[lis[i - 1]]$
14 **return** lis;

3.2 Approximation-Algorithm

Consider the *Core-Algorithm* described in Algorithm 1 that computes the longest increasing subsequence of the input sequence S in the error-free case. The algorithm processes the input elements one by one, maintaining the longest increasing subsequence found so far. In particular, it maintains a parameter k and an array L, such that k is the length of the longest increasing subsequence found so far and L contains an entry for each length 1 to k, such that $L[i]$ stores the smallest element processed so far that can be at the end of an increasing subsequence of length i.

– The first element is placed to $L[1]$ and k is set to 1.
– Each subsequent element x is placed to $L[j + 1]$, such that j is the largest position where $y = L[j]$ is smaller than x.
– If x is placed to $L[k + 1]$, then k is updated to $k + 1$.
– Whenever a new element x is placed, put a pointer *prec* from x to the element in $y = L[j]$, that, by construction, has a lower value than x.
– In the end, follow these pointers from the top element of the last pile to recover the longest increasing subsequence (in reverse order).

An entry $L[j]$ basically represents the increasing sequence of length j that ends with the smallest possible element processed so far. When an element x is inserted into some position $L[j + 1]$ this means that it is appended to the sequence represented by $L[j]$. Hence, x either increases the longest increasing sequence so far (case $j = k$) or the sequence $L[j + 1]$ gets replaced by this new sequence (case $x < L[j + 1]$).

Our *Approximation-Algorithm*, as described in Algorithm 2, is obtained by modifying the Core-Algorithm such that it works in our error model.

Algorithm 2. *Approximation-Algorithm*$(S = \langle s_1, \ldots, s_n \rangle)$

1 $S^{apx} \longleftarrow$ approximately sort S as shown in [17] ;
2 $d \longleftarrow c \cdot \log n$ // $\exists c$ s.t. w.h.p. disl$(s) \leq c \cdot \log n$ [17];
3 $L[1] \longleftarrow s_1$;
4 $k \longleftarrow 1$;
5 **foreach** $i = 2, \ldots, n$ **do**
6     $x \longleftarrow s_i$;
7   **if** $\mathrm{pos}(x, S^{apx}) < \mathrm{pos}(L[1], S^{apx})$ **then** $L[1] \longleftarrow x$;
8   **else**
9        $j \longleftarrow \max\{j \leq k \colon \mathrm{pos}(L[j], S^{apx}) < \mathrm{pos}(x, S^{apx})\}$;
10   **if** $\mathrm{pos}(L[j], S^{apx}) + 2d \leq \mathrm{pos}(x, S^{apx})$ **then**
11   **if** $j = k$ **then** $k \longleftarrow k + 1$;
12          $L[j + 1] \longleftarrow x$;
13          $prec[x] \longleftarrow L[j]$;

14 $lis[1] \longleftarrow L[k]$;
15 **foreach** $i = 2, \ldots, k$ **do**
16    $lis[i] \longleftarrow prec[lis[i - 1]]$;

17 **return** lis;

- We first approximately sort (using the algorithm from [17], see also Theorem 4 in the current paper) the elements of S to obtain S^{apx}, and we redefine the comparison outcomes based on this total order, i.e., the result of a comparison between two elements now corresponds to their relative order in S^{apx}.
- To compute a suitable subsequence, we change the algorithm so that it remembers the longest $2d$-distant in S^{apx} subsequences instead of the longest increasing subsequences. This implies that an element x is only appended to an (intermediate) subsequence that ends with element y if $\mathrm{pos}(y, S^{apx}) + 2d < \mathrm{pos}(x, S^{apx})$.

For easier analysis, we introduce some additional notation. We call one execution of the lines 5 to 13 of Algorithm 2 an *iteration*, and enumerate them such that element s_i is considered in iteration i. We also say that line 4 corresponds to the first iteration. Furthermore, we denote by L_t and k_t the state and the value of L and k after the t-th iteration, respectively, and for any $j \leq k_t$, we call the subsequence $\langle L_t[j], prec[L_t[j]], prec[prec[L_t[j]]], \ldots \rangle$ with length j the *implied* sequence of $L_t[j]$.

Lemma 7. *For every $t \leq n$, after the t-th iteration of our Approximation-Algorithm, every implied sequence is a subsequence of S that is $2d$-distant in S^{apx}. Moreover, $\langle L_t[1], \ldots, L_t[k_t] \rangle$ is also $2d$-distant in S^{apx}.*

Proof. For any t and $j \leq k_t$, let $S' = \langle s'_1, \ldots, s'_m \rangle$ be the implied sequence of $L_t[j]$. Observe that to every element $s'_i \in S'$, such that $i > 1$, the algorithm has assigned s'_{i-1} as its predecessor. Since the predecessor of any element can only have been processed in an earlier iteration, S' is a subsequence of S.

It follows by induction, that the condition on line 10 in Algorithm 2 ensures that S' is $2d$-distant in S^{apx}: It is trivial to see for $t = 1$, thus, assume that every implied sequence before the t-th iteration is $2d$-distant in S^{apx}. If s_t is inserted into $L[j]$ (nothing changes in the other case), the implied sequence of $L_t[j]$ is equal to s_t appended to the implied sequence of $L_{t-1}[j-1]$ (if it exists). By hypothesis and the condition on line 10, $L_t[j]$ is still $2d$-distant, and since the other implied sequences do not change, the claim also holds after iteration t.

That $\langle L_t[1], \ldots, L_t[k_t] \rangle$ is $2d$-distant in S^{apx} also follows by induction: If $L[j]$ changes (thus $L[j']$ does not change for all $j' \neq j$), then by hypothesis and the conditions in lines 7, 9, and 10, $\mathrm{pos}(L_{t-1}[j-1], S^{apx}) + 2d \leq \mathrm{pos}(L_t[j], S^{apx}) < \mathrm{pos}(L_{t-1}[j], S^{apx}) \leq \mathrm{pos}(L_{t-1}[j+1], S^{apx}) - 2d$ (for all those entries that exist).
□

Lemma 8. *Let $S' = \langle s_1', \ldots, s_m' \rangle$ be the sequence that our Approximation-Algorithm returns. Then, S' is a longest subsequence of S that is $2d$-distant in S^{apx}.*

Proof. Lemma 7 implies that S' is a subsequence of S and $2d$-distant in S^{apx}. Let $S^* = \langle s_1^*, \ldots, s_{m^*}^* \rangle$ be a longest subsequence of S that is $2d$-distant in S^{apx}. We now show that $|S'| \geq |S^*|$. In particular, we show by induction that after iteration t_i^*, $\mathrm{pos}(L_{t_i^*}[i], S^{apx}) \leq \mathrm{pos}(s_i^*, S^{apx})$. For the base case, consider iteration t_1^*, where s_1^* is processed. Either s_1^* gets inserted into some position $j \geq 1$, i.e., $L_{t_1^*}[j] = s_1^*$, or not. If it gets inserted, then by conditions in lines 7 or 9 in Algorithm 2, $\mathrm{pos}(L_{t_1^*}[1], S^{apx}) \leq \mathrm{pos}(s_1^*, S^{apx})$. If not, then it must hold that $L_{t_1^*}[1] = L_{t_1^*-1}[1]$ and thus $\mathrm{pos}(L_{t_1^*}[1], S^{apx}) < \mathrm{pos}(s_1^*, S^{apx})$.

For the step case, consider iteration t_{i+1}^*, where s_{i+1}^* is processed, and observe that the value of k only increases during the algorithm, and for any $t' < t$ and $j \leq k_{t'}$ it holds that $\mathrm{pos}(L_{t'}[j], S^{apx}) \geq \mathrm{pos}(L_t[j], S^{apx})$. Therefore, and by induction hypothesis and the assumption that S^* is $2d$-distant in S^{apx}, $\mathrm{pos}(L_{t_{i+1}^*-1}[i], S^{apx}) + 2d \leq \mathrm{pos}(s_{i+1}^*, S^{apx})$. And Lemma 7 implies, $\mathrm{pos}(L_{t_{i+1}^*-1}[i], S^{apx}) + 2d \leq \mathrm{pos}(L_{t_{i+1}^*-1}[i+1], S^{apx})$. Thus, if s_{i+1}^* does not get inserted, it is because $\mathrm{pos}(L_{t_{i+1}^*-1}[i+1], S^{apx}) < \mathrm{pos}(s_{i+1}^*, S^{apx})$, and if it gets inserted, it will be in some position $j \geq i+1$. In any case, the hypothesis also holds after the iteration iteration t_{i+1}^*, which means that S' has indeed maximum length.   □

3.3   Proof of Theorem 1

We now prove the initially stated Theorem 1, which for convenience, we restate here:

Theorem 1 (Upper Bounds). *For any sequence S that contains n distinct elements, our Approximation-Algorithm computes an $O(\log n)$-approximation of the longest increasing sequence of S, in $O(n \log n)$ time, with probability at least $1 - \frac{1}{n}$.*

Proof. Let $d \in O(\log n)$ according to Theorem 4, such that with probability $1 - \frac{1}{n}$, the maximum dislocation in S^{apx} is at most d. If this is true, by Lemmata 5–8, our Approximation-Algorithm returns a subsequence S' of S that is increasing, and that has length at least $\frac{LIS(S)}{2d} \in \Omega\left(\frac{LIS(S)}{\log n}\right)$.

The running time consists of the initial sorting, which by Theorem 4 takes $O(n \log n)$ time[1], and the n iterations of the algorithm, which take $O(\log n)$ time each if binary search is used to implement line 10. The final construction of the output takes $O(k)$ time, where $k \leq LIS(S) \leq n$ is the length of the approximation.   □

4   Lower Bound on the Approximation Factor

We continue this paper with a lower bound on the approximation factor, that implies that the upper bound we showed in Theorem 1 is tight up to constant factors. In particular, we prove Theorem 2, which we restate here:

Theorem 2 (Lower Bound – Approximation Factor). *There exists a collection of sequences S (permutations of length n) and a probability distribution on S, such that no algorithm can return an $O(\log n)$-approximation (for some suitable hidden constant that depends on p) of the longest increasing subsequence with probability $1 - \frac{1}{n}$.*

Our proof can be seen as a generalization of the lower bound on the maximum dislocation for sorting (see proof of Theorem 9 in [15]), where it is shown that two elements whose ranks differ by less than $O(\log n)$ are likely to be indistinguishable by any algorithm, and hence to appear in the wrong relative order. Intuitively, the argument there is as follows: consider the sorted sequence and the sequence obtained by swapping two elements, and assume that the comparison outcomes on these sequences look identically. It turns out that the probability of this happening is larger than $\frac{1}{n}$, whenever the rank difference is smaller than $O(\log n)$, since only a small number of comparison outcomes must differ.

This is not enough in our case, since an algorithm could simply ignore such two elements. For instance, consider an increasing sequence of c adjacent elements. If the first and the last element are swapped, the algorithm could simply return the subsequence without these two elements and be almost optimal. A first idea to fix this problem could be to consider the case, where one observes the whole increasing sequence to be reversed. However, to have this happen with probability larger than $\frac{1}{n}$, c needs to be smaller than $O(\sqrt{\log n})$, thus implying a weaker lower bound.

Instead, we shall use a collection of similar sequences (more than two), such that if an algorithm *succeeds* on one of these sequences it must fail on another one.

---

[1] By modifying this algorithm so that it returns also the mapping from each element in S to its position in S^{apx} we can obtain the new comparison results in the same time.

Proof. We say that an algorithm *succeeds* if it returns a $(c \log n)$-approximation for any constant $c < \frac{1}{2 \log \frac{1-p}{p}}$, otherwise we say it *fails*. We shall first define our collection $\mathcal{S}$ of similar sequences. Let $\eta := \lceil \frac{\log n}{2 \log \frac{1-p}{p}} \rceil$. Let S^* denote the sequence, in which the largest η elements appear first in increasing order and then the remaining elements appear in decreasing order,

$$S^* := \langle n - \eta + 1, \ldots, n - 1, n, \ n - \eta, \ldots, 1 \rangle.$$

Furthermore, for $1 \leq i < \eta$, let $S_{(i)}$ be the sequence obtained from S^* when the largest element is moved to position i,

$$S_{(i)} := \langle n - \eta + 1, \ldots, n - \eta + (i - 1), n, n - \eta + i, \ldots, n - 1, \ n - \eta, \ldots, 1 \rangle.$$

Now, let $\mathcal{S} := \{S^*, S_{(1)}, S_{(2)}, \ldots, S_{(\eta-1)}\}$ (note that basically $S^* = S_{(\eta)}$) and let $\mathcal{P}$ be the uniform distribution over $\mathcal{S}$. We will show (proof by contradiction) that no algorithm succeeds on this pair $(\mathcal{S}, \mathcal{P})$ with probability at least $1 - \frac{1}{n}$.

Assume towards a contradiction that algorithm A succeeds with high probability on a sequence S' chosen uniformly at random from $\mathcal{S}$, i.e.,

$$\Pr(A(S') \text{ succeeds}) = \sum_{i=1}^{\eta} \Pr(A(S_{(i)}) \text{ succeeds}) \cdot \Pr(S' = S_{(i)}) \geq 1 - \frac{1}{n}.$$

This implies that

$$P := \Pr(A(S^*) \text{ succeeds}) \geq 1 - \frac{\eta}{n}, \tag{2}$$

since by hypothesis and assuming the case where the algorithm succeeds on all the other input sequences (i.e., best case for the algorithm, worst case for the proof), $\frac{P}{\eta} + \frac{\eta-1}{\eta} \geq 1 - \frac{1}{n}$ resolves to (2).

Let $C \in \{\prec, \succ\}^{\binom{n}{2}}$, then $A(S, C)$ means that algorithm A runs on sequence S and observes comparison outcomes C. Now, consider the set of all comparison outcomes that the algorithm can observe and let $\mathcal{C} := \{C \in \{\prec, \succ\}^{\binom{n}{2}} : A(S^*, C) \text{ succeeds}\}$ denote the set of all possible comparison outcomes for which A succeeds on input S^*. We define $R(S) \in \{\prec, \succ\}^{\binom{n}{2}}$ to be the random variable corresponding to the comparison outcomes as they would be observed by the algorithm when the input sequence is S. Then, the probability that $A(S^*)$ succeeds is expressed by the total probabilities of the events that A observes comparison outcomes in $\mathcal{C}$,

$$P = \Pr(A(S^*) \text{ succeeds}) = \sum_{C \in \mathcal{C}} \Pr(R(S^*) = C). \tag{3}$$

Before we continue the proof, we shall first show the following lemma.

Lemma 9. $\forall S \in \mathcal{S} \setminus \{S^*\}$ *and* $C \in \{\prec, \succ\}^{\binom{n}{2}}$, $\Pr(R(S) = C) > \Pr(R(S^*) = C) \cdot \left(\frac{p}{1-p}\right)^{\eta}$.

Proof. Consider $S^* = \langle s_1^*, \ldots, s_n^* \rangle$ and C and let $E(S^*, C)$ be the set of wrong comparison results, i.e., the set of pairs (s_i^*, s_j^*) with $i < j$ such that either $s_i^* < s_j^*$ and $c_{(i-1)n+j} = $ "$\succ$" (i.e., $s_i^* \succ s_j^*$) or $s_i^* > s_j^*$ and $c_{(i-1)n+j} = $ "$\prec$". Thus,

$$\Pr(R(S^*) = C) = (1-p)^{\binom{n}{2} - |E(S^*, C)|} \cdot p^{|E(S^*, C)|} = (1-p)^{\binom{n}{2}} \cdot \left(\frac{p}{1-p} \right)^{|E(S^*, C)|}.$$

Now consider $S = S_{(k)} = \langle s_1, \ldots, s_n \rangle$ and observe that only the relative order of the pairs (s_k, s_j) with $k < j \leq \eta$, changed compared to S^*. This implies that there can be at most $\eta - k < \eta$ additional wrong comparison results, i.e., $|E(S, C)| < |E(S^*, C)| + \eta$. Therefore, and since $\frac{p}{1-p} \leq 1$,

$$\Pr(R(S) = C) = (1-p)^{\binom{n}{2}} \cdot \left(\frac{p}{1-p} \right)^{|E(S, C)|}$$

$$> (1-p)^{\binom{n}{2}} \cdot \left(\frac{p}{1-p} \right)^{|E(S^*, C)| + \eta} = \Pr(R(S^*) = C) \cdot \left(\frac{p}{1-p} \right)^{\eta}.$$

$\square$

Continuation of the Proof of Theorem 2. Now notice that in order to succeed, A needs to return at least two of the first η elements in S^*. Therefore, we can map every $C \in \mathcal{C}$ to a (not necessarily unique) sequence of $\mathcal{S}$ as follows: for each $C \in \mathcal{C}$, let i_C be the position of the first element that $A(S^*, C)$ returns and let $S(C) := S_{(i_C)}$. (Note that $i_C < \eta$ as otherwise A does not return at least two elements of the first η elements in S^*.) For each $S \in \mathcal{S} \setminus \{S^*\}$,

$$\Pr(A(S) \text{ fails}) \geq \sum_{C \in \mathcal{C}:\, S = S(C)} \Pr(R(S) = C)$$

$$> \sum_{C \in \mathcal{C}:\, S = S(C)} \Pr(R(S^*) = C) \cdot \left(\frac{p}{1-p} \right)^{\eta}.$$

And as a consequence, for $S' \in \mathcal{S}$ chosen uniformly at random,

$$\Pr(A(S') \text{ fails}) \geq \sum_{S \in \mathcal{S} \setminus \{S^*\}} \Pr(S' = S) \cdot \Pr(A(S) \text{ fails})$$

$$> \sum_{S \in \mathcal{S} \setminus \{S^*\}} \frac{1}{\eta} \sum_{C \in \mathcal{C}:\, S = S(C)} \Pr(R(S^*) = C) \cdot \left(\frac{p}{1-p} \right)^{\eta}$$

$$= \frac{1}{\eta} \left(\frac{p}{1-p} \right)^{\eta} \sum_{S \in \mathcal{S} \setminus \{S^*\}} \sum_{C \in \mathcal{C}:\, S = S(C)} \Pr(R(S^*) = C)$$

$$\geq \frac{1}{\eta} \left(\frac{p}{1-p} \right)^{\eta} \sum_{C \in \mathcal{C}} \Pr(R(S^*) = C) \geq \frac{1}{\eta} \left(\frac{p}{1-p} \right)^{\eta} \left(1 - \frac{\eta}{n} \right),$$

where from line 3 to line 4 we use that every instance of comparison results is mapped to exactly one sequence, and on the last line we use Eqs. (2) and (3).

Now, observe that for n large enough, $\left(1 - \frac{\eta}{n}\right) > \frac{1}{2}$ and that, by our choice of η, $\left(\frac{p}{1-p}\right)^\eta \geq \frac{1}{\sqrt{n}}$. Therefore,

$$\Pr(A(S') \text{ fails}) > \frac{2 \log \frac{1-p}{p}}{\log n} \cdot \frac{1}{\sqrt{n}} \cdot \frac{1}{2} > \frac{1}{n}.$$

To conclude the proof, note that this is a contradiction to our assumption that A succeeds with high probability. □

The lower bound shown in Theorem 2 holds for all deterministic algorithms, but can be expanded to also hold for probabilistic algorithms as explained in the following remark.

Remark 10. To make the lower bound on the approximation factor work also for any randomized algorithm A, we can turn A into a deterministic version by fixing a sequence $\lambda \in \{0,1\}^t$ random bits that can be used by the algorithm. Thus, for the resulting deterministic algorithm A_λ, the lower bound holds. Let p_λ be the probability to generate the sequence λ of random bits. To lower bound the probability that $A(S')$ fails, where S' is chosen uniformly at random from $\mathcal{S}$, one simply needs to sum over all λ the probabilities that A_λ fails multiplied by p_λ, i.e., $\Pr(A(S') \text{ fails}) = \sum_{\lambda \in \{0,1\}^t} \Pr(A_\lambda(S') \text{ fails}) \cdot p_\lambda \geq \frac{1}{n} \sum_{\lambda \in \{0,1\}^t} p_\lambda = \frac{1}{n}$.

5   Lower Bound on the Running Time

We complement this paper by showing that the running time of our Approximation-Algorithm is asymptotically optimal. In [13], it is shown that (in the error-free model) computing the longest increasing subsequence is at least as hard as sorting. We will use this proof to informally show Theorem 3 which we restate here:

Theorem 3 (Lower Bound – Running Time). *Any* $\log n$-*approximation algorithm for longest increasing subsequence requires* $\Omega(n \log n)$ *comparisons, even if no errors occur.*

The proof techniques of the lower bound in [13] are as follows: Assume that we are in the error-free case. Consider the easier problem of deciding on a given sequence S of n distinct elements whether $|LIS(S)| < k$, and consider the comparison tree of an algorithm A with leaves that tell as an answer to this question either "yes" or "no". Without loss of generality, assume that no useless comparisons are made on a root to a leaf path (i.e., no comparison twice and no comparisons whose outcome is predictable by the outcomes of previous comparisons).

Every leaf ℓ can be associated with a partial order implied by a set of linear orderings on S that are consistent with the transitive closure of the comparisons performed on the path from the root to ℓ. If the answer in a leaf is "yes", this implies that there are no k elements of S that are pairwise incomparable in this partial order (i.e., the relative order of every pair is neither tested in any comparison on the path, nor implied by other comparisons), as otherwise, these elements could possibly form an increasing sequence of length k. Such a subset of elements is called *antichain*, while a *chain* is a subset of elements that are linearly ordered. An important property of chains and antichains used in the proof is based on the so-called Dillworth theorem:

Lemma 4 (Lemma 3.1 in [13]). *In any finite partial order, the elements can be partitioned into m chains, where m is the size of the largest antichain.*

This implies that in a "yes"-leaf, the elements can be partitioned into less than k chains, since there is no antichain of size k. Furthermore, given such a partition into (less than) k chains, the elements can be sorted with $n \log k + O(n)$ comparisons, think for instance of natural merge sort:

Lemma 5 (Lemma 3.3 in [13]). *If a linear order is partitioned into k chains, then this linear order can be algorithmically restored with at most $n\lceil \log k \rceil$ comparisons.*

In order to lower bound the number of comparisons needed to end in a "yes"-leaf, algorithm A can be extended to A^* as follows: whenever A concludes to be in a "yes"-leaf, A^* continues to completely sort the elements of S (which requires no more than $n \log k + O(n)$ further comparisons). Let $S(n, k)$ denote the number of linear orderings of the elements in S that end in a "yes"-leaf, i.e., the number of linear orderings such that the longest increasing subsequence in S is strictly smaller than k. Then,

$$S(n, k) \geq n! \left(1 - \frac{\binom{n}{k}}{k!} \right),$$

since there are $n!$ different linear orderings and $\binom{n}{k}$ possible subsequences of size k each increasing with probability $1/k!$. The comparison tree corresponding to A^* has thus at least $S(n, k)$ leaves, and therefore must perform at least $\log S(n, k)$ comparisons in its worst case. Therefore and by Lemma 5, algorithm A must perform at least

$$\log S(n, k) - n \log k - O(n)$$

comparisons in its worst case to end up in a "yes"-leaf, which is $\Omega(n \log n)$ when choosing $k = 3 \cdot \sqrt{n}$, since in this case $0 < \binom{n}{k}/k! < 1$, and therefore $S(n, k) \sim n!$ (see also Theorem 3.5 in [13]).

We can use the above proof techniques to show that every algorithm, that computes a $\log n$-approximation on longest increasing subsequence must perform at least $\Omega(n \log n)$ comparisons.

Proof (of Theorem 3). Let B be an $\log n$-approximation algorithm for $LIS(S)$ under our error model (i.e., we can always simulate our error model in the error-free case) and consider a relaxation of the problem of determining whether $|LIS(S)|$ is smaller than $k \log n$. In this relaxation we require the answer to be "yes" (resp. "no") if $|LIS(S)| < k$ (resp. $|LIS(S)| \geq k \log n$), while we do not impose any restriction on the range $k \leq |LIS(S)| < k \log n$.

It is clear that algorithm B can be used to solve this relaxed problem without increasing the number of needed comparisons. Therefore, the associated comparison tree must reach a leaf corresponding to answer "yes" for all linear orderings on the elements in S that contain no increasing subsequence of length k, while the largest antichain in any such an ordering is smaller than $k \log n$. This implies, by using Lemmata 4 and 5, that algorithm B^* (now in the error-free case) needs at least $n \log(k \log n) + O(n)$ further comparisons in the worst case to sort the elements in S, and B thus needs at least

$$\log S(n, k) - n \log(k \log n) - O(n)$$

comparisons in the worst case to end in a "yes"-leaf, which is in $\Omega(n \log n)$ if we set again $k = 3 \cdot n^{1/2}$. This step follows since by Stirling's approximation $k! \geq (k/e)^k$, and thus

$$S(n, 3 \cdot n^{1/2}) \geq n! \left(1 - \frac{n!}{(n - 3 \cdot n^{1/2})!(3 \cdot n^{1/2})!(3 \cdot n^{1/2})!}\right)$$

$$\geq n! \left(1 - \frac{n^{3 \cdot n^{1/2}}}{\left(\frac{9n}{e^2}\right)^{3 \cdot n^{1/2}}}\right)$$

$$\geq n! \left(1 - 1.2^{-3 \cdot n^{1/2}}\right),$$

which is larger than $1/2 \cdot n!$ for $n \geq 2$. Therefore,

$$\log S(n, 3 \cdot n^{1/2}) \geq \log \frac{n!}{2} \geq n \log n - O(n),$$

while

$$n \log(3 \cdot n^{1/2} \log n) \leq \frac{1}{2} n \log n + O(n \log \log n).$$

$\square$

Finally, we can conclude that our Approximation-Algorithm performs in asymptotically optimal time, since we can always simulate our error model in the error-free case.

6   Conclusion

Although a logarithmic approximation ratio might not seem very exciting at first glance, it turns out that this is the best one that can be obtained in the presence of persistent comparison errors. In this respect, it is interesting to see that there exist such simple recipes to compute a logarithmic approximation.

We have seen in the very beginning one recipe that can use as a black box any algorithm that computes a longest increasing sequence if no comparison errors happen. And we have seen afterwards another recipe that dives into such an algorithm and changes the rule of when to add (or append) an element to a previously computed increasing subsequence. Note that this approach also works for the very similar *patience sort* algorithm. As indicated earlier, our Approximation-Algorithm has the advantage, that it performs much better than $O(\log n)$-approximate on many input sequences and is even optimal in the case where the longest increasing subsequence is already $2d$-distant in S^{apx}, whereas this is not necessarily true when using the black-box recipe. Moreover, it is easy to observe that the Approximation-Algorithm is never worse than the other.

Finally, we would like to explain how the upper bound on the approximation factor can be generalized. Our Approximation-Algorithm actually succeeds whenever the approximately sorted sequence has maximum dislocation at most d. This implies that the result can be parametrized and also used in other models with comparison comparison errors.

- Whenever one can obtain a total order with maximum dislocation d, the Approximation-Algorithm is $2d$-approximative.

Consider for instance the so-called *threshold*-model [1,14,18], where comparisons between numbers that differ by more than some threshold τ are always correct, while those between numbers that differ by less than τ can fail persistently (with some probability possibly depending on the difference or even adversarially). If the input sequence S is a permutation of the numbers $\{1\ldots,n\}$, running Quicksort in this error model yields a sequence with maximum dislocation 2τ (see [18]). Thus, our Approximation-Algorithm finds a 4τ-approximation of the longest increasing subsequence in S.

References

1. Ajtai, M., Feldman, V., Hassidim, A., Nelson, J.: Sorting and selection with imprecise comparisons. ACM Trans. Algorithms **12**(2), 19 (2016)
2. Aldous, D., Diaconis, P.: Longest increasing subsequences: from patience sorting to the Baik-Deift-Johansson theorem. Bull. Am. Math. Soc. **36**(4), 413–432 (1999)
3. Alonso, L., Chassaing, P., Gillet, F., Janson, S., Reingold, E.M., Schott, R.: Quicksort with unreliable comparisons: a probabilistic analysis. Comb., Probab. Comput. **13**(4–5), 419–449 (2004)
4. Bachmat, E., Berend, D., Sapir, L., Skiena, S., Stolyarov, N.: Analysis of aeroplane boarding via spacetime geometry and random matrix theory. J. Phys. A: Math. Gen. **39**(29), L453 (2006). http://stacks.iop.org/0305-4470/39/i=29/a=L01
5. Baik, J., Deift, P., Johansson, K.: On the distribution of the length of the longest increasing subsequence of random permutations. J. Am. Math. Soc. **12**(4), 1119–1178 (1999)
6. Bespamyatnikh, S., Segal, M.: Enumerating longest increasing subsequences and patience sorting. Inf. Process. Lett. **76**(1–2), 7–11 (2000). https://doi.org/10.1016/S0020-0190(00)00124-1

7. Braverman, M., Mossel, E.: Noisy sorting without resampling. In: Proceedings of the Nineteenth Annual ACM-SIAM Symposium on Discrete Algorithms, SODA 2008, San Francisco, California, USA, 20–22 January 2008, pp. 268–276 (2008). http://dl.acm.org/citation.cfm?id=1347082.1347112

8. Chandramouli, B., Goldstein, J.: Patience is a virtue: revisiting merge and sort on modern processors. In: International Conference on Management of Data, SIGMOD 2014, Snowbird, UT, USA, 22–27 June 2014, pp. 731–742 (2014). https://doi.org/10.1145/2588555.2593662

9. Crochemore, M., Porat, E.: Fast computation of a longest increasing subsequence and application. Inf. Comput. **208**(9), 1054–1059 (2010). https://doi.org/10.1016/j.ic.2010.04.003

10. Damaschke, P.: The solution space of sorting with recurring comparison faults. In: Proceedings of the 27th International Workshop on Combinatorial Algorithms, IWOCA 2016, Helsinki, Finland, 17–19 August 2016, pp. 397–408 (2016). https://doi.org/10.1007/978-3-319-44543-4_31

11. Delcher, A.L., Kasif, S., Fleischmann, R.D., Peterson, J., White, O., Salzberg, S.L.: Alignment of whole genomes. Nucleic Acids Res. **27**(11), 2369–2376 (1999). https://doi.org/10.1093/nar/27.11.2369

12. Feige, U., Raghavan, P., Peleg, D., Upfal, E.: Computing with noisy information. SIAM J. Comput. **23**(5), 1001–1018 (1994). https://doi.org/10.1137/S0097539791195877

13. Fredman, M.L.: On computing the length of longest increasing subsequences. Discret. Math. **11**(1), 29–35 (1975). https://doi.org/10.1016/0012-365X(75)90103-X

14. Funke, S., Mehlhorn, K., Näher, S.: Structural filtering: a paradigm for efficient and exact geometric programs. Comput. Geom. **31**(3), 179–194 (2005)

15. Geissmann, B., Leucci, S., Liu, C., Penna, P.: Sorting with recurrent comparison errors. In: 28th International Symposium on Algorithms and Computation, ISAAC 2017, Phuket, Thailand, 9–12 December 2017, pp. 38:1–38:12 (2017). https://doi.org/10.4230/LIPIcs.ISAAC.2017.38

16. Geissmann, B., Leucci, S., Liu, C., Penna, P.: Optimal dislocation with persistent errors in subquadratic time. In: 35th Symposium on Theoretical Aspects of Computer Science, STACS 2018, Caen, France, 28 February to 3 March 2018, pp. 36:1–36:13 (2018). https://doi.org/10.4230/LIPIcs.STACS.2018.36

17. Geissmann, B., Leucci, S., Liu, C., Penna, P.: Optimal sorting with persistent comparison errors. ArXiv e-prints, April 2018

18. Geissmann, B., Penna, P.: Inversions from sorting with distance-based errors. In: SOFSEM 2018: Theory and Practice of Computer Science - Proceedings of 44th International Conference on Current Trends in Theory and Practice of Computer Science, Krems, Austria, 29 January–2 February 2018, pp. 508–522 (2018). https://doi.org/10.1007/978-3-319-73117-9_36

19. Hadjicostas, P., Lakshmanan, K.B.: Bubble sort with erroneous comparisons. Australas. J. Comb. **31**, 85–106 (2005). https://www.scopus.com/inward/record.uri?eid=2-s2.0-79551499675&partnerID=40&md5=caf06fd3542cd9364588bfa79c433629

20. Hadjicostas, P., Lakshmanan, K.B.: Measures of disorder and straight insertion sort with erroneous comparisons. Ars Comb. **98**, 259–288 (2011). https://www.scopus.com/inward/record.uri?eid=2-s2.0-79551490783&partnerID=40&md5=7bfbc2e4c824bba2e3dbe1289807d896

21. Hadjicostas, P., Lakshmanan, K.B.: Recursive merge sort with erroneous comparisons. Discret. Appl. Math. **159**(14), 1398–1417 (2011). https://doi.org/10.1016/j.dam.2011.05.010

22. Klein, R., Penninger, R., Sohler, C., Woodruff, D.P.: Tolerant algorithms. In: Demetrescu, C., Halldórsson, M.M. (eds.) ESA 2011. LNCS, vol. 6942, pp. 736–747. Springer, Heidelberg (2011). https://doi.org/10.1007/978-3-642-23719-5_62
23. Masek, W.J., Paterson, M.S.: A faster algorithm computing string edit distances. J. Comput. Syst. Sci. **20**(1), 18–31 (1980). https://doi.org/10.1016/0022-0000(80)90002-1. http://www.sciencedirect.com/science/article/pii/0022000080900021
24. Potts, C.N., Shmoys, D.B., Williamson, D.P.: Permutation vs. non-permutation flow shop schedules. Oper. Res. Lett. **10**(5), 281–284 (1991)
25. Yang, I., Huang, C., Chao, K.: A fast algorithm for computing a longest common increasing subsequence. Inf. Process. Lett. **93**(5), 249–253 (2005). https://doi.org/10.1016/j.ipl.2004.10.014
26. Zhang, H.: Alignment of BLAST high-scoring segment pairs based on the longest increasing subsequence algorithm. Bioinformatics **19**(11), 1391–1396 (2003). https://doi.org/10.1093/bioinformatics/btg168

Cut Sparsifiers for Balanced Digraphs

Motoki Ikeda[✉] and Shin-ichi Tanigawa

The University of Tokyo, Tokyo, Japan
{motoki_ikeda,tanigawa}@mist.i.u-tokyo.ac.jp

Abstract. In this paper we consider a cut sparsification problem for digraphs parametrized by balancedness. A weighted digraph $D = (V, E)$ is said to be α-balanced if the total weight of the edges from U to $V \setminus U$ is at most α times the total weight of the edges from $V \setminus U$ to U for any $U \subseteq V$. Based on the combinatorial cut-sparsification framework by Fung et al. (2011), we show that for any α-balanced weighted digraph D with n vertices and m edges there is a weighted subdigraph D' with $O(\alpha\epsilon^{-2}n \log n \log(nW))$ edges that $(1+\epsilon)$-cut-approximates D, where W is the maximum weight of an edge in D. We also show how to compute such a cut sparsifier in $O(m \log \alpha + \alpha^3 n \log W \operatorname{poly}(\log n))$ time with high probability.

Applying our sparsifier as a preprocessing, the running time of the minimum cut approximation algorithm by Ene et al. (2016) is improved to $O(m \log \alpha + \alpha^3 \epsilon^{-4} n \operatorname{poly}(\log n))$ for an α-balanced digraph with n vertices and m edges.

Keywords: Cut sparsification · Balanced digraph
Minimum cut problem

1 Introduction

Graph sparsification is one of the fundamental tools for developing efficient graph algorithms. The seminal work of Karger [9] and Benczúr and Karger [1,2] showed that for any positively weighted undirected graph G with n vertices and m edges, there is a weighted subgraph G' with $O(\epsilon^{-2}n \log n)$ edges such that the size of each cut is within $(1 \pm \epsilon)$ factor of the original cut size. Such a sparse subgraph is called a *cut sparsifier*. They also gave an $O(m \log^3 n)$ time algorithm for constructing a cut sparsifier with high probability, and demonstrated applications to several cut and flow problems. Later, Spielman and Teng [15] introduced a generalized notion, a *spectral sparsifier*, that sparsifies G keeping the spectral of the Laplacian, and have broadened applications to solving linear systems. Since the work of [15], various improved spectral sparsifiers and efficient algorithms have been developed.

This successful line of research is only for undirected graphs, and despite its obvious importance, there has been little progress for digraphs. Cohen et

This work is supported by JST CREST (JPMJCR1402).

L. Epstein and T. Erlebach (Eds.): WAOA 2018, LNCS 11312, pp. 277–294, 2018.
https://doi.org/10.1007/978-3-030-04693-4_17

al. [4] recently introduced a new notion of spectral sparsifiers based on a scaled norm, and they showed the existence of sparsifiers with $O(\epsilon^{-2}n\mathrm{poly}(\log n))$ edges for any strongly connected digraphs. However, unlike the undirected case, their spectral sparsifier does not imply a cut sparsifier. In fact there are digraphs which do not admit cut sparsifiers with sub-quadratic size (see [4]). This is a typical reason why there is no counterpart theory for digraphs.

Although we cannot hope for a perfect theory for digraphs, there is a natural question; for which class of digraphs can we construct good cut sparsifiers? In this paper we study this problem by focusing on balanced graphs. Balancedness is a new notion introduced by Ene et al. [5] for expressing the ratio of the incoming and out-going cut sizes. More formally, for $\alpha \geq 1$, a digraph $D = (V, E)$ is called α-balanced if

$$\delta^+(U; D) \leq \alpha\delta^-(U; D)$$

holds for any $U \subseteq V$, where $\delta^+(U; D)$ (resp., $\delta^-(U; D)$) denotes the sum of the weights of the edges from U to $V \setminus U$ (resp., from $V \setminus U$ to U). The *imbalance* b_D of D is defined to be the infimum of α such that D is α-balanced. Note that $b_D = 1$ if and only if D is Eulerian.

The main contribution of this paper is to show the existence of cut sparsifiers whose sizes are parametrized by b_D. We show that for any weighted digraph D with n vertices and m edges, there is a weighted subdigraph D' with $O(b_D\epsilon^{-2}n \log n \log(nW))$ edges such that

$$(1 - \epsilon)\delta^+(U; D) \leq \delta^+(U; D') \leq (1 + \epsilon)\delta^+(U; D) \text{ for all } U \subseteq V,$$

where W is the maximum weight of an edge in D. We further show how to obtain such a cut sparsifier in $O(m \log b_D + b_D^3 n \log W \mathrm{poly}(\log n))$ time with high probability.

Our result on the existence of cut sparsifiers is actually a direct application of a result on undirected cut sparsifiers. Although the main focus of the research for undirected graphs has been shifted to spectral sparsifiers, still interesting questions remain even for cut sparsifiers. One such a question is to understand which graph parameter can be used as a sampling parameter in a sampling-type algorithm. Fung et al. [6] gave a general framework to solve this question for undirected graphs. In this paper we exploit the power of their remarkable combinatorial approach; we show that the proof of the main result in [6] can be applied even to digraphs without any substantial modification.

As is always the case with cut sparsifiers, our result can be used as a pre-processing of algorithms for any cut problem. One interesting example is the minimum cut problem of balanced digraphs studied by Ene et al. [5]. Ene et al. [5] gave an algorithm to find a $(1 + \epsilon)$-approximate minimum cut (and a $(1-\epsilon)$-approximate maximum flow) of a digraph D that runs in $O(mb_D^2\epsilon^{-2}\log^c n)$ time for some constant c. (Here the current best c is 45, see [14].) Using our sparsifier at a preprocessing phase, we obtain an algorithm that runs in $O(m \log b_D + b_D^3\epsilon^{-4}n\,\mathrm{poly}(\log n))$ time. This is a substantial improvement if b_D

is not too large. (Note that an exact algorithm in [11] is faster than that of Ene et al. [5] if $b_D = \Omega(n^{1/4})$.)

The paper is organized as follows. In Sect. 2 we show the existence of cut sparsifiers for balanced digraphs, and in Sect. 3 we give an efficient algorithm for constructing those sparsifiers. In Sect. 4 we explain an application to the minimum cut problem. In Sect. 5 we give a short remark on the number of cut projections in α-balanced digraphs.

Throughout the paper we consider a digraph $D = (V, E)$ or an undirected graph $G = (V, E)$ with n vertices, m edges, and each edge weight is a positive integer. As defined above, for $U \subseteq V$, $\delta^+(U; D)$ (resp., $\delta^-(U; D)$) denotes the sum of the weights of the edges from U to $V \setminus U$ (resp., from $V \setminus U$ to U). In an undirected graph G, we use $\delta(U; G)$ to denote the sum of the weights of the edges between U and $V \setminus U$. The (local) *edge connectivity* $\kappa(e; G)$ of $e = \{u, v\}$ in G is defined by $\kappa(e; G) = \kappa_e = \min\{\delta(U; G) \mid U \subseteq V, u \in U, v \notin U\}$. Similarly, the edge connectivity $\kappa(e; D)$ of $e = (u, v)$ in D is defined by $\kappa(e; D) = \min\{\delta^+(U; D) \mid U \subseteq V, u \in U, v \notin U\}$.

2 Digraph Sparsification

In this section, we give cut sparsifications for digraphs based on the result by Fung et al. [6]. Let us first give the following formal definition.

Definition 1. *Let* $D = (V, E)$ *be a digraph. A digraph* $D' = (V, E')$ ϵ-*cut-approximates* D, *which is often abbreviated as* $D' \in (1 \pm \epsilon)D$, *if for all* $U \subseteq V$,

$$(1 - \epsilon)\delta^+(U; D) \leq \delta^+(U; D') \leq (1 + \epsilon)\delta^+(U; D).$$

A sparse subgraph that ϵ-cut-approximates the original graph is called a cut sparsifier.

As is in the ordinary sparsification framework, our algorithm is a random sampling algorithm. More specifically, we use the *compression* of each edge, first introduced by Benczúr and Karger [1,2], where each edge e is sampled with probability p_e and the sampled edge is given a weight $1/p_e$. The sampling probability is determined by a graph parameter λ_e for each edge e. The original algorithm by Benczúr and Karger [1,2] uses the strong connectivity of each edge e for λ_e, which is defined to be the largest k for which a k-edge-connected subgraph containing the edge exists. Fung et al. [6] showed that it is possible to construct a cut sparsifier using edge connectivity, effective resistance, or Nagamochi-Ibaraki index (defined in Sect. 2.1).

A formal description of the *compression* for digraphs is given in Algorithm 1.

We now analyze the quality of the output D_ϵ. Following the analysis by Fung et al. [6], we consider a partition $F_0, F_1, \ldots, F_\Lambda$ of the edge set E of D defined by

$$F_i := \{e \in E \mid 2^i \leq \lambda_e < 2^{i+1}\}$$

where $\Lambda = \lfloor \lg(\max_{e \in E} \lambda_e) \rfloor$.

Algorithm 1. Compress($D, \lambda, \gamma, d, \epsilon$)

Input: A weighted simple digraph $D = (V, E, w)$ with weight $\omega : E \to \mathbb{Z}_+$, an edge
   parameter $\lambda : E \to \mathbb{Z}_+$, a constant $\gamma, d \in \mathbb{R}_+$, and $\epsilon \in (0, 1)$
Output: A cut sparsifier $D_\epsilon = (V, F, u)$
1: $C \leftarrow 43(d + 7)$
2: $\rho \leftarrow C\gamma \ln n/\epsilon^2$
3: $F \leftarrow \emptyset$
4: **for** each $e \in E$ **do**
5:    $p_e \leftarrow \min\{\rho/\lambda_e, 1\}$
6:    Generate a random number X_e from a binomial distribution $B(w_e, p_e)$
7:    **if** $X_e > 0$ **then**
8:       Add edge e to F and set $u_e = X_e/p_e$
9:    **end if**
10: **end for**
11: **return** $D_\epsilon = (V, F, u)$

We say that a family $G_0, \ldots, G_\Lambda$ of weighted undirected graphs *covers* D if
for each i and for each $(u, v) \in F_i$, the weight of $\{u, v\}$ in G_i is greater than or
equal to the sum of the weights of (u, v) and (v, u) in F_i. Such a cover is said to
be a γ-*certificate*[1] if the following two properties are satisfied:

(Connectivity) For each $i \geq 0$ and each edge $(u, v) \in F_i$, $\kappa(\{u, v\}; G_i) \geq 2^{i-1}$.
(Overlapped) For any $U \subseteq V$, $\sum_{i=0}^{\Lambda} \delta(U; G_i) \leq \gamma \cdot \delta^+(U; D)$.

Given γ-certificates, the following theorem states the existence of cut sparsifiers.

Theorem 1. *Let D be a weighted digraph, and λ_e be a positive integer for
each $e \in E$. Suppose that there exists a γ-certificate family of weighted
undirected graphs that covers D. Then, $D_\epsilon = $ Compress$(D, \lambda, \gamma, d, \epsilon)$ contains
$O(\frac{\gamma \log n}{\epsilon^2} \sum_{e \in E} \frac{w_e}{\lambda_e})$ edges in expectation, and $D_\epsilon \in (1 \pm \epsilon)D$ with probability at
least $1 - 1/n^d$.*

Proof. The theorem follows from the following more general statement, Theorem 2, by observing that each undirected graph G_i is considered as an Eulerian
digraph if we regard each undirected edge as two parallel directed edges of both
directions.   □

We can apply the above definition of a covering family and a γ-certificate
to a family of weighted digraphs $D_0, \ldots, D_\Lambda$. Formally, a family $D_0, \ldots, D_\Lambda$ of
weighted digraphs *covers* D if for each i and for each $(u, v) \in F_i$, the weight of
(u, v) in D_i is greater than or equal to the weight of (u, v) in F_i. A cover is said
to be a γ-*certificate* if $\kappa((u, v); D_i) \geq 2^{i-1}$ holds for each $i \geq 0$ and each edge
$(u, v) \in F_i$, and $\sum_{i=0}^{\Lambda} \delta^+(U; D_i) \leq \gamma \cdot \delta^+(U; D)$ for any $U \subseteq V$.
   Theorem 1 still holds if G_i is substituted by an Eulerian digraph D_i.

---

[1] This is a simplified and adapted notion of the (π, α)-certificate introduced by Fung
et al. [6].

Theorem 2. *Let D be a weighted digraph, and λ_e be a positive integer for each $e \in E$. Suppose that there exists a γ-certificate family of weighted Eulerian digraphs that covers D. Then, $D_\epsilon = \mathsf{Compress}(D, \lambda, \gamma, d, \epsilon)$ contains $O(\frac{\gamma \log n}{\epsilon^2} \sum_{e \in E} \frac{w_e}{\lambda_e})$ edges in expectation, and $D_\epsilon \in (1 \pm \epsilon)D$ with probability at least $1 - 1/n^d$.*

Theorem 2 is a proper extension of Theorem 1. The proof of Theorem 2 is an adaptation of that of Fung et al. [6], but for completeness we give a formal proof in Appendix B.

2.1   Compression Using NI Indexes

In the following two subsections, we shall show how to set up parameter λ_e to apply Theorem 1.

Nagamochi and Ibaraki [12,13] showed how to compute a sparse certificate for the k-connectivity of undirected graphs. Motivated by their work, Fung et al. [6] introduced the following simplified variant of the local connectivity.

Definition 2 (NI forest, NI index [6]). *Let G be an undirected graph with integer-valued edge weight, and let $\tilde{G}$ be the multigraph obtained from G by replacing each edge e with weight w_e by w_e parallel edges. A sequence of edge-disjoint spanning forests $T_1, T_2, \ldots$ of $\tilde{G}$ is said to be an NI forest packing if T_i is a spanning forest on the edges left in $\tilde{G}$ after removing those in $T_1, T_2, \ldots, T_{i-1}$. An edge with weight w_e in G must appear in w_e contiguous forests. The NI index of edge e in G, denoted ℓ_e, is the index of the last NI forest in which e appears.*

Let $D = (V, E)$ be an α-balanced digraph, and G be the undirected graph obtained from D by ignoring the direction. For each edge $e \in E$, we set $\lambda_e = \ell_e$, where ℓ_e is the NI index of e in G. It turns out that the compression using this parameter gives a good sparsifier. To see this we need to construct a family $G_0, \ldots, G_\Lambda$ of undirected graphs with the properties as given in Theorem 1.

Let $T_1, T_2, \ldots, T_k$ be an NI forest packing of G. We define a weighted undirected graph H_i to be the union of $T_{2^{i-1}}, T_{2^{i-1}+1}, \ldots, T_{2^i-1}$ (i.e., the weight of $\{u, v\}$ is the number of appearances of edge $\{u, v\}$ in $T_{2^{i-1}}, T_{2^{i-1}+1}, \ldots, T_{2^i-1}$.) We then define $G_i = (V, E_i)$ such that the weight of $\{u, v\}$ is the sum of the weight of $\{u, v\}$ in H_i and the weights of (u, v) and (v, u) in F_i for every pair $u, v \in V$ (and E_i is defined to be the set of pairs of vertices with nonzero weight).

Lemma 1. *A family G_i of undirected graphs defined above is a $2(1 + \alpha)$-certificate covering D.*

Proof. Clearly the family covers D.

To see the connectivity, recall first that $\lambda_e \geq 2^i$ for any $e = (u, v) \in F_i$. Hence u and v are connected in each of $T_{2^{i-1}}, T_{2^{i-1}+1}, \ldots, T_{2^i-1}$ by the definition of NI forest packing. Therefore $\kappa(\{u, v\}; G_i) \geq 2^{i-1}$.

To evaluate the overlapping, note that for any $i \neq j$, $F_i \cap F_j = \emptyset$ and the edge set of H_i is disjoint from that of H_j. Hence the sum of the weights of

$\{u,v\}$ over G_i is at most two times the weight of $\{u,v\}$ in G. Thus for each $U \subseteq V$ we get $\sum_i \delta(U;G_i) \leq 2\delta(U;G) \leq 2(1+\alpha)\delta^+(U,D)$, and it is $2(1+\alpha)$-overlapped.   □

We can now apply Theorem 1.

Theorem 3. *Let D be a weighted digraph, and D_ϵ = Compress$(D, \ell, 2(1 + b_D), d, \epsilon)$. Then, D_ϵ contains $O(b_D \epsilon^{-2} n \log n \log(nW))$ edges in expectation, and $D_\epsilon \in (1 \pm \epsilon)D$ with probability at least $1 - 1/n^d$, where W is the maximum weight of an edge in D.*

Proof. By Lemma 1, there always exists a $2(1+b_D)$-certificate covering D. Thus by Theorem 1, we have a weighted subgraph D_ϵ with $O(\rho \sum_e w_e/\ell_e)$ edges and $D_\epsilon \in (1 \pm \epsilon)D$ with probability at least $1 - 1/n^d$. It was shown by Fung et al. [6] that $\sum_{e \in E} w_e/\ell_e = O(n \log(nW))$. Therefore D_ϵ has the properties in the statement.   □

2.2 Compression Using Edge Connectivities

If we use the local edge connectivity, we have a slightly better sparsifier. But computing the local edge connectivities is more expensive than computing the NI indexes.

Let $D = (V, E)$ be an α-balanced digraph, and G be the undirected graph obtained from D by ignoring the direction. For an edge $e = (u, v)$ in D, we consider the local edge connectivity κ_e of $\{u, v\}$ in G. We consider the compression by setting $\lambda_e = \kappa_e$ for each $e \in E$. We need to construct a family $G_0, \ldots, G_\Lambda$ of undirected graphs with the properties as given in Theorem 1.

Let $T_1, \ldots, T_k$ be an NI forest packing of G. We define a weighted undirected graph H_i to be the union of $T_1, T_2, \ldots, T_{2^{i-1}-1}$ for $i \leq \lg n$, the union of $T_{2^{i-1}-\lg n}, T_{2^{i-1}-\lg n+1}, \ldots, T_{2^{i+1}-1}$ for $i \geq \lg n + 1$. We then define $G_i = (V, E_i)$ such that the weight of $\{u, v\}$ is the sum of the weight of $\{u, v\}$ in H_i and the weights of (u, v) and (v, u) in F_i for every pair $u, v \in V$ (and E_i is defined to be the set of pairs of vertices with nonzero weight).

Lemma 2 (Fung et al. [6]). *Let $T_1, T_2, \ldots$ be an NI forest packing of an undirected graph $G = (V, E)$. For any pair of vertices $u, v \in V$ and for any $i \geq 1$, $\kappa(u, v; T_1 \cup T_2 \cup \cdots \cup T_i) \geq \min\{\kappa_{uv}, i\}$.*

Lemma 3. *A family G_i of undirected graphs defined above is a $(1+\alpha)(3+\lg n)$-certificate covering D.*

Proof. Clearly the family covers D.

To see the connectivity, recall first that $\lambda_e \geq 2^i$ for any $e = (u, v) \in F_i$. Hence, for $i \leq \lg n$, it holds that $\kappa(\{u, v\}; H_i) \geq 2^{i-1} - 1$ by Lemma 2, and $\kappa(\{u, v\}; G_i) \geq 2^{i-1}$. For $i \geq \lg n + 1$, it holds that $\kappa(\{u, v\}; T_1 \cup \cdots \cup T_{2^{i+1}-1}) \geq 2^i$ by Lemma 2. Since there are at most 2^{i-1} edges in $T_1, T_2, \ldots, T_{2^{i-1}-\lg n-1}$, we have $\kappa(\{u, v\}; G_i) \geq 2^{i-1}$.

To evaluate the overlapping, note that for any $i \neq j$, $F_i \cap F_j = \emptyset$ and each edge of G appears in H_i for at most $2 + \lg n$ different values of i. Hence the sum of the weights of $\{u, v\}$ over G_i is at most $3 + \lg n$ times the weight of $\{u, v\}$ in G. Thus for each $U \subseteq V$ we get $\sum_i \delta(U; G_i) \leq (3 + \lg n)\delta(U; G) \leq (1 + \alpha)(3 + \lg n)\delta^+(U; D)$, and it is $(1 + \alpha)(3 + \lg n)$-overlapped. $\qquad\square$

We can now apply Theorem 1.

Theorem 4. *Let D be a weighted digraph, and $D_\epsilon = \mathsf{Compress}(D, \kappa, (1+b_D)(3+ \lg n), d, \epsilon)$. Then, D_ϵ contains $O(b_D\epsilon^{-2}n\log^2 n)$ edges in expectation, and $D_\epsilon \in (1 \pm \epsilon)D$ with probability at least $1 - 1/n^d$.*

Proof. By Lemma 3, there always exists a $(1 + b_D)(3 + \lg n)$-certificate covering D. Thus by Theorem 1, we have a weighted subgraph D_ϵ with $O(\rho\sum_e w_e/\kappa_e)$ edges and $D_\epsilon \in (1 \pm \epsilon)D$ with probability at least $1 - 1/n^d$. It is known [6] that $\sum_{e \in E} w_e/\kappa_e \leq n - 1$. Therefore D_ϵ has the properties in the statement. $\qquad\square$

We can also apply the analysis to the compression algorithm using the local edge connectivity of digraphs (rather than that of the underlying undirected graphs) as sampling parameter λ_e. However, the resulting edge density is no better than that in Theorem 4.

3   Digraph Sparsification Algorithm

In this section we give an efficient implementation of $\mathsf{Compress}$ based on the NI index. For this, we compute the NI index of a weighted graph before calling $\mathsf{Compress}$. It is implicit in the work by Nagamochi and Ibaraki [13] that the NI index of a weighted graph can be computed in $O(m + n\log n)$ time. The generation of a random variable from a binomial distribution $B(w_e, p_e)$ can be done in $O(w_ep_e)$ time (see e.g. [8]). Therefore, $\mathsf{Compress}(D, \ell, 2(1 + \alpha), \epsilon)$ can be implemented in $O(m + \sum_e w_ep_e)$ time if we know that D is α-balanced in advance. Here $\sum_e w_ep_e = O(\alpha\epsilon^{-2}n\log n\log(nW))$ is the expected number of the edges in the sparsifier, and we may always assume that it is $O(m)$ since otherwise we can simply return D as a better sparsifier. Hence the total running time is $O(m)$. To apply the algorithm to any digraph D, we need to (approximately) compute the imbalance of D. For this, the following result is known.

Lemma 4 (Ene et al. [5, Lemma 2.9]). *Given a weighted digraph D and α such that D is α-balanced, there is an algorithm $\mathsf{ApproxBal}(D, \alpha, \epsilon_0)$ that outputs $(1 + \epsilon_0)$-approximate b_D in $O(m\alpha^2\epsilon_0^{-2}\mathrm{poly}(\log n))$ time.*

By simply calling the algorithm in Lemma 4, we obtain an $O(mb_D^2\mathrm{poly}(\log n))$ time algorithm for constructing a cut sparsifier for a digraph D. In this section we shall present an improved implementation by first showing the following.

Lemma 5. *Given a weighted digraph D, there is an algorithm that outputs α with $b_D \leq \alpha \leq 27b_D$ with probability at least $1 - 1/n^d$ in $O(m\log b_D + b_D^3n\log W\,\mathrm{poly}(\log n))$ time, where W is the maximum weight of an edge in D.*

In the algorithm stated in Lemma 5, we use the following two algorithms as subroutines.

- ApproxBal(H, α, ϵ_0): Given $\alpha \in \mathbb{Z}_+, \epsilon_0 \in \mathbb{R}_+$, and an α-balanced digraph H with n vertices and m edges, output b with $b_H \leq b \leq (1 + \epsilon_0)b_H$ in $O(m\alpha^2\epsilon_0^{-2}\,\mathrm{poly}(\log n))$ time.
- Sparsify(H, α, ϵ_0): Given $\alpha \in \mathbb{Z}_+, \epsilon_0 \in \mathbb{R}_+$, and an α-balanced digraph H with n vertices and m edges, output H' with $O(\alpha\epsilon_0^{-2}n \log n \log(nW))$ edges that $(1 + \epsilon_0)$-cut-approximates H with probability at least $1 - 1/n^{d+1}$ in $O(m)$ time.

Note that in these subroutines we are required to know that the input is α-balanced in advance.

Combining these two subroutines, we consider Algorithm 2 to compute the imbalance approximately. Here D^{-1} denotes the digraph obtained from D by reversing the direction of each edge, and αD denotes the weighted digraph in which the weight of each edge is α times of the original weight.

Algorithm 2. An algorithm to approximate imbalance

Input: A weighted digraph $D = (V, E, w)$
1: $\epsilon_0 \leftarrow 0.1$, $\epsilon_1 \leftarrow 2(1 + \epsilon_0)/(1 - \epsilon_0)$, $\alpha \leftarrow 1$
2: **while** $\alpha \leq n$ **do**
3:     $D_\alpha \leftarrow D \cup \alpha D^{-1}$
4:     $H_\alpha \leftarrow$ Sparsify$(D_\alpha, \alpha, \epsilon_0)$
5:     $b_\alpha \leftarrow$ ApproxBal$(H_\alpha, \alpha(1 + \epsilon_0)/(1 - \epsilon_0), \epsilon_0)$
6:     **if** $b_\alpha \leq (\alpha + \alpha^{-1})/2\epsilon_1$ **then**
7:         **return** α
8:     **else**
9:         $\alpha \leftarrow 2\alpha$
10:     **end if**
11: **end while**
12: Output "α is larger than n"

We show that Algorithm 2 outputs a constant-factor-approximation of b_D. We first remark that, since there are at most $\log n$ iterations, all Sparsify$(D_\alpha, \alpha, \epsilon_0)$ outputs a cut sparsifier with probability at least $1 - 1/n^d$.

Lemma 6. *For any $\alpha \in \mathbb{Z}_+$, $D_\alpha = D \cup \alpha D^{-1}$ satisfies*

$$b_{D_\alpha} = \frac{1 + \alpha b_D}{\alpha + b_D} \leq \alpha.$$

Proof. For any nonempty subset $U \subsetneq V$,

$$\frac{\delta^-(U; D_\alpha)}{\delta^+(U; D_\alpha)} = \frac{\delta^-(U; D) + \alpha\delta^+(U; D)}{\delta^+(U; D) + \alpha\delta^-(U; D)} = \frac{1 + \alpha\beta(U)}{\beta(U) + \alpha}$$

where $\beta(U) := \delta^+(U; D)/\delta^-(U; D)$. For $a \geq 1$, a function $f(x) = (1 + ax)/(a + x) = a - (a^2 - 1)/(a + x)$ is monotonically increasing. Hence, b_{D_α} is given by U that maximizes $\beta(U)$, implying the first equation in the statement.

The second inequality simply follows by observing

$$b_{D_\alpha} = \alpha - \frac{\alpha^2 - 1}{\alpha + b_D} \leq \alpha.$$

$\square$

Lemma 7. *Let* $H_\alpha = \mathsf{Sparsify}(D_\alpha, \alpha, \epsilon_0)$ *and* $b_\alpha = \mathsf{ApproxBal}(H_\alpha, \alpha(1+\epsilon_0)/(1-\epsilon_0), \epsilon_0)$. *Then with probability at least* $1 - 1/n^{d+1}$,

$$\frac{1 - \epsilon_0}{1 + \epsilon_0} b_{D_\alpha} \leq b_\alpha \leq \frac{(1 + \epsilon_0)^2}{1 - \epsilon_0} b_{D_\alpha}.$$

Proof. From Lemma 6, $\mathsf{Sparsify}(D_\alpha, \alpha, \epsilon_0)$ correctly outputs a cut sparsifier with probability at least $1 - 1/n^{d+1}$. Hence,

$$\frac{1 - \epsilon_0}{1 + \epsilon_0} b_{D_\alpha} \leq b_{H_\alpha} \leq \frac{1 + \epsilon_0}{1 - \epsilon_0} b_{D_\alpha}.$$

By Lemma 6 this in particular implies $b_{H_\alpha} \leq \alpha(1 + \epsilon_0)/(1 - \epsilon_0)$, and therefore $\mathsf{ApproxBal}(H_\alpha, \alpha(1 + \epsilon_0)/(1 - \epsilon_0), \epsilon_0)$ correctly outputs a $(1 + \epsilon_0)$-approximate of b_{H_α}, i.e., $b_{H_\alpha} \leq b_\alpha \leq (1 + \epsilon_0)b_{H_\alpha}$. Therefore we obtain the relation in the statement. $\square$

Lemma 8. *Let* $b_\alpha = \mathsf{ApproxBal}(H_\alpha, \alpha(1 + \epsilon_0)/(1 - \epsilon_0), \epsilon_0)$, *and suppose that* $b_\alpha \leq (\alpha + \alpha^{-1})/2\epsilon_1$ *where* $\epsilon_1 = 2(1 + \epsilon_0)/(1 - \epsilon_0)$. *Then* $\alpha \geq b_D$ *with probability at least* $1 - 1/n^{d+1}$.

Proof. If $b_\alpha \leq (\alpha + \alpha^{-1})/2\epsilon_1$,

$$\frac{1 - \epsilon_0}{1 + \epsilon_0} \cdot \frac{1 + \alpha b_D}{\alpha + b_D} \leq \frac{(1 - \epsilon_0)}{2(1 + \epsilon_0)} \cdot \frac{1}{2}\left(\alpha + \frac{1}{\alpha}\right) \tag{1}$$

holds from Lemmas 6 and 7. Then (1) is equivalent to

$$0 \leq \alpha^3 - 3\alpha^2 b_D - 3\alpha + b_D = \alpha(\alpha - b_D)(\alpha - 1) - (2b_D - 1)\alpha^2 - b_D(\alpha - 1) - 3\alpha.$$

Since $\alpha \geq 1$ and $b_D \geq 1$, it is necessary that $\alpha \geq b_D$. $\square$

Lemma 9. *Let* $b_\alpha = \mathsf{ApproxBal}(H_\alpha, \alpha(1 + \epsilon_0)/(1 - \epsilon_0), \epsilon_0)$ *and* $\Delta = 4\epsilon_1(1 + \epsilon_0)^2/(1 - \epsilon_0)$. *If* $\alpha \geq \Delta b_D$, *then* $b_\alpha \leq (\alpha + \alpha^{-1})/2\epsilon_1$ *with probability at least* $1 - 1/n^{d+1}$.

Proof.

$$
\begin{aligned}
b_\alpha &\le \frac{(1+\epsilon_0)^2}{1-\epsilon_0} b_{D_\alpha} && \text{(by Lemma 7)} \\
&= \frac{(1+\epsilon_0)^2}{1-\epsilon_0} \cdot \frac{1+\alpha b_D}{\alpha + b_D} && \text{(by Lemma 6)} \\
&\le \frac{(1+\epsilon_0)^2}{1-\epsilon_0} \left(\frac{1}{\alpha + b_D} + \frac{\alpha}{\Delta} \right) && \text{(by } \alpha \ge \Delta b_D) \\
&\le \frac{(1+\epsilon_0)^2}{1-\epsilon_0} \cdot \frac{2\alpha}{\Delta} && \left(\text{by } \frac{1}{\alpha + b_D} \le \frac{1}{2} < b_D \le \frac{\alpha}{\Delta} \right) \\
&\le \frac{(1+\epsilon_0)^2}{1-\epsilon_0} \cdot \frac{2}{\Delta} \left(\alpha + \frac{1}{\alpha} \right) \\
&= \frac{1}{2\epsilon_1} \left(\alpha + \frac{1}{\alpha} \right)
\end{aligned}
$$

$\square$

We are now ready prove Lemma 5.

Proof (of Lemma 5). Let α^* be the output. By Lemma 8 we have $b_D \le \alpha^*$. By Lemma 9 and Line 6 of Algorithm 2, it holds that α in the second to last loop $(= \alpha^*/2)$ is at most Δb_D. Thus, by the definition of Δ and ϵ_1,

$$
\alpha^* \le 2\Delta b_D = \frac{16(1+\epsilon_0)^3}{(1-\epsilon_0)^2}.
$$

When we take $\epsilon_0 = 0.1$, we have $\alpha^* \le 27 b_D$.

The time complexity can be obtained by replacing m of the time complexity of ApproxBal with the edge size of the output of Sparsify. $\square$

Now, by using Algorithm 2 to compute the imbalance of a given digraph, we have the following computational result for cut sparsifiers.

Theorem 5. *Given a weighted digraph D and ϵ, there is an algorithm that outputs a cut sparsifier with $O(b_D \epsilon^{-2} n \log n \log(nW))$ edges in expectation with probability at least $1 - 1/n^d$ in time $O(m \log b_D + b_D^3 n \log W \operatorname{poly}(\log n))$, where W is the maximum weight of an edge in D.*

4 Minimum Cut Problem

Ene et al. [5] show the following algorithm.

Theorem 6 (Ene et al. [5]). *Given a weighted digraph D, a source s, a sink t, and ϵ_0 with $0 < \epsilon_0 < 1$, there is an algorithm that outputs a $(1+\epsilon_0)$-approximate minimum s-t cut in time $O(m b_D^2 \epsilon_0^{-2} \operatorname{poly}(\log n))$.*

When computing a $(1 + \epsilon_0)$-approximate minimum s-t cut, there is a simple trick to suppose that the edge weight is integer-valued and the maximum weight value is at most $O(m^2/\epsilon_0)$ [3]. Hence by using the cut sparsifier in Theorem 5 at a preprocessing phase in the algorithm in Theorem 6, we have the following.

Theorem 7. *Given a weighted digraph D, a source s, a sink t, and ϵ_0 with $0 < \epsilon_0 < 1$, there is an algorithm that outputs a $(1 + \epsilon_0)$-approximate minimum s-t cut with probability at least $1 - 1/n^d$ in $O(m \log b_D + b_D^3 \epsilon_0^{-4} n \operatorname{poly}(\log n))$ time.*

5   Bound of the Number of Cut Projections in Balanced Digraphs

It was shown by Karger and Stein [10] that the number of cuts of size at most β times the minimum cut size is bounded by $n^{2\beta}$ for any undirected graph with n vertices. This was generalized by Fung et al. [6] in the form of cut projections (defined below for digraphs), and was crucially used in the analysis of cut sparsifiers. On the other hand, there is a family of digraphs for which the number of the minimum cuts exponentially grows in n, and this is a critical difference between undirected and directed graphs. In view of this, in this section we shall give a new bound on the number of cut projections in terms of imbalance.

Definition 3 (Fung et al. [6]). *An edge is said to be k-heavy if its connectivity is at least k; otherwise, it is said to be k-light. The k-projection of an edge set is the set of k-heavy edges in it.*

The following theorem, a natural extension of the theorem by Fung et al. [6], is a key tool in the proof of Theorem 2.

Theorem 8. *Let λ be the weight of a minimum weight cut in a digraph D. Then, for any integer $k \geq \lambda$ and any real number $\beta \geq 1$, the number of k-projections of cuts with size at most βk is at most $2n^{2\beta b_D}$.*

Suppose that D is an Eulerian digraph, i.e., $b_D = 1$. One natural idea to prove Theorem 8 for D is to apply the undirected version of Fung et al. [6] to the underlying undirected graphs. Specifically for counting the number of cuts of size k, we may count the number of cuts of size $2k$ in the underlying undirected graph since in the Eulerian digraph $D = (V, E)$ we have $\delta^+(U; D) = \delta^-(U; D)$ for any $U \subseteq V$. This simple approach does not seem to work. Consider, for example, the graph in Fig. 1, which consists of n disjoint pairs of strongly connected graphs of two vertices. The corresponding undirected graph consists of n pairs of vertices with two multiple edges. Consider counting the number of cuts of size n in this digraph. This corresponds to picking one vertex from each component, and hence we have 2^n choices. On the other hand, in the underlying graph, there is only one cut of size $2n$, that is, the whole edge set.

We may guess such a gap does not occur in strongly connected digraphs; however we cannot ignore disconnected digraphs in order to count the number of k-projections stated in Theorem 8; even if a given digraph is strongly connected, it can be disconnected after removing k-light edges.

Nevertheless we can apply the proof of the undirected counterpart by Fung et al. [6]. One critical ingredient in the proof by Fung et al. [6] is Mader's splitting-off theorem, whose directed counterpart does not hold in general. Fortunately, Jackson [7] already pointed out an extension of Mader's theorem to Eulerian digraphs, and this extension enables us to apply the proof of Fung et al. [6] to Eulerian digraphs. Extending the result from Eulerian digraphs to general digraphs using the imbalance parameter is done by a simple counting argument. See Appendix A for a formal proof.

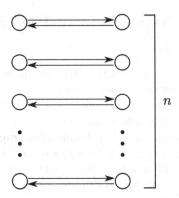

Fig. 1. An example that shows the simple approach for counting does not seem to work.

A   Proof of Theorem 8

Let $D = (V, E)$ be a strongly connected digraph. Since the edge weight is integer-valued, in the following discussion we may assume that D is an unweighted multigraph. For $U \subseteq V$, we define $C^+(U; D)$ (resp., $C^-(U; D)$) be the set of edges from U to $V \setminus U$ (resp., from $V \setminus U$ to U). We also define $P(k, \beta; D)$ to be the set of the k-projections of cuts with size at most βk in D. Our goal is to prove $|P(k, \beta; D)| \leq 2n^{2\beta b_D}$.

We first consider the case when D is Eulerian.

Lemma 10. *Let λ be the weight of a minimum weight cut in an Eulerian digraph D. Then, for any integer $k \geq \lambda$ and any real number $\beta \geq 1$, $|P(k, \beta, D)| \leq 2n^{2\beta}$.*

To prove this, we introduce the splitting-off operation.

Definition 4. *The* splitting-off *operation replaces a pair of edges (u, v) and (v, w) with the edge (u, w), and is said to be* admissible *if it does not change the edge connectivity k_{st} between any two vertices $s, t \neq v$. It is well-known that splitting-off operation never increases the size of any cut.*

The complete splitting-off *operation at a vertex v repeatedly performs admissible splitting-off operations on the edges incident on v until v becomes an isolated vertex, and then removes v.*

Lemma 11 (Jackson [7]). *Let v be a non isolated vertex of an Eulerian digraph D. Then, there exists a complete splitting-off operation at v.*

The proof of Lemma 10 is done by analyzing the following algorithm, Algorithm 3, which is identical to that given by Fung et al. [6]. In Algorithm 3, a vertex v is said to be k-*heavy* if there exists an k-heavy edge incident on v; otherwise, it is said to be k-*light*.

Algorithm 3 performs a set of iterations. In each iteration, it performs complete splitting-off at all k-light vertices in D, contracts an edge selected uniformly at random, and removes all self-loops. The iterations terminate when at most $\lceil 2\beta \rceil$ vertices are left in the graph. At this point, the algorithm outputs the k-projection of a cut selected uniformly at random. Note that a complete splitting-off adds new edges to D. All new edges are treated as k-light irrespective of their connectivity. Therefore, the k-projection of a cut that is output by the algorithm does not include any new edge.

Algorithm 3. An algorithm for proving bound on cut projections

---

Input: An Eulerian digraph $D = (V, E)$, an integer $k \geq \lambda$ where λ is the weight of a minimum weight cut in D, and a real number $\beta \geq 1$
1: **while** there are more than $\lceil 2\beta \rceil$ vertices remaining **do**
2:     **while** there exists a k-light vertex v in D **do**
3:         Perform a complete splitting-off at v
4:     **end while**
5:     Pick an edge e uniformly at random
6:     Contract e and remove all self-loops
7: **end while**
8: **return** the k-projection of a cut selected uniformly at random

---

Lemma 10 follows from the following.

Lemma 12. *Let F be the k-projection of a cut with size at most βk. Then, Algorithm 3 outputs F with probability at least $n^{-2\beta}/2$.*

Indeed, if Lemma 12 holds, the probability that a k-projection of a cut with size at most βk is returned is at least $n^{-2\beta}/2$ times the number of such k-projections. Thus we get $|P(k, \beta, D)| \leq 2n^{2\beta}$.

We shall now prove Lemma 12. Algorithm 3 changes a graph to a different graph by complete splitting-offs, edge-contractions, and removals of self-loops. Let $D_i = (V_i, E_i)$ $(i = 0, \ldots, M)$ be the graphs which we consider during the algorithm. To prove Lemma 12, we consider the following properties:

- (I1) D_i is Eulerian.
- (I2) There exists a subset $U_i \subseteq V_i$ such that $\mathrm{pr}_k(C^+(U_i; D_i); D) = F$ and $\delta^+(U_i; D_i) \leq \beta k$, where $\mathrm{pr}_k(C; D)$ is the k-projection of C in D.
- (I3) If $e \in E_i$ is not an edge added by complete splitting-offs and $\kappa(e; D) \geq k$, then $\kappa(e; D_i) \geq k$.

Clearly, $D = D_0$ has the properties (I1)–(I3). Since the removal of a self-loop does not affect any cut set, (I1)–(I3) are preserved. For a complete splitting-off operation,

- (I1) is preserved from the definition of splitting-off.
- (I2) is preserved since we only split-off at a k-light vertex and a splitting-off never increases the size of any cut.
- (I3) is preserved since we only split-off at a k-light vertex and the splitting-offs are admissible.

Lemma 13. *Let D_{i+1} be the result of a contraction of an edge $f = (w, x)$ chosen from D_i uniformly at random. Suppose that D_i has the properties (I1)–(I3). Then, D_{i+1} has the properties (I1)–(I3) with probability at least $1 - 2\beta/|V_i|$.*

Proof. Clearly, (I1) is preserved. For (I3), since a contraction does not create new cuts, the edge connectivity of an uncontracted edge cannot decrease. Now we consider the probability that D_{i+1} has (I2). (I2) is preserved if $C^+(U_i; D_i) \cup C^-(U_i; D_i)$ does not contain f, and,

$$\Pr[f \notin C^+(U_i; D_i) \cup C^-(U_i; D_i)] = 1 - \frac{|C^+(U_i; D_i) \cup C^-(U_i; D_i)|}{|E_i|}$$

$$= 1 - \frac{2\delta^+(U_i; D_i)}{|E_i|}.$$

Since every vertex in D_i is k-heavy, the outdegree of each vertex is at least k. Therefore, we have

$$|E_i| = \sum_{v \in V_i} \delta^+(v; D_i) \geq k|V_i|,$$

and

$$\Pr[f \notin C^+(U_i; D_i) \cup C^-(U_i; D_i)] \geq 1 - \frac{2\delta^+(U_i; D_i)}{k|V_i|} \geq 1 - \frac{2\beta}{|V_i|}.$$

$$\square$$

We use a following technical lemma.

Lemma 14 (Karger [9, p. 42, ll.27–28]). *For any real number $\beta \geq 1$ and any positive integer $n > 2\beta$,*

$$\frac{n!}{\Gamma(n - 2\beta + 1)} < n^{2\beta}.$$

Now we are ready to prove Lemma 12.

We assume that $n > 2\beta$; otherwise there is nothing to prove. Let N be the number of contractions, and suppose that the ith contraction transforms D_{j_i} into D_{j_i+1}. The output is F if $D_0, D_1, \ldots, D_M$ satisfies (I1)–(I3) and the algorithm selects the cut defined by U_M in Line 8. Let R be $\lceil 2\beta \rceil$. Then,

$$\Pr[\text{Algorithm 3 outputs } F]$$

$$\geq \left(1 - \frac{2\beta}{|V_{j_1}|}\right)\left(1 - \frac{2\beta}{|V_{j_2}|}\right) \cdots \left(1 - \frac{2\beta}{|V_{j_N}|}\right) 2^{-|V_M|}$$

$$= \left(1 - \frac{2\beta}{n}\right)\left(1 - \frac{2\beta}{n-1}\right) \cdots \left(1 - \frac{2\beta}{R+1}\right) 2^{-R}$$

$$= \frac{n - 2\beta}{n} \cdot \frac{n - 1 - 2\beta}{n-1} \cdots \frac{R + 1 - 2\beta}{R+1} \cdot 2^{-R}$$

$$= \frac{\Gamma(n - 2\beta + 1)}{\Gamma(R - 2\beta + 1)} \cdot \frac{R!}{n!} \cdot 2^{-R} \geq \frac{\Gamma(n - 2\beta + 1)}{2 \cdot n!} > n^{-2\beta}/2,$$

where the second last inequality follows from $2^{R-1} \leq R!$ and $0 < \Gamma(x) \leq 1$ for $1 \leq x \leq 2$, and the last inequality follows from Lemma 14. This completes the proof of Lemma 12, and hence Lemma 10.

Proof (of Theorem 8). Let $D' = D \cup D^{-1}$. For any $U \subseteq V$, it follows from the definition of imbalance that

$$(1 + b_D^{-1})\delta^+(U; D) \leq \delta^+(U; D') \leq (1 + b_D)\delta^+(U; D). \tag{2}$$

From the first inequality of (2), for any $u, v \in V$,

$$\begin{aligned}
\kappa((u, v); D') &= \min\left\{\delta^+(U; D') \mid U \subseteq V, u \in U, v \notin V\right\} \\
&\geq \min\left\{(1 + b_D^{-1})\delta^+(U; D) \mid U \subseteq V, u \in U, v \notin V\right\} \\
&= (1 + b_D^{-1})\kappa((u, v); D).
\end{aligned}$$

Hence for k-heavy edge e in D,

$$\kappa(e; D') \geq (1 + b_D^{-1})\kappa(e; D) \geq (1 + b_D^{-1})k.$$

Furthermore, from the second inequality of (2), if $U \subseteq V$ satisfies $\delta^+(U; D) \leq \beta k$, then $\delta^+(U; D') \leq \beta(1 + b_D)k$. Thus,

$$\begin{aligned}
|P(k, \beta; D)| &= |\{\mathrm{pr}_k(C^+(U; D); D) \mid U \subseteq V,\ \delta^+(U; D) \leq \beta k\}| \\
&\leq |\{\mathrm{pr}_{(1+b_D^{-1})k}(C^+(U; D'); D') \mid U \subseteq V,\ \delta^+(U; D') \leq \beta(1 + b_D)k\}|.
\end{aligned}$$

Note that the last formula is $|P((1 + b_D^{-1})k, \beta b_D; D')|$. Thus, by Lemma 10, we have $|P(k, \beta; D)| \leq |P((1 + b_D^{-1})k, \beta b_D; D')| \leq 2n^{2\beta b_D}$. $\qquad\square$

B   Proof of Theorem 2

The proof is again an adaptation of that for the undirected counterpart [6].

We prepare some notations. Recall that $C^+(U; D)$ is the set of edges from U to $V \setminus U$. For $U \subseteq V$, we define $F_i^{(U)} = F_i \cap C^+(U; D)$, and $f_i^{(U)} = |F_i^{(U)}|$. Let $\widehat{f_i^{(U)}}$ be the sum of weight over edges in $F_i^{(U)}$ that appear in D_ϵ. It holds that $\mathbb{E}[\widehat{f_i^{(U)}}] = f_i^{(U)}$.

The following Chernoff bound will be used.

Lemma 15 (Fung et al. [6]). *Let $X_1, X_2, \ldots, X_n$ be n independent random variables such that X_i takes value $1/p_i$ with probability p_i and 0 otherwise. Then, for any p such that $p \leq p_i$ for each i, any $\epsilon \in (0, 1)$, and any $N \geq n$,*

$$\Pr\left[\left|\sum_{i=1}^{n} X_i - n\right| > \epsilon N\right] < 2e^{-0.38\epsilon^2 pN}.$$

The following lemma is a key to prove Theorem 2.

Lemma 16. *Let $D_0, \ldots, D_\Lambda$ be a γ-certificate family of weighted Eulerian digraphs that covers D, and $i \in \{0, 1, \ldots, \Lambda\}$. Then with probability at least $1 - 1/n^{d+2}$, any $U \subseteq V$ satisfies*

$$|f_i^{(U)} - \widehat{f_i^{(U)}}| \leq \frac{\epsilon}{2} \max\left\{ \frac{\delta^+(U; D_i)}{\gamma}, f_i^{(U)} \right\}. \tag{3}$$

Proof. If $f_i^{(U)} = 0$, (3) holds with probability one. So we only consider U such that $f_i^{(U)} > 0$. By the connectivity condition of γ-certificates, we have $\delta^+(U; D_i) \geq 2^{i-1}$ for any such U. Then we partition subsets of V into $\mathcal{U}_{ij}$ ($j \geq 0$) based on $\delta^+(U; D_i)$:

$$\mathcal{U}_{ij} = \{U \subseteq V \mid f_i^{(U)} > 0, \ 2^{i+j-1} \leq \delta^+(U; D_i) \leq 2^{i+j} - 1\}.$$

In order to analyze the worst situation, we may assume that each edge is sampled with probability strictly less than one, i.e, $p_e = \frac{\rho}{\lambda_e}$. We claim the following:

Each $U \in \mathcal{U}_{ij}$ satisfies (3) with probability at least $1 - 2n^{-(d+7)2^j}$. (4)

To see this, recall that $\lambda_e < 2^{i+1}$ for each $e \in F_i^{(U)}$. Hence

$$p_e = \frac{\rho}{\lambda_e} \geq \frac{\rho}{2^{i+1}}.$$

Therefore by Lemma 15, we have

$$\Pr\left[|f_i^{(U)} - \widehat{f_i^{(U)}}| > \left(\frac{\epsilon}{2}\right) \max\left\{ \frac{\delta^+(U; D_i)}{\gamma}, f_i^{(U)} \right\}\right]$$

$$< 2\exp\left(-0.38 \frac{\epsilon^2}{2^2} \frac{\rho}{2^{i+1}} \max\left\{ \frac{\delta^+(U; D_i)}{\gamma}, f_i^{(U)} \right\}\right)$$

$$\leq 2\exp\left(-0.38 \frac{\epsilon^2}{2^2} \frac{\rho}{2^{i+1}} \frac{\delta^+(U; D_i)}{\gamma}\right).$$

Using $\delta^+(U; D_i) \geq 2^{i+j-1}$ and $\rho = C\gamma \ln n/\epsilon^2$ with $C = 43(d+7)$, the last term is bounded by $2n^{-(d+7)2^j}$.

By (4) and the union bound, the failure probability of (3) is at most

$$\sum_{j \geq 0} |\{F_i^{(U)} \mid U \in \mathcal{U}_{ij}\}| \cdot 2n^{-(d+7)2^j}. \tag{5}$$

To bound $|\{F_i^{(U)} \mid U \in \mathcal{U}_{ij}\}|$ we use Theorem 8. By the connectivity condition of γ-certificates,

$$|\{F_i^{(U)} \mid U \in \mathcal{U}_{ij}\}| \leq |\{F_i^{(U)} \mid \delta^+(U; D_i) \leq 2^{i-1} \cdot 2^{j+1} - 1\}|$$
$$\leq |P(2^{i-1}, 2^{j+1}; D_i)| \leq 2n^{4 \cdot 2^j}. \tag{6}$$

By (5) and (6) the failure probability of (3) is at most

$$\sum_{j\geq 0} 4n^{-(d+3)2^j} \leq \frac{4n^{-(d+3)}}{1-n^{-(d+3)}} \leq \frac{1}{n^{d+2}}.$$

□

Proof (of Theorem 2). In the graph D, there are at most n^2 pairs of vertices, so the number of distinct λ_e is at most n^2. Hence the number of nonempty F_i is at most n^2. Using union bound over these values of i, we can conclude that (3) is satisfied for all i and U with probability at least $1 - 1/n^d$. Thus, from the triangle inequality, we have

$$|\delta^+(U;D) - \delta^+(U;D_\epsilon)| = \left|\sum_{i=0}^{\Lambda} f_i^{(U)} - \sum_{i=0}^{\Lambda} \widehat{f_i^{(U)}}\right| \leq \sum_{i=0}^{\Lambda} |f_i^{(U)} - \widehat{f_i^{(U)}}|$$

$$\leq \frac{\epsilon}{2} \sum_{i=0}^{\Lambda} \max\left\{\frac{\delta^+(U;D_i)}{\gamma}, f_i^{(U)}\right\} \leq \frac{\epsilon}{2}\left(\sum_{i=0}^{\Lambda} \frac{\delta^+(U;D_i)}{\gamma} + \sum_{i=0}^{\Lambda} f_i^{(U)}\right)$$

$$\leq \epsilon \cdot \delta^+(U;D)$$

since γ-overlapped property holds and $\sum_{i=0}^{\Lambda} f_i^{(U)} = \delta^+(U;D)$. Hence we conclude that $D_\epsilon \in (1 \pm \epsilon)D$.

Finally, observe that the expected number of edges in D_ϵ is $\sum_e(1 - (1 - p_e)^{w_e}) \leq \sum_e w_e p_e = O(\sum_e \rho w_e/\lambda_e)$. This completes the proof.   □

References

1. Benczúr, A.A., Karger, D.R.: Approximating s-t minimum cuts in $\tilde{O}(n^2)$ time. In: Proceedings of the 28th Annual ACM Symposium on Theory of Computing, pp. 47–55 (1996)
2. Benczúr, A.A., Karger, D.R.: Randomized approximation schemes for cuts and flows in capacitated graphs. SIAM J. Comput. **44**(2), 290–319 (2015)
3. Christiano, P., Kelner, J.A., Madry, A., Spielman, D.A., Teng, S.H.: Electrical flows, Laplacian systems, and faster approximation of maximum flow in undirected graphs. In: Proceedings of the 43rd Annual ACM Symposium on Theory of Computing, pp. 273–281 (2011)
4. Cohen, M.B., et al.: Almost-linear-time algorithms for Markov chains and new spectral primitives for directed graphs. In: Proceedings of the 49th Annual ACM Symposium on Theory of Computing, pp. 410–419 (2017). A full version is available at http://arxiv.org/abs/1611.00755
5. Ene, A., Miller, G., Pachocki, J., Sidford, A.: Routing under balance. In: Proceedings of the 48th Annual ACM Symposium on Theory of Computing, pp. 598–611 (2016)
6. Fung, W.S., Hariharan, R., Harvey, N.J.A., Panigrahi, D.: A general framework for graph sparsification. In: Proceedings of the 43rd Annual ACM Symposium on Theory of Computing, pp. 71–80 (2011). A full version is available at https://www.cs.ubc.ca/~nickhar/Publications/Sparsifier/Sparsifier-Long.pdf

7. Jackson, B.: Some remarks on arc-connectivity, vertex splitting, and orientation in graphs and digraphs. J. Graph Theory **12**(3), 429–436 (1988)
8. Kachitvichyanukul, V., Schmeiser, B.W.: Binomial random variate generation. Commun. ACM **31**(2), 216–222 (1988)
9. Karger, D.R.: Random sampling in cut, flow, and network design problems. Math. Oper. Res. **24**(2), 383–413 (1999)
10. Karger, D.R., Stein, C.: A new approach to the minimum cut problem. J. ACM **43**(4), 601–640 (1996)
11. Lee, Y.T., Sidford, A.: Path finding methods for linear programming: solving linear programs in $\tilde{O}(\sqrt{\text{rank}})$ iterations and faster algorithms for maximum flow. In: Proceedings of IEEE 55th Annual Symposium on Foundations of Computer Science, pp. 424–433 (2014)
12. Nagamochi, H., Ibaraki, T.: Computing edge-connectivity in multigraphs and capacitated graphs. SIAM J. Discret. Math. **5**(1), 54–64 (1992)
13. Nagamochi, H., Ibaraki, T.: A linear-time algorithm for finding a sparse k-connected spanning subgraph of a k-connected graph. Algorithmica **7**(1), 583–596 (1992)
14. Peng, R.: Approximate undirected maximum flows in $O(m \, \text{polylog}(n))$ time. In: Proceedings of the 27th Annual ACM-SIAM Symposium on Discrete Algorithms, pp. 1862–1867 (2016)
15. Spielman, D.A., Teng, S.H.: Spectral sparsification of graphs. SIAM J. Comput. **40**(4), 981–1025 (2011)

Reconfiguration of Graphs with Connectivity Constraints

Nicolas Bousquet[1][✉] and Arnaud Mary[2]

[1] CNRS, G-SCOP, Grenoble-INP, Univ. Grenoble-Alpes, Grenoble, France
`nicolas.bousquet@grenoble-inp.fr`
[2] LBBE, Université Claude Bernard Lyon 1, Lyon, France
`arnaud.mary@univ-lyon1.fr`

Abstract. A graph G realizes the degree sequence S if the degrees of its vertices is S. Hakimi [5] gave a necessary and sufficient condition to guarantee that there exists a connected multigraph realizing S. Taylor [13] later proved that any connected multigraph can be transformed into any other via a sequence of flips (maintaining connectivity at any step). A flip consists in replacing two edges ab and cd by the diagonals ac and bd. In this paper, we study a generalization of this problem. A set of subsets of vertices $\mathcal{CC}$ is *nested* if for every $C, C' \in \mathcal{CC}$ either $C \cap C' = \emptyset$ or one is included in the other. We are interested in multigraphs realizing a degree sequence S and such that all the sets of a nested collection $\mathcal{CC}$ induce connected subgraphs. Such constraints naturally appear in tandem mass spectrometry.

We show that it is possible to decide in polynomial if there exists a graph realizing S where all the sets in $\mathcal{CC}$ induce connected subgraphs. Moreover, we prove that all such graphs can be obtained via a sequence of flips such that all the intermediate graphs also realize S and where all the sets of $\mathcal{CC}$ induce connected subgraphs. Our proof is algorithmic and provides a polynomial time approximation algorithm on the shortest sequence of flips between two graphs whose ratio depends on the depth of the nested partition.

1 Introduction

Let $G = (V, E)$ be a graph where V denotes the set of vertices and E the set of edges. All along the paper, unless otherwise specified, all the graphs are loop-free but may admit multiple edges. Reconfiguration problems consist in finding a step by step transformation between two solutions of a given problem such that all intermediate states are also solutions. Reconfiguration problems arise in many different fields (e.g. graph theory [2,3], statistical physics [10], combinatorial games [8], chemistry [12] and peer-to-peer networks [4]) and received a considerable attention in the last few years. For a complete overview of the reconfiguration field, the reader is referred to the recent surveys of van den Heuvel [9] and Nishimura [11]. In this paper we consider the reconfiguration of graphs with a fixed degree sequence and its applications to cheminformatics.

L. Epstein and T. Erlebach (Eds.): WAOA 2018, LNCS 11312, pp. 295–309, 2018.
https://doi.org/10.1007/978-3-030-04693-4_18

The *degree sequence* of a graph G is the sequence of the degrees of its vertices in non-increasing order. Given a non-increasing sequence of integers $S = \{d_1, \ldots, d_n\}$, a graph $G = (V, E)$ with $V = \{v_1, \ldots, v_n\}$ *realizes* S if $d(v_i) = d_i$ for all $i \leq n$.

In the fifties, mathematicians tried to find conditions that guarantee that given a sequence of integers $S = \{d_1, \ldots, d_n\}$, there exists a graph realizing S. Senior [12] gave necessary and sufficient conditions for the case of connected (multi)graphs. Havel [7] proposed a polynomial time algorithm that outputs a simple loop-free graph realizing S if such a graph exists or returns no otherwise. Hakimi [5] re-discovered the results of both Senior and Havel and also proposed an algorithm that outputs a connected loop-free graph realizing S if such a graph exists or returns no otherwise.

A *flip* (also called *swap* or *switch* in the literature) on two edges ab and cd consists in deleting the edges ab and cd and creating the edges ac and bd (or ad and bc)[1]. The flip operation that transforms the edges ab and cd into the edges ac and bd is denoted $(ab, cd) \to (ac, bd)$. When the target edges are not important we will simply say that we flip the edges ab and cd.

Let $S = \{d_1, \ldots, d_n\}$ be a non-increasing sequence and let G and H be two graphs on n vertices $v_1, \ldots, v_n$ realizing S. The graph G can be *transformed* into H if there is a sequence of flips that transforms G into H. Since flips do not modify the degree sequence, the intermediate graphs also realize S. Let $\mathcal{G}(S)$ be the graph whose vertices are loop-free multigraphs realizing S and such that two vertices G and H of $\mathcal{G}(S)$ are adjacent if G can be transformed into H via a single flip. Since the flip operation is reversible, the graph $\mathcal{G}(S)$ is an undirected graph called the *reconfiguration graph of S*. Note that there exists a sequence of flips between any pair of graphs realizing S if and only if the graph $\mathcal{G}(S)$ is connected. In [6], Hakimi proved the following:

Theorem 1 (Hakimi [6]). *Let S be a non-increasing sequence. If the graph $\mathcal{G}(S)$ is not empty, it is connected.*

A connected reconfiguration graph has some interesting consequences for sampling or enumerating solutions. For instance, it implies that all the solutions can be enumerated with polynomial delay (as long as we get one of them). An enumeration algorithm is an algorithm that lists without repetition all the solutions of a given problem. An algorithm solves an enumeration problem *with polynomial delay*, if the delay between two consecutive outputs is bounded by a polynomial of the input size. Any reconfiguration problem such that the reconfiguration graph is connected and the number of local operations (in our case flips) is polynomial admits a polynomial delay enumeration algorithm. So Theorem 1 ensures that there exists an algorithm that enumerates with polynomial delay all the graphs realizing S. Note however that the space needed by this algorithm might be exponential. As far as we know, the existence of a polynomial delay algorithm with polynomial space to generate all the graphs realizing S is open.

---

[1] In the case of multigraphs, we simply decrease by one the multiplicities of edges ab and cd and increase by one the ones of ac and bd.

One can wonder if the reconfiguration graph is still connected if only we consider graphs with additional properties. For a graph property Π, let us denote by $\mathcal{G}(S, \Pi)$ the subgraph of $\mathcal{G}(S)$ induced by the graphs realizing S with the property Π. If we respectively denote by $\mathscr{C}$ and $\mathscr{S}$ the property of being connected and simple, Taylor proved in [13] that $\mathcal{G}(S, \mathscr{C})$, $\mathcal{G}(S, \mathscr{S})$ and $\mathcal{G}(S, \mathscr{C} \wedge \mathscr{S})$ are connected ($\wedge$ stands for "and"). Let G, H be two graphs of $\mathcal{G}(S, \Pi)$. A sequence of flips *transforms* G *into* H in $\mathcal{G}(S, \Pi)$ if the sequence of flips transforms G into H and all the intermediate graphs also have the property Π. Note that it is equivalent to find a path between G and H in $\mathcal{G}(S, \Pi)$.

Applications to Mass Spectrometry. Mass spectrometry is a technique used to measure the mass-to-charge (m/z) ratio of molecules. The process results in a m/z-spectrum, whose deconvolution provides a histogram with the quantity of each complex. Given this histogram, chemists can determine how many atoms of each type compose the molecule (i.e. the chemical formula of the molecule). With this chemical formula, we want to understand the structure of the molecule. Two question naturally arise: (i) Can we find a molecule structure satisfying this formula? (ii) Can we find all of them? Two molecules with the same chemical formula are called *structural isomers.*

The problem of determining the structure of a molecule given its chemical formula, can be formulated as a combinatorial problem. Let $v_1, \ldots, v_n$ be the n atoms of the molecules. The degree of each atom v_i is its valence. The questions then become: (i) Can we find a connected loop-free (multi-)graph on vertices $v_1, \ldots, v_n$ for which the degree of each v_i is equal to the valence of its corresponding atom? (ii) If yes, can we generate all of them? As we have already seen, Hakimi [6] and Taylor [13] answered positively to both questions: we can enumerate with polynomial delay all the graphs in $\mathcal{G}(S, \mathscr{C})$.

In the last few years, with the development of tandem mass spectrometry, we get more information on the structure of the original molecule. With this technology, we can again break the molecule into several fragments which in turn can be broken into other fragments...etc... For each produced fragment of this subdivision, we can determine its atoms constitution. Finally, we can obtain a tree of fragments, where each fragment corresponds to a part of the molecule that have to be connected. Rephrased in terms of graphs, it means that instead of simply knowing that the whole graph is connected, we are given a collection of subsets of vertices that have to induce connected subgraphs. Since the number of graphs realizing a degree sequence is usually large, this additional information can drastically reduce the number of possible molecules.

Our Results. A collection $\mathcal{CC}$ of subsets of vertices is *nested* if for every pair C_i, C_j in $\mathcal{CC}$, either $C_i \cap C_j = \emptyset$ or one is included in the other. The *height* d of a nested partition $\mathcal{CC}$ is the maximum number of sets $C_{i_1}, \ldots, C_{i_d}$ in $\mathcal{CC}$ such that $C_{i_1} \subsetneq C_{i_2} \subsetneq \ldots \subsetneq C_{i_d}$.

Let $S = \{d_1, \ldots, d_n\}$ be a degree sequence and $\mathcal{CC}$ be a nested collection such that $V \in \mathcal{CC}$. Let us denote by $\mathcal{G}(S, \mathcal{CC})$ the subgraph of $\mathcal{G}(S)$ induced by the graphs such that $G[C]$ is connected for C in CC. We study the three following questions:

(i) Is it possible to find in polynomial time a graph G in $\mathcal{G}(S, CC)$ if such a graph exists?
(ii) Is $\mathcal{G}(S, CC)$ a connected subgraph of $\mathcal{G}(S)$?
(iii) If yes, is it possible to find or approximate a shortest transformation between two graphs of $\mathcal{G}(S, CC)$?

In Sect. 4, we answer positively to (i). We actually provide a necessary and sufficient condition for a graph to be realizable and then prove that this characterization can actually be turned into an algorithm.

In Sect. 5, we answer to both (ii) and (iii). We show that, given two graphs G, H in $\mathcal{G}(S, CC)$, there always exists a transformation between G and H in $\mathcal{G}(S, CC)$. To prove it, we exhibit an algorithm that finds a sequence of at most $(2d + 1)\delta(G, H)$ flips transforming G into H, where d is the height of the nested partition and $\delta(G, H)$ is the size of the *symmetric difference* (see paragraph Notations for a formal definition). Since the length of a minimum transformation between G and H is at least $\delta(G, H)/4$ (a flip decreases by at most four the size of the symmetric difference), we get an $(8d + 4)$-approximation of the shortest sequence. Note that it also provides as an immediate corollary a polynomial delay algorithm to enumerate all the graph in $\mathcal{G}(S, CC)$.

Theorem 2. *Let CC be a nested collection of subsets that contains V. The graph $\mathcal{G}(S, CC)$ is connected and the distance between any pair of graphs G and H in $\mathcal{G}(S, CC)$ is at most $(2d + 1)\delta(G, H)$.*

Moreover, there is a polynomial time algorithm that, given $G, H \in \mathcal{G}(S, CC)$, computes a sequence of flips transforming G into H in $\mathcal{G}(S, CC)$ of length at most $(8d + 4)OPT$ where OPT denotes the length of a shortest sequence.

We moreover show that finding a shortest sequence of flips between two graphs in $\mathcal{G}(S, \mathscr{C})$ is NP-complete. It in particular implies as an immediate corollary that it is NP-complete for two graphs in $\mathcal{G}(S, CC)$. Due to space restriction, the proof of this result is not included in this short version. The proof follows the scheme of the NP-hardness proof of Will [14] in the case of simple graphs.

In order to prove Theorem 2, we need as a black-box an approximation algorithm of the shortest transformation between two graphs in $\mathcal{G}(S, \mathscr{C})$. In Sect. 2, we give an algorithm that provides a transformation from G into H in $\mathcal{G}(S, \mathscr{C})$ of size at most four times the optimal one for any pair of graphs G, H in $\mathcal{G}(S, \mathscr{C})$. This result also provides an alternative proof of the result of Taylor [13]. Most of the proofs are not included in this extended abstract, we refer the reader to a complete version for all the details[2].

Notations.
All along the paper, we consider unoriented loop-free multigraphs. Given two graphs G and H on the same vertex set V, we denote by $G \Delta H$ their symmetric

---

[2] Available online at https://arxiv.org/abs/1809.05443.

difference i.e. the (multi)set of edges such that e appears in $G\Delta H$ with multiplicity $r > 0$ if the difference between the multiplicities of e in G and in H is equal to r or $-r$. We denote by $\delta(G, H)$ the size of $G\Delta H$. By abuse of notations, we often assimilate $G\Delta H$ to the graph $G = (V, G\Delta H)$. We denote by $G - H$ the (multi)set of edges such that e appears in $G - H$ with multiplicity $r > 0$ if the difference between the multiplicities of e in G and in H is equal to r. For simple graphs it simply corresponds to the edges which are in G and not in H. We also assimilate $G - H$ to the graph $(V, G - H)$. Note that $\delta(G, H)$ is twice the number of edges in $G - H$. Finally let $G \cap H$ be the set of edges containing e is an edge with multiplicity r if the minimum multiplicity of e in G and H is exactly r. As for $G - H$ and $G\Delta H$, we assimilate $G \cap H$ to $(V, G \cap H)$. Note that $E(G) = E(G \cap H) \cup E(G - H)$.

2   4-Approximation for Connected Graphs

Let S be a non-increasing degree sequence. In [13], Taylor proved that $\mathcal{G}(S, \mathscr{C})$ is connected. However, his proof does not immediately provide an approximation algorithm of the shortest transformation between pairs of graphs in $\mathcal{G}(S, \mathscr{C})$. In this section we give an alternative proof of the result of Taylor that provides a 4-approximation algorithm of the shortest transformation between any pair of graphs in $\mathcal{G}(S, \mathscr{C})$.

Theorem 3. *Let S be a non-increasing degree sequence and G and H in $\mathcal{G}(S, \mathscr{C})$. We can find in polynomial time a sequence of at most $\delta(G, H)$ flips transforming G into H in $\mathcal{G}(S, \mathscr{C})$.*

Moreover this transformation never flips any edge that is already in both G and H[3].

The size of a shortest transformation is at least $\delta(G, H)/4$ since at most two edges of G can be flipped on edges of H at every step. So Theorem 3 provides a 4-approximation algorithm. The remaining of this section is devoted to prove the following lemma whose iterated application immediately implies Theorem 3. We say that a flip *maintains connectivity* (of a connected graph G) if the resulting graph after the flip is still connected. A sequence of flips maintains connectivity if all intermediate graphs are connected.

Lemma 1. *Let S be a non-increasing degree sequence and G and H in $\mathcal{G}(S, \mathscr{C})$. There exists a sequence of at most two flips in maintaining connectivity that decreases by at least two the size of the symmetric difference. Moreover the sequence never flips any edge that is already in both G and H.*

Proof. In order to prove it, let us first prove that there exist cases where we can easily decrease the symmetric difference in one step. Due to space restriction the proof of the next claim is omitted.

---

[3] We say that an edge e in $G \cap H$ is never flipped if the multiplicity of e at any intermediate step never goes below the multiplicity of e in $G \cap H$.

Claim. If an edge of $G - H$ is contained in a cycle of G^4, then there exists a flip maintaining the connectivity that decreases by at least two the size of the symmetric difference.

If the claim can be applied, Lemma 1 holds. So we can assume that all the edges of $G - H$ do not belong to a cycle of G, i.e. all of them are bridges. Let G' be the graph obtained from G by contracting all the connected components of $G \cap H$ into a single vertex. And there is an edge $S_1 S_2$ in G' if there is an edge $u_1 u_2$ of G with u_1 in S_1 and u_2 in S_2. We say that $v_1 v_2$ *corresponds to* $S_1 S_2$ in G'. Note that the claim ensures that there is a bijection between the edges of $G - H$ and the edges of G'. Since no edge of $G - H$ is contained in a cycle of G, the graph G' is a tree (without multiedges). In that case, we can prove that it is always possible to find a sequence of at most two flips that decreases the symmetric difference by at least 2. A formal proof of this fact is proposed in the full version of the paper.

3   Tree of the Fragments

Let V be a vertex set and S be a degree sequence of size $|V|$. Let $\mathcal{CC}$ be a nested collection of subsets of V such that V and all the singletons belong to $\mathcal{CC}$. Singletons are included in $\mathcal{CC}$ for convenience since their addition does not change the graph $\mathcal{G}(S, \mathcal{CC})$. Indeed, a single vertex induces a connected subgraph.

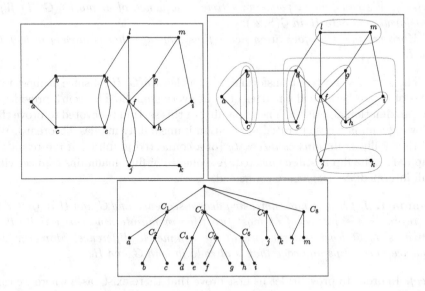

Fig. 1. A graph G, a nested partition $\mathcal{CC}$ and the tree of the fragments of $\mathcal{CC}$.

---

[4] Two parallel edges form a cycle of length 2.

Let G and H be two graphs in $\mathcal{G}(S, \mathcal{CC})$. Let $e = uv$ be an edge with multiplicity r_1 in G and r_2 in H and let $r = \min(r_1, r_2)$. Then r copies of e are *good* and the others are bad. (In case of simple graph an edge e of G is good if it is also an edge of H and it is bad otherwise). A flip in G is *correct* if it maintains the connectivity of $G[C]$ for any $C \in \mathcal{CC}$. A flip *does not modify a good edge* e if the multiplicity of e after the flip is still at least the multiplicity of e in $G \cap H$.

Let $\mathcal{CC}$ be a nested collection. The *tree of the fragments* T is the tree whose nodes are labeled by elements of $\mathcal{CC}$ and there is an arc from C_1 to C_2 if $C_2 \subseteq C_1$ and there does not exist $C \in \mathcal{CC}$ distinct from C_1 and C_2 such that $C_2 \subseteq C \subseteq C_1$. In other words, T is the tree rooted at V corresponding to the nested partition of $\mathcal{CC}$ (see Fig. 1 for an illustration). By abuse of notation and when no confusion is possible, C will denote both the node of T and the corresponding set in $\mathcal{CC}$. Since $\mathcal{CC}$ contains all the sets of size one, the leaves of T are the vertices of V. Since $\mathcal{CC}$ is nested, T is well-defined and is a tree. Given a node C of the tree of the fragments and $v \in V$, $v \in C$ if and only if the leaf labeled with v is a leaf of the subtree rooted at C. We denote by $G[C]$ the subgraph of G induced by the vertices in C.

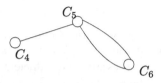

Fig. 2. The graph $G^{ch}(C_3)$

Given a node C, we denote by $ch(C)$ the children of C in T. Let C be an internal node of T and $G \in \mathcal{G}(S, \mathcal{CC})$. We denote by $G^{ch}(C)$, the graph with vertex set $ch(C)$ where C' and C'' in $V(G^{ch}(C))$ are adjacent with multiplicity k if there exist exactly k edges with one endpoint in C' and one endpoint in C'' in G. See Fig. 2 for an illustration.

Note that there is a natural bijection between edges of G and edges of $G^{ch}(C)$. Indeed, for any edge e of G, there exists a unique $C \in \mathcal{CC}$ in which an edge is created in $G^{ch}(C)$ because of e. The following remark follows from that observation.

Remark 1. Let G be a graph in $\mathcal{G}(S, \mathcal{CC})$.

1. For every $C \in T$, $|E(G[C])| = \bigcup_{C' \subseteq C} |E(G^{ch}(C'))|$;
2. Let $C \in T$ and let $e \in E(G^{ch}(C))$, then $G^{ch}(C') - e$ is connected for every $C' \in T$, $C' \neq C$.

The second point holds since e has no impact on $G^{ch}(C')$ for $C' \neq C$. Let us first prove the following lemma that will be used all along the proof.

Lemma 2. *Let G be a graph in $\mathcal{G}(S)$.*

– If $G[C]$ is connected, then $G^{ch}(C)$ is connected.
– Let C be a node of T and T' be the subtree rooted at C. If $G^{ch}(C')$ is connected for every node C' of T' then $G[C']$ is connected for every node C' of T'.

The second point will be applied in several situations since manipulating $G^{ch}(C)$ is often simpler than manipulating $G[C]$. Note that when C is the root, the conclusion of the second point ensures that G is in $\mathcal{G}(S, \mathcal{CC})$.

Let V_1, V_2 be a partition of V. Then $E_G(V_1, V_2)$ denotes the set of edges of G with one endpoint in V_1 and one endpoint in V_2.

Lemma 3. *Let G, H be two graphs of $\mathcal{G}(S, \mathcal{CC})$ and let $C \in \mathcal{CC}$. If $|E_G(C, V \setminus C)| < |E_H(C, V \setminus C)|$, then there exists $e \in G[C] - H[C]$ such that $G[C'] - e$ is connected for every node C' of T.*

Note that when we say *"there exists $e \in G[C] - H[C]$"* in the statement of Lemma 3, the edge e might exist in both G and H, but in that case the multiplicity of e in H has to be strictly smaller than the multiplicity of e in G.

4   Realizability

Let $V = \{v_1, \ldots, v_n\}$ be a set of vertices and let $\mathcal{CC}$ be a nested partition containing V and all the singletons. Let T be the tree of the fragments of $\mathcal{CC}$. In this section, we provide a necessary and sufficient condition on the degree sequence $S = \{d_1, \ldots, d_n\}$ to be realizable by a loop-free multigraph such that C induces a connected subgraph for every $C \in \mathcal{CC}$. This characterization generalizes the ones of [5] and [12] for loop-free multigraphs since in this case $\mathcal{CC} = \{V\}$. We end the section by explaining how this characterization can be turned into a polynomial time algorithm.

Let C be a node of T. A graph G' is *coherent on C* if it is defined on C and:

– for every $v_i \in C$, $d_{G'}(v_i) \leq d_i$ and,
– for every $C' \in \mathcal{CC}$ such that $C' \subseteq C$, $G'[C']$ is connected.

Let G' be a graph *coherent* on C. The *degree-deficit* of C (for G') is equal to $\sum_{v_i \in C}(d_i - d_{G'}(v_i))$. In other words, the degree-deficit represents the amount of endpoints of edges "missing" to complete the degree of the vertices of C. Note that if a graph G defined on V realizes T, then the degree-deficit of C is the number of edges with one endpoint in C and one endpoint in $V \setminus C$. Moreover for any node C, the graph $G[C]$ is coherent on C. Let us start with a straightforward remark.

Remark 2. Let C' and C'' be two disjoint sets in $\mathcal{CC}$. Let G' be a graph defined on $C' \cup C''$. If the degree-deficits of both C' and C'' are positive, then there exists $u \in C'$ and $v \in C''$ such that the edge uv can be added in G' without violating any degree constraint.

We now define $\ell(C)$ and $u(C)$. We will then prove that, given a graph G in $\mathcal{G}(S, \mathcal{CC})$, they correspond to respectively the minimum and the maximum degree-deficit of C for any graph G' coherent on C. For any node C in $\mathcal{CC}$, we define

$$u(C) := \sum_{v_i \in C} d_i - (2|C| - 2)$$

and

$$\ell(C) = \begin{cases} d_i & \text{if } C \text{ is the leaf } v_i \\ \varphi(C) & \text{if } \varphi(C) \geq 0 \\ 0 & \text{if } \varphi(C) < 0 \text{ and } \varphi(C) \text{ is even} \\ 1 & \text{if } \varphi(C) < 0 \text{ and } \varphi(C) \text{ is odd} \end{cases}$$

where

$$\varphi(C) = \max_{C_j \in ch(C)} \left(\ell(C_j) - \sum_{\substack{C_i \in ch(C) \\ C_i \neq argmax\{\ell(C_j)\}}} u(C_i) \right).$$

Lemma 4. *Let S be a degree sequence and $\mathcal{CC}$ be a nested partition. Let G in $\mathcal{G}(S, \mathcal{CC})$. Then for every node C of the tree of the fragments, the degree-deficit $s(C)$ of C satisfies:*

$$\ell(C) \leq s(C) \leq u(C)$$

Moreover, $\ell(C)$ and $u(C)$ are even if and only if $\sum_{v_i \in C} d_i$ is even.

Proof. The number of edges of a connected subgraph on n vertices is at least $n-1$. Since $G[C]$ is connected, there are at least $|C|-1$ edges with both endpoints in C and then we have $s(C) \leq u(C) = \sum_{v_i \in C} d_i - (2|C| - 2)$. Since we removed an even value from $\sum_{v_i \in C} d_i$, $u(C)$ is even if and only if $\sum_{v_i \in C} d_i$ is even.

Let us now prove that $s(C) \geq \ell(C)$. We prove it by induction bottom-up from the leaves. If C is a leaf, then $C = \{v_i\}$ and there are exactly d_i edges between v_i and its complement in G. Thus $s(C) = E(C, V \setminus C) = \ell(C)$ and the parity of $s(C)$ is indeed the one of $\ell(C)$.

Now, let C be an internal node. By induction hypothesis, for every child C_i of C, the parities of $\ell(C_i)$ and $u(C_i)$ are the parity of $\sum_{v_i \in C_i} d_i$. Thus, by definition of $\ell(C)$, the parity of $\ell(C)$ is the parity of $\sum_{v_i \in C} d_i$. So, in particular, if $\varphi(C) \leq 1$, the conclusion holds since the degree-deficit cannot be negative. So we can assume that $\varphi(C) \geq 2$. Let us denote by $C_1, \ldots, C_r$ the children of C and we can assume w.l.o.g. that C_1 is the child of C satisfying $\varphi(C) = \ell(C_1) - \sum_{i \geq 2} u(C_i)$. Let $N := \sum_{i \geq 2} u(C_i)$. The first part of the proof ensures that the degree-deficit of $\cup_{i \geq 2} C_i$ is at most N. Moreover, by induction, the degree-deficit of C_1 is at least $\ell(C_1)$. So the maximum number of edges between C_1 and $\cup_{i \geq 2} C_i$ in a graph satisfying all the constraints is N (since $\ell(C_1) > N$). So the degree-deficit of C is at least $\ell(C_1) - N$ which completes the proof. $\square$

The goal of this section consists in proving the following theorem that ensures that it suffices to look at the values $u(C)$ and $\ell(V)$ to determine if a graph is realizable.

Theorem 4. *Let S be a degree sequence and $\mathcal{CC}$ be a nested partition containing V. There exists a graph in $\mathcal{G}(S, \mathcal{CC})$ if and only if:*

1. *For every internal node C of the tree of the fragments, $u(C) \geq 1$.*
2. *$\ell(V) = 0$ and $u(V) \geq 0$.*

Let G be a graph in $\mathcal{G}(S, \mathcal{CC})$. Then the degree-deficit of V equals 0. So by Lemma 4 applied on V, $\ell(V) = 0$ and $u(V) \geq 0$ is necessary. Moreover, since G is connected, at least one edge has to have one endpoint in C and one endpoint in $V \setminus C$ for every $C \subsetneq V$. Thus the degree-deficit of C is at least one for every $C \in \mathcal{CC}$, $C \neq V$, and then Lemma 4 ensures that the first condition is necessary. To prove Theorem 4, we have to show that these two conditions are sufficient. The sufficiency is an immediate corollary of the next lemma applied to V with $s = \ell(V) = 0$.

Lemma 5. *Let S be a degree sequence and $\mathcal{CC}$ be a nested partition. Let C be a node of the tree of the fragments T such that $u(C) \geq 0$ and, for every $C' \subsetneq C$, $u(C') \geq 1$. For every s such that $\ell(C) \leq s \leq u(C)$ and such that s has the same parity as $\ell(C)$ and $u(C)$, there exists a graph G' coherent on C with degree-deficit s.*

In order to prove Lemma 5, we need to show the following lemma as an intermediate step. Its proof is not included in the extended abstract.

Lemma 6. *Let S be a degree sequence and $\mathcal{CC}$ be a nested partition. Let C be a node of the tree of the fragments T such that $u(C) \geq 0$ and such that, for every $C' \subsetneq C$ we have $u(C') \geq 1$. Then there exists graph G' coherent on C with degree-deficit $u(C)$.*

Using Lemma 6, we can now prove Lemma 5 via decreasing induction. Due to space restriction, the proof is not included in this extended abstract. The proof consists in proving that if no edge can be added to decrease the degree-deficit of C, then graph structure is constrained. The core of the proof consists in showing that we can slightly modify this structure, either immediately or by induction (on subsets of C) in such a way an edge can be added in the subgraph induced by C without violating any constraint.

Lemma 7. *The proof of Theorem 4 can be turned into a polynomial time algorithm.*

5   Connectivity of $\mathcal{G}(S, \mathcal{CC})$ and Approximation Algorithm

Well-Structured Subtrees and Extensions. A subtree T' of the tree of the fragment T is *well-structured* if

(i) it contains the root, and

(ii) if u, v are two children of $w \in T'$ then either both u and v are in T' or none of them is in T' (see Fig. 3 for an illustration).

A set $C \in \mathcal{CC}$ is in a well-structured subtree T' if the node labeled by C is in T'. Given a well-structured subtree T', we denote by $\mathcal{CC}_{T'}$ the subset of $\mathcal{CC}$ corresponding to the *inner* nodes of T', i.e. all the nodes of T' but the leaves. Note that the root of T and the whole tree T are well-structured.

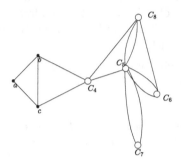

Fig. 3. The graph $G^s(T')$ where the leaves of the well-structured subtree T' are $a, b, c, C_4, C_5, C_6, C_7$. The graph G and the tree T are the ones of Fig. 1.

The graph *inherited from a well-structured subtree* $T' \subseteq T$ denoted by $G^s(T')$ is the graph where the vertex set is the set of leaves of T' and where there is an edge between X and Y with multiplicity α if there are α edges with one endpoint in X and one endpoint in Y in G. Note that $G^s(T)$ is the graph G. Let $C \in \mathcal{CC}_{T'}$. We denote by $G^s(T')[C]$ the subgraph of $G^s(T')$ induced by the leaves of the subtree of T' rooted at C. A flip in $G^s(T')$ is T'-*correct* if it maintains the connectivity of $G^s(T')[C]$ for any $C \in \mathcal{CC}_{T'}$. Note that a correct flip is a T-correct flip.

For the sake of readability, vertices of $G^s(T')$ will be denoted with capital letters and vertices of the original graph G will be denoted with lower case letters. Due to space restriction, the proof of the following lemma is not included in this extended abstract.

Lemma 8. *Let S be a degree sequence and $\mathcal{CC}$ be a nested partition. Let G and H be two graphs in $\mathcal{G}(S, \mathcal{CC})$. Let T' be a well-structured subtree of the tree of the fragments T and let $(AB, CD) \to (AC, BD)$ be a T'-correct flip.*

We can find in polynomial time a flip $(ab, cd) \to (ac, bd)$ of G where a, b, c and d are respectively in $G[A], G[B], G[C]$ and $G[D]$ which is correct for G. Moreover if AB and CD are in $G^s(T') - H^s(T')$ then we can assume that ab and cd are in $G - H$.

Lemma 8 permits to work with the graphs $G^s(T')$ and $H^s(T')$ and ensures that if we make a flip on one of these graphs, it can be simulated by a flip in the original graph. A natural question immediately arises, how can we ensure that the graph G obtained after the flip still satisfies all the constraints? The next lemma will permit to answer this question.

Let G and H be two graphs in $\mathcal{G}(S, \mathcal{CC})$. Let T' be a well-structured subtree of the tree of the fragments T. We can extend the notion of good and bad edges to the graphs $G^s(T')$ and $H^s(T')$. (It is *good* if it is in both graphs and *bad* otherwise).

Let T be the tree of the fragments and T_1 be a well-structured subtree. The tree T_2 is an *extension* of T_1 on *extension node* C if C is a leaf of T_1 and T_2 is T_1 plus all the children of C in T. The set of children $\mathcal{X}$ of C is then called the set of *special vertices* of $G^s(T_2)$. Any flip between two edges of $G^s(T_2)$ with at least one endpoint in $\mathcal{X}$ which:

- maintains the connectivity of $G^s(T_2)[\mathcal{X}]$ and,
- does not create any edge in $G^s(T_2) \setminus \mathcal{X}$.

is called a *special flip*.

Lemma 9. *Let S be a degree sequence and $\mathcal{CC}$ be a nested partition. Let G in $\mathcal{G}(S, \mathcal{CC})$. Let T_2 be an extension of a well-structured subtree T_1. Any special flip is T_2-correct.*

Two graphs G and H agree on well-structured subtree T' if $G^s(T') = H^s(T')$. Note that if G and H agree on T then $G = H$.

The connectivity of $\mathcal{G}(S, \mathcal{CC})$ and the approximation algorithm will follow from the next two lemmas. Before stating them formally, let us briefly describe the ideas of the proof. We will start with a graph the well-structured subtree T' reduced to the root. This well-structured subtree will grow little by little during the proof until $T' = T$. Our goal consists in transforming $G^s(T')$ into $H^s(T')$ via special flips. Lemma 9 ensures that this sequence of flips is T'-correct and Lemma 8 ensures that this sequence can be adapted into a sequence of flips for G that are correct. However, in order to be able to find such a transformation we first need to transform the graphs in such a way the two graphs $G^s(T')$ and $H^s(T')$ have the same degree sequence. Lemma 10 will ensure that it is possible to assume it. Then Lemma 11 will guarantee that a sequence of special flips permits to transform $G^s(T')$ into $H^s(T')$. So we finally obtain two graphs (still denoted by G and H for convenience) such that $G^s(T') = H^s(T')$. In that case, we perform an extension on a leaf of the subtree T' and repeat the process until $T' = T$. At the end of this last step, we get $G = H$ since $G = G^s(T') = H^s(T') = H$.

Lemma 10. *Let S be a degree sequence and $\mathcal{CC}$ be a nested partition. Let T_2 be an extension of a well-structured subtree T_1 on extension node C. Let G, H be two graphs of $\mathcal{G}(S, \mathcal{CC})$ that agree on T_1.*

We can find in polynomial time a sequence of correct flips that transform G into G' and H into H' in such a way $G'^s(T_2)$ and $H'^s(T_2)$ have the same degree sequence and G', H' still agree on T_1. The number of flips in the sequence is at most $\delta(G^s(T_2), H^s(T_2))/2$ and no flip modifies a good edge. Moreover we have $\delta(G'^s(T_2), H'^s(T_2)) \leq \delta(G^s(T_2), H^s(T_2))$.

The proof, not included in this extended abstract, is based on applications of Lemma 3 which permits to equilibrate degrees. The most technical part consists in proving that all the constraints are still satisfied and that the symmetric difference is not increasing.

Lemma 11. *Let T_2 be an extension of a well-structured subtree T_1 on extension node C. Let G, H be two graphs of $\mathcal{G}(S, \mathcal{CC})$ that agree on T_1 and such that $G^s(T_2)$ and $H^s(T_2)$ have the same degree sequence.*

We can find in polynomial time a sequence of at most $\delta(G^s(T_2), H^s(T_2))$ special flips transforming $G^s(T_2)$ into $H^s(T_2)$ which only flips bad edges.

Note that Lemma 9 then ensures that this sequence of special flips is a sequence of correct flips. The proof of Lemma 11 consists in creating an auxiliary graph such that all the possible flips maintaining connectivity correspond to special flips. The construction of such a graph is possible since G and H agree on T_1. We can finally use Theorem 3 to conclude that there exists a transformation from $G^s(T_2)$ to $H^s(T_2)$).

The algorithm. Let us now present the algorithm to compute a sequence of correct flips that transform G into H:

Procedure 1. Find a sequence of flip that transforms G into H

1: Compute the tree of the fragments T rooted at r.
2: $T - 1 \leftarrow$ Root of T.
3: $G_1 \leftarrow G, H_1 \leftarrow H$.
4: **while** $T_1 \neq T$ **do**
5:     Let C be a leaf of T_1 which is an internal node of T.
6:     Let $_2$ be well-structured subtree inherited from T' on extension node C.
7:     Transform G_1 and H_1 into G_2 and H_2 via a sequence of T_2-correct flips in such a way they have the same degree sequence on $V(G^s(T_2))$ using Lemma 10.
8:     Transform G_2 into G_3 via a sequence of T_2-correct flips in such a way the two graphs G_3 and H_2 agree on T_2 using Lemma 11.
9:     $G_1 \leftarrow G_3, H_1 \leftarrow H_2, T_1 \leftarrow T_2$.
10: **end while**

Let us first prove the correctness of the algorithm. Lemma 10 ensures that it is possible to transform in polynomial time the graph the graphs G_1 and H_1 in such a way all the connectivity constraints are still satisfied and the two resultats graphs G_2 and H_2 have the same degree sequence on T_2. Moreover G_2 and H_2 still agree on T_1. So the step of line 7 can be performed in polynomial time. Moreover, Lemma 11 ensures that there exists a transformation using only T_2-correct flips that transform $G_2^s(T_2)$ into $H_2^s(T_2)$. Lemmas 8 and 9 ensure that this sequence can be transformed into a sequence of correct flips that transforms G_2 into G_3 and such that $G_2^s(T_2) = H_2^s(T_2)$. So the step of line 8 can be performed (in polynomial time). When the algorithm stops, the graph G_1 and H_1 agree on T and then the two graphs are the same.

Theorem 5. *Algorithm 1 provides a sequence of flips of length at most $(2d + 1)\delta(G, H)$ transforming G into H in $\mathcal{G}(S, CC)$. In particular, it provides a $(8d + 4)$-approximation algorithm of the shortest sequence.*

6   Conclusion and Open Problems

In Sect. 2, we provide a 4-approximation algorithm to transform a connected multigraph into another. In this paper, we were simply interested in the existence of a constant approximation algorithm in order to obtain an approximation algorithm for the generalized problem. We did not make any attempts to optimize our bound. It is likely that a more careful analysis provides a better approximation ratio. In particular, if one can prove that there always exists a flip that creates a good edge (without breaking ones) that does not disconnect the graph, then we would immediately obtain a 2-approximation algorithm. Recently, Bereg and Ito [1] provide a 3/2-approximation algorithm to transform a multigraph into another based on the symmetric circuit partition, beating the trivial 2-approximation algorithm in that case. A similar technique might be useful for significantly improve the approximation ratio for connected graphs.

The approximation ratio of Theorem 2 depends on the depth of the tree of the fragment. Can we avoid this dependency and simply find an approximation algorithm that does not depend of the depth? Note that we did not try to optimize the constant in front of the linear function of d in the approximation ratio that is probably easily improvable.

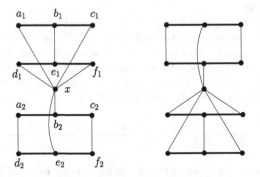

Fig. 4. It is impossible to transform the left graph into the right graph when the set of subsets that have to induce connected subgraphs is the set of thick edges plus the sets $\{a_1, b_1, d_1, x\}, \{a_1, b_1, e_1, x\}, \{b_1, c_1, f_1, x\}, \{b_1, c_1, e_1, x\}, \{a_2, b_2, d_2, x\}, \{a_2, b_2, e_2, x\}, \{b_2, c_2, f_2, x\}$ and $\{b_2, c_2, e_2, x\}$.

In practice, instead of one tree of the fragments, we are often given a collection of trees of the fragments instead of one. It means that subsets of vertices that have to be connected might intersect and not be contained one in the other. One can wonder if it is still true that the reconfiguration graph is connected in this setting? Unfortunately the answer to this question is negative, for instance in the graph provided in Fig. 4.

If yes, does it always exist a transformation that is linear in the size of the symmetric difference and can we approximate it in polynomial time?

Acknowledgments. The authors want to thank the anonymous reviewers of WAOA for their careful reading of the paper which permits to significantly improve its quality.

References

1. Bereg, S., Ito, H.: Transforming graphs with the same graphic sequence. J. Inf. Process. **25**, 627–633 (2017). https://doi.org/10.2197/ipsjjip.25.627
2. Bonamy, M., Bousquet, N.: Recoloring graphs via tree decompositions. Eur. J. Comb. **69**, 200–213 (2018). https://doi.org/10.1016/j.ejc.2017.10.010
3. Cereceda, L., van den Heuvel, J., Johnson, M.: Finding paths between 3-colorings. J. Graph Theory **67**(1), 69–82 (2011). https://doi.org/10.1002/jgt.20514
4. Cooper, C., Dyer, M., Greenhill, C.: Sampling regular graphs and a peer-to-peer network. Comb. Probab. Comput. **16**(4), 557–593 (2007)
5. Hakimi, S.L.: On realizability of a set of integers as degrees of the vertices of a linear graph. I. J. Soc. Ind. Appl. Math. **10**(3), 496–506 (1962). http://www.jstor.org/stable/2098746
6. Hakimi, S.L.: On realizability of a set of integers as degrees of the vertices of a linear graph II. Uniqueness. J. Soc. Ind. Appl. Math. **11**(1), 135–147 (1963). http://www.jstor.org/stable/2098770
7. Havel, V.: A remark on the existence of finite graphs. Casopis Pest. Mat. **80**, 477–480 (1955)
8. Hearn, R.A., Demaine, E.: PSPACE-completeness of sliding-block puzzles and other problems through the nondeterministic constraint logic model of computation. Theor. Comput. Sci. **343**(1–2), 72–96 (2005)
9. van den Heuvel, J.: The Complexity of change. In: Blackburn, S.R., Gerke, S., Wildon, M. (eds.) Part of London Mathematical Society Lecture Note Series, p. 409 (2013)
10. Mohar, B., Salas, J.: On the non-ergodicity of the Swendsen-Wang-Kotecký algorithm on the Kagomé lattice. J. Stat. Mech. Theory Exp. **2010**(05), P05016 (2010)
11. Nishimura, N.: Introduction to reconfiguration (2017). Preprint
12. Senior, J.: Partitions and their representative graphs. Am. J. Math. **73**(3), 663–689 (1951)
13. Taylor, R.: Contrained switchings in graphs. In: McAvaney, K.L. (ed.) Combinatorial Mathematics VIII. LNM, vol. 884, pp. 314–336. Springer, Heidelberg (1981). https://doi.org/10.1007/BFb0091828
14. Will, T.G.: Switching distance between graphs with the same degrees. SIAM J. Discrete Math. **12**(3), 298–306 (1999). https://doi.org/10.1137/S0895480197331156

The Itinerant List Update Problem

Neil Olver[1,2], Kirk Pruhs[3], Kevin Schewior[4,5(✉)], René Sitters[1,2],
and Leen Stougie[6]

[1] Department of Econometrics and Operations Research,
Vrije Universiteit Amsterdam, Amsterdam, The Netherlands
{n.olver,r.a.sitters,l.stougie}@vu.nl
[2] CWI, Amsterdam, The Netherlands
[3] Computer Science Department, University of Pittsburgh, Pittsburgh, USA
kirk@cs.pitt.edu
[4] Institut für Informatik, Technische Universität München, Munich, Germany
[5] Département d'Informatique, École Normale Supérieure Paris,
PSL University, Paris, France
kschewior@gmail.com
[6] CWI & INRIA-Erable, Amsterdam, The Netherlands

Abstract. We introduce the *itinerant list update problem (ILU)*, which is a relaxation of the classic list update problem in which the pointer no longer has to return to a home location after each request. The motivation to introduce ILU arises from the fact that it naturally models the problem of track memory management in Domain Wall Memory. Both online and offline versions of ILU arise, depending on specifics of this application.

First, we show that ILU is essentially equivalent to a dynamic variation of the classical minimum linear arrangement problem (MLA), which we call DMLA. Both ILU and DMLA are very natural, but do not appear to have been studied before. In this work, we focus on the offline ILU and DMLA problems. We then give an $O(\log^2 n)$-approximation algorithm for these problems. While the approach is based on well-known divide-and-conquer approaches for the standard MLA problem, the dynamic nature of these problems introduces substantial new difficulties. We also show an $\Omega(\log n)$ lower bound on the competitive ratio for any randomized online algorithm for ILU. This shows that online ILU is harder than online LU, for which $O(1)$-competitive algorithms, like Move-To-Front, are known.

1 Introduction

We introduce a variation of the classical list update problem, which we call the *itinerant list update problem (ILU)*. The setting consists of n (data) items, that

N. Olver—Supported in part by an NWO Veni grant.
K. Pruhs—Supported in part by NSF grants CCF-1421508 and CCF-1535755, and an IBM Faculty Award.
K. Schewior—Supported by DFG grant GRK 1408, Conicyt Grant PII 20150140, and DAAD PRIME program.

L. Epstein and T. Erlebach (Eds.): WAOA 2018, LNCS 11312, pp. 310–326, 2018.
https://doi.org/10.1007/978-3-030-04693-4_19

without loss of generality we will assume are the integers $[n] = \{1, \ldots, n\}$, stored linearly in n locations on a track (tape). The track has a single read/write head. Requests for these items arrive over time. In response to the arrival of a request for an item x, the algorithm can perform an arbitrary sequence of the following unit cost operations:

Move: Move the head to the left, or to the right, one position.
Swap: Swap the item pointed to by the head with the adjacent item on the left, or the adjacent item on the right.

In order to be a feasible response, at some point in this response sequence, the tape head must point to the position holding x. The objective is to minimize the total cost over all requests. In the offline version of ILU, the request sequence is known in advance, and in the online problem, only after the previous request has been serviced.

Our motivation for introducing ILU is that it captures the problem of dynamic memory management of a single track of Domain Wall Memory (DWM). Here, dynamic means that the physical memory location where a data item is stored may change over the execution of the application. DWM technology is discussed in more detail in the full version of the paper, but for our algorithmic purposes it is sufficient to know that conceptually, a track of DWM can be viewed as a tape with a read/write head. At least in the near term, it is envisioned that DWM will be deployed close to the processor in the memory hierarchy, and used as scratchpad memory instead of cache memory, so the stored data there would not have a copy in a lower level of the memory hierarchy [6,7,10,13,14,16]. If the application is an embedded application, where the sequence of memory accesses is (essentially) known before execution, then dynamic memory management can be handled at compile time, and thus is an offline problem [7]. If the sequence of memory accesses is not known before execution, then dynamic memory management would be handled by the operating system at run time, and the problem is online. Additionally in this case, at run time there would need to be an auxiliary data structure translating virtual memory addresses to physical memory addresses. We abstract away this issue (which is independent of the memory technology), and model these two settings by the offline and online ILU problems.

1.1 Relationship of ILU to List Update and Minimum Linear Arrangement

The main difference between ILU and the standard list update problem (LU) is that in LU there is an additional feasibility constraint. At the end of each response sequence, the head has to return to a fixed home position. If the head has a home position, there is a simple $O(1)$-approximation, which is also online: The Move-To-Front (MTF) policy, which moves the last-accessed item to the home position (and moves intermediate items one position further from the home) can be shown to be $O(1)$-competitive by simple modifications to the analysis of MTF in [19]. There, the home position is the first position, and costs are defined somewhat differently.

However, the natural adaptations of MTF for online ILU are all $\tilde{\Omega}(n)$-competitive, see the full version of the paper. These lower bounds hint at an additional difficulty of online ILU relative to the standard list update problem. In both problems it seems natural for the online algorithm to aggregate recently accessed items together. However, in the standard list update problem it is obvious where to aggregate these items, near the home location, while in ILU, it seems unclear where these items should be aggregated.

In the standard formulation of the list update problem [2,19], MTF is 2-competitive, which is optimal for deterministic algorithms [19]. The optimal competitive ratio for randomized algorithms (against an oblivious adversary) is between 1.5 [20] and 1.6 [1]. The offline version of the list update problem is shown to be NP-hard in [3,4], and there is an exact algorithm with running time $O(2^n n! m)$, where m is the number of requests [18]. For a survey of many further results related to the list update problem, see [2,12].

If items can only be reordered once at the beginning (so the memory management is not allowed to be dynamic), then offline ILU is essentially the classical minimum linear arrangement problem (MLA) [11]. In MLA, the input is an edge-weighted graph G with n vertices. The output is an embedding of the vertices of G into a track with n locations. The objective is to minimize the sum over the edges of the weight of the edge times the distance between the endpoints of the edge in the track. Here in the ILU application, the weight of an edge (x, y) roughly corresponds to the number of times that item y is requested immediately after item x is requested. We will make the connection between ILU and MLA more precise shortly.

Hansen [11] gave a polynomial-time $O(\log^2 n)$-approximation algorithm for MLA. This algorithm is a divide-and-conquer algorithm where the divide step computes a balanced cut (say using [15]) to determine a partition of the items into two sets, where all the items in the first set will eventually be embedded to the left of all the items in the second set. As noted by Rao and Richa [17], this same algorithmic design technique can be used to obtain approximation algorithms with similar approximation ratios for the minimum containing interval graph problem, and the minimum storage-time product problem. The algorithm by Feige and Lee [9], which achieves the currently best known approximation guarantee of $O(\sqrt{\log n} \log \log n)$ for these problems, combines rounding techniques for semidefinite programs [5] and spreading-metric techniques [17].

1.2   Our Results

We have already mentioned the connection of ILU to the minimum linear arrangement problem; we now make the connection more precise by defining the dynamic minimum linear arrangement (DMLA) problem. The setting for DMLA is the same as the setting for ILU: a linear track of items $[n]$. A sequence of graphs $H_1, H_2, \ldots$ arrives over time, with the vertex set $V(H_t) = [n]$ for each time t. In response to the graph H_t, the algorithm can first perform an arbitrary sequence of swaps of adjacent items on the track; each such swap has a cost of 1. After this, the *service cost* for H_t is (as in MLA) the sum over the edges of the distance between the current positions of the endpoints in the track. The objective is to minimize the overall cost due to both swaps and service costs. Note that in DMLA there is no concept of a track head, and swaps can be made anywhere on the track. Once again, DMLA has both an online and offline version. The standard MLA problem is essentially a special case of the offline DMLA problem in which all of the many arriving graphs are identical (so there is nothing to be gained from reordering the track).

We use DMLA1 to refer to the DMLA problem restricted to instances where each request graph H_t has a single edge. We show the following.

Theorem 1. *Consider offline ILU, DMLA, and DMLA1. If there is a polynomial-time $f(n)$-approximation algorithm for one of these problems, there are polynomial-time $O(f(n))$-approximation algorithms for the two other problems as well, as long as $f(n) = O(\text{polylog } n)$.*

The proof will be provided in the full version of the paper. It involves a somewhat intricate sequence of reductions.

As our aim is a polylogarithmic approximation to ILU, we can henceforth restrict our attention to DMLA1. Our main theorem is the following.

Theorem 2. *There is a polynomial-time $O(\log^2 n)$-approximation algorithm for offline DMLA1, implying the same for DMLA and ILU.*

As the DMLA problem generalizes the standard MLA problem, it is natural to suspect that the divide and conquer algorithmic design approaches used in, e.g., [9,11], might be applicable. It turns out that the dynamic nature of the problem introduces major new difficulties. We discuss these difficulties, and how we succeed in bypassing them, in Sect. 2.1. We believe that our more sophisticated algorithm design and analysis techniques may also be useful for other linear arrangement problems, where the simpler techniques used for MLA, the minimum containing interval graph problem and the minimum storage-time product problem are also not sufficient [9,11,17].

We now turn to online ILU. We have already seen that it seems much harder than the classical list update problem. This is confirmed by the following theorem, proved in Sect. 3.

Theorem 3. *The competitive ratio of any randomized online ILU algorithm against an oblivious adversary is $\Omega(\log n)$.*

In the construction, the algorithm only gradually "learns" that certain items should be close to each other to handle the requests cheaply. To profitably aggregate these items, however, the algorithm would need to know where to aggregate them, which requires more global information. This manifests the difficulties encountered when trying to adapt MTF.

It remains a very interesting and challenging open problem to give a polylog-competitive algorithm for online ILU. The reductions in the full version of the paper show that online ILU, DMLA and DMLA1 are also equivalent, and it suffices to give a polylog-competitive algorithm for online DMLA1. We anticipate that the insights obtained in the analysis of the approximation algorithm will be crucial in making progress.

2   Approximation Algorithm

We now prove Theorem 2, by giving an $O(\log^2 n)$-approximation algorithm for DMLA1; by Theorem 1 this implies the same for ILU and DMLA. The design of our algorithm is described in Sect. 2.2, and its analysis in Sect. 2.3. We first give a technical overview of our algorithm and analysis.

2.1   Overview

As the starting point for our algorithm for DMLA1 was the divide-and-conquer algorithm for MLA by Hansen [11], we start by discussing this algorithm. The MLA algorithm finds an approximate minimum balanced cut of the input graph G into a "left" side and a "right" side (the balance is randomly selected). The algorithm then recurses on the subgraph of G induced by the "left" vertices, and on the subgraph of G induced by the "right" vertices, This recursion is "simple", in the sense that the subproblems are just smaller instances of the MLA problem. This recursive process constructs a laminar family[1] of subsets of the vertices of G, with each set labeled left or right, and from which the ordering can be obtained in the natural way. One issue that must be addressed in the analysis is ruling out the possibility that choosing a high-cost balanced cut at the root can drastically reduce costs at lower levels of the recursion, so that taking a low-cost balanced cut at the root is already an unfixable mistake. In [11] this is handled by showing that the MLA problem has the following *subadditivity property*: the optimal cost for the left subinstance plus the optimal cost for the right subinstance is at most the optimal cost for the original instance. This subadditivity property makes it straightforward to observe that a c-approximate algorithm for minimum balanced cut implies a $c \cdot h$-approximate algorithm for MLA, where $h = \Theta(\log n)$ is the height of the recursion tree.

In order to adapt this algorithm and analysis from MLA to DMLA1, the first question is to determine what should play the role of the input graph. Our algorithm operates on a time-expanded graph G, defined in Definition 1 (see Fig. 1), that contains a vertex (x, t) for every item x and time t. A cut in G can again be interpreted as dividing the nodes into a left and right part, but now in a dynamic way: an item x might be in the left side of the cut at some time t_1, but on the right side at another time t_2. "Consistency edges" of the form $\{(x, t-1), (x, t)\}$ play a role in encoding swap costs; the edge contributes to the cut if item x switches sides between times $t-1$ and t. However, we now

---

[1] A family $\mathcal{F} \subseteq 2^S$ of sets over some ground set S is called *laminar* if, for all $F_1, F_2 \in \mathcal{F}$, we have $F_1 \cap F_2 = \emptyset$, $F_1 \subseteq F_2$, or $F_1 \supseteq F_2$.

encounter a significant complication: a balanced cut of G does not suffice. One instead needs a cut that is balanced *at each time*; in other words, a constant fraction of the items should be to the "left" at any given time t. This is crucial for the same reason as in MLA; to ensure, in essence, that the expected distortion between the original line metric and the random tree metric described by the laminar family is not too large.

Before describing how our algorithm finds per-time balanced cuts, we note a major hurdle. Firstly, we cannot hope for a "simple" recursion. Consider the left side of some per-time balanced cut; this will typically have some items that enter and leave this set over time. So from the left subproblem's point of view, items are arriving and leaving over time. It is tempting to try to define a generalization of DMLA1 in which items are allowed to enter and leave. However, we failed to find a formulation of such a problem that (a) we could approximately solve in an efficient manner, and (b) has the subadditivity property which is so critical to the MLA analysis. Rather than surmounting this hurdle, we more or less bypass it, as we will now describe.

Let us return to the issue of finding per-time balanced cuts. Our algorithm proceeds as follows. First, compute a balanced cut of G; if this is sufficiently cheap, it is easily argued (by virtue of the consistency edges) that the cut is in fact per-time balanced. Otherwise, we find a balanced cut of the subgraph corresponding to those vertices up to some time r, where r is chosen as large as possible but such that the balanced cut is cheap, and hence per-time balanced. Again our algorithm then recurses on the left and right subgraphs of the vertices of G up to time r, but then also recurses on the subgraph consisting of vertices with times after r. This however means that we make no effort whatsoever to prevent a complete reordering between times r and $r+1$; there may be a complete "reshuffle" of the items, and this is not captured in the cost of any of the balanced cuts.

So the final major hurdle is to bound the cost of these reshufflings, which can occur in all levels of the recursion. Because we don't know how to show that DMLA1 has an appropriate subadditivity property, we cannot charge these reshuffling costs locally. This is unlike in the analysis of the MLA algorithm, where all charging is done locally. Instead, we charge, in a somewhat delicate way, the cost of reshuffling at a given level to cuts higher up in the laminar family.

Finally we have to relate the expected reduced cost of the algorithm to the cost of the optimum. This is broadly similar to the MLA algorithm analysis, with some extra technical work to handle the dynamic aspect. In the end we obtain an $O(\log^2 n)$ approximation factor. As in the analysis of the MLA algorithm, we lose one log factor in the approximation of the balanced cut, and one log factor that is really the height of the recursion tree.

2.2   Algorithm Design

We begin by introducing some needed notation.

For a graph $G = (V, E)$ and a subset $W \subseteq V$, $G[W]$ denotes the subgraph induced by W, and $E[W]$ the set of edges in $G[W]$. For a set $S \subseteq V$, $\delta(S)$ denotes the set of edges crossing the cut S. We also use the less standard notation $\delta_W(S)$ to denote $\delta(S) \cap E[W]$. The *balance* of a cut S in G is simply $|S|/|V|$. Furthermore, by T we refer to the total number of requests. The following "time-expanded graph" will be used throughout the algorithm.

Definition 1. $G = (V, E)$ *is defined as follows:*

- *There is a vertex (x, t) for each item $x \in [n]$ and each time t, where $t \in \{0, 1, 2, \ldots, T\}$. We call the set of nodes at time t layer t, and denote it by L_t.*
- *For each time $t \in [T]$, and the single edge $\{x, y\} \in H_t$, there is an edge $e_t := \{(x, t), (y, t)\}$. We call these request edges.*
- *For each item x and $t \in [T]$ there is an edge $\{(x, t-1), (x, t)\}$. We will call these consistency edges.*

Let E^{r} and E^{c} denote the set of request and consistency edges respectively.

Note that the cost of any solution is certainly at least T, since each request incurs a cost of at least 1. Thus we can afford to return the items to their original order after every n^2 requests, at only a constant-factor increase in cost. This splits up the instance into completely independent sub-instances with at most n^2 requests each, and so we may assume that $T \leq n^2$. Next, we prove this formally.

Lemma 1. *If there is a polynomial-time $f(n)$-approximation algorithm A for DMLA1 with $T \leq n^2$, then there is a polynomial-time $O(f(n))$-approximation algorithm B for DMLA1 in general.*

Proof. Given some DMLA1 instance I with one edge at a time and without restrictions on T, Algorithm B first cuts I into contiguous sub-instances $I_1', I_2', \ldots, I_k'$ of n^2 requests each (possibly except for the last one). Then B calls A on each of these sub-instances and then connects these solutions up by moving back to the initial order before each new sub-instance, at a total additional cost of at most $(k-1)n^2$.

Then we have

$$\mathrm{cost}^B(I) \leq \sum_{i=1}^{k} \mathrm{cost}^A(I_i') + (k-1)n^2$$

$$\leq f(n) \cdot \sum_{i=1}^{k} \mathrm{cost}^{\mathrm{OPT}}(I_i') + (k-1)n^2$$

$$\leq f(n) \cdot (\mathrm{cost}^{\mathrm{OPT}}(I) + (k-1)n^2) + (k-1)n^2,$$

$$\leq O(f(n)) \cdot \mathrm{cost}^{\mathrm{OPT}}(I).$$

In the second-to-last step we use that the optimal solution for I can be transformed into optimal solutions for $I_1', I_2', \ldots, I_k'$ by moving to the identical orders, again at a total additional cost of at most $(k-1)n^2$. In the last step, we use that $\mathrm{cost}^{\mathrm{OPT}}(I)$ is at least the number of requests in I and $f(n) \geq 1$. $\qquad\square$

This is important when applying approximation algorithms to G (or subgraphs of it) whose approximation guarantees depend on the size of the input graph. We proceed with further definitions.

Definition 2. *For any $W \subseteq V$, let*

$$t_{\min}(W) := \min\{t : (x,t) \in W \text{ for some } x \in [n]\},$$
$$t_{\max}(W) := \max\{t : (x,t) \in W \text{ for some } x \in [n]\}.$$

We say an item x is present *in W if $(x,t) \in W$ for some $t \in \{0,1,\dots,T\}$. We say x is* permanent *in W if $(x,t) \in W$ for all $t_{\min}(W) \leq t \leq t_{\max}(W)$; all other items present in W are called* temporary *in W. Let $\alpha(W)$ be the number of items present in W, and $\beta(W)$ the number of temporary items in W. For any $t_{\min}(W) \leq r \leq t_{\max}(W)$, let $W_{(r)} = \{(x,t) \in W : t \leq r\}$. By* layer t *of W, we refer to the set $L_t \cap W$.*

Algorithm Description. The first stage of the algorithm will recursively and randomly hierarchically partition G. The output of this first stage will be described by a laminar family $\mathcal{L}$ on V, with each set $S \in \mathcal{L}$ labeled either *left* or *right*.

So let W be a subset of V, representing the vertex set of a subproblem. Throughout, c will denote a positive constant chosen sufficiently small; $c = 1/100$ suffices. We assume that

$$\beta(W) \leq c\alpha(W); \tag{1}$$

this is of course true when $W = V$ (because $\beta(V) = 0$), and we will ensure that it holds for each subproblem that we create. If $\alpha(W) < 16/c$, we will terminate, and this subproblem will be a leaf of the laminar family constructed. So asume $\alpha(W) \geq 16/c$ from now on. The algorithm chooses κ uniformly from the interval $[\frac{1}{2} - c, \frac{1}{2} + c]$, which is used as the balance parameter for a certain balanced cut problem. The problem differs depending on whether $t_{\min}(W) = 0$ or $t_{\min}(W) > 0$, since the initial ordering is fixed. If $t_{\min}(W) > 0$, define $\bar{G}_{(r)} = G[W_{(r)}]$. If $t_{\min}(W) = 0$, define $\bar{G}_{(r)}$ as the graph obtained from $G[W_{(r)}]$ by choosing z so that the set $A = \{(x,0) \in W : x \leq z\}$ has cardinality $\kappa\alpha(W)$, and then contracting A into single node s, and the nodes $\{(x,0) \in W : x > z\}$ into a single node t. Let $\bar{W}_{(r)}$ be the vertex set of $\bar{G}_{(r)}$.

We now compute a cut S_r in $\bar{G}_{(r)}$ with

$$|S_r| \in [(\kappa - 8c)|\bar{W}_{(r)}|, (\kappa + 8c)|\bar{W}_{(r)}|], \tag{2}$$

in such a way that $|\delta_{W_{(r)}}(S_r)| = O(\log|\bar{W}_{(r)}| \cdot |\delta_{\bar{W}_{(r)}}(S_r^*)|)$, where S_r^* is the minimum cut with

$$|S_r^*| \in [(\kappa - 4c)|\bar{W}_{(r)}|, (\kappa + 4c)|\bar{W}_{(r)}|]. \tag{3}$$

Note that the intervals in (2) and (3) are non-empty because $\alpha(W) \geq 16/c$, implying that both S_r^* and S_r exist; further $O(\log|\bar{W}_{(r)}|) \subseteq O(\log n)$ because

$T \le n^2$. Bicriteria approximation algorithms to balanced cut required to compute some $|S_r|$ as above are well known [15,21]. In the case $t_{\min}(W) = 0$, we ensure that S_r is chosen so that $s \in S_r$, by replacing S_r with $W_{(r)} \setminus S_r$ if necessary. Note that if $|\delta_{\bar{W}_{(r)}}(S_r)| \le c\alpha(W)$ (which will be the case of interest), S_r will separate s and t. This is because, if $s, t \in S_r$ ($s, t \notin S_r$ analogously), $|\delta_{\bar{W}_{(r)}}(S_r)|$ is at least the number of items x permanent in W for which there is a $(x, \tau) \in W_{(r)} \setminus S_r$. Using the balance requirement (2), we can lower bound this quantity by $(\frac{1}{2} - 9 \cdot c)\alpha(W) - \beta(W) \ge (\frac{1}{2} - 10 \cdot c)\alpha(W)$, which exceeds $c\alpha(W)$ for c sufficiently small. We can interpret S_r as a cut in $G[W_{(r)}]$ by uncontracting s and t.

Roughly, the plan now is to pick r^* as large as possible such that $|\delta_{W_{(r^*)}}(S_{r^*})|$ is not too big; small enough so that we can be sure that at each time t between $t_{\min}(W)$ and r^*, $S_{r^*} \cap L_t$ has size roughly $\kappa\alpha(W)$. However, some additional care is needed, since we also would like that $|\delta_{W_{(r^*)}}(S_{r^*})|$ is not too small—unless $r^* = t_{\max}(W)$. This is needed so that the edges in $\delta_{W_{(r^*)}}(S_{r^*})$ can be charged to later.

Thus, we proceed as follows. We define the cuts $\bar{S}_r$ inductively by: $\bar{S}_{t_{\min}(W)} = S_{t_{\min}(W)}$, and for $r > t_{\min}(W)$, $\bar{S}_r$ is either S_r, or an *extension* S' of $\bar{S}_{r-1}$ to layer r, whichever is cheaper. This extension is obtained, roughly speaking, by duplicating the layer $r-1$ nodes of $\bar{S}_{r-1}$, i.e., taking all (x, r) for which $(x, r-1)$ is in $\bar{S}_{r-1}$. But in order to ensure that S' is sufficiently balanced, we adjust this so that $|S' \cap L_r| \in [(\kappa - 8c)|W \cap L_r|, (\kappa + 8c)|W \cap L_r|]$, by adding or removing an arbitrary set of items of minimum cardinality that is sufficient to satisfy this requirement. Then clearly S' satisfies (2), since inductively $\bar{S}_{r-1}$ satisfied it for $\bar{W}_{(r-1)}$. Note that if q items are added or removed, then $|\delta_{W_{(r)}}(S')| \le |\delta_{W_{(r-1)}}(\bar{S}_{r-1})| + q + 1$; here we use that H_r consists of only a single edge.

Our algorithm sets r^* to be the maximum $r \le t_{\max}(W)$ such that

$$\beta(W_{(r)}) + |\delta_{W_{(r)}}(\bar{S}_r)| \le \tfrac{1}{4}c\alpha(W). \tag{4}$$

We argue that $r = t_{\min}(W)$ always fulfills this inequality, so that r^* does always exist: In this case, $\beta(W_{(r)}) = 0$ and $|\delta_{W_{(r)}}(\bar{S}_r)| \le 1$. To see the latter, distinguish two cases. If $t_{\min}(W) = 0$, then the corresponding layer has been contracted into two nodes, there is only one possible balanced cut, and it can be cut only by a single request edge. If on the other hand $t_{\min}(W) > 0$, then there is a balanced cut of cost 0. Since $\alpha(W) \ge 16/c$ by assumption, the inequality holds.

For convenience let $S^* := \bar{S}_{r^*}$ and $W^* := W_{(r^*)}$. We illustrate this in Fig. 1. We now note a property that will be required later in the analysis.

Property 1. If $r^* < t_{\max}(W)$, then $\beta(W) + |\delta_{W^*}(S^*)| = \Omega(\alpha(W))$.

Proof. Let S' be the extension of S^* that was considered by the algorithm at time $r^* + 1$, and let q be the number of items that needed to be added to or removed from the duplication of layer r^* of S^* in order to ensure $|S' \cap L_{r^*+1}| \in [(\kappa - 8c)|W \cap L_{r^*+1}|, (\kappa + 8c)|W \cap L_{r^*+1}|]$. We bound q: Since S^* fulfills (2), there is a layer r' such that $|S^* \cap L_{r'}| \in [(\kappa - 8c)|W \cap L_{r'}|, (\kappa + 8c)|W \cap L_{r'}|]$.

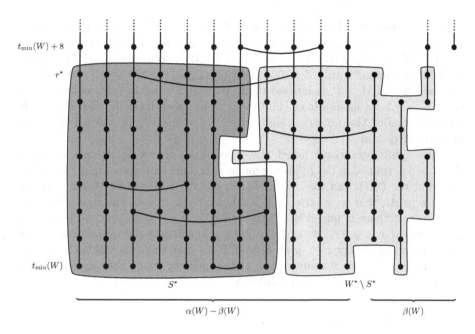

$t_{\min}(W) + 8$

r^*

$t_{\min}(W)$

S^*

$W^* \setminus S^*$

$\underbrace{\hspace{5cm}}_{\alpha(W) - \beta(W)}$ $\underbrace{\hspace{2cm}}_{\beta(W)}$

Fig. 1. An example of $G[W]$, S^*, and $W^* \setminus S^*$. For the sake of illustration, we do not require S^* to satisfy the inequalities with the same constants as in our algorithm.

Now note that, when going from layer r' of S^* to layer $r^* + 1$, each item that we need to add or remove in order to restore the balance requirement is due to an edge in $E[W^*]$ being cut by S^* or an item temporary in W leaving or entering. So we have $q \leq |\delta_{W^*}(S^*)| + \beta(W)$.

Then $|\delta_{W_{(r^*+1)}}(S')| \leq |\delta_{W^*}(S^*)| + q + 1 \leq 2|\delta_{W^*}(S^*)| + \beta(W) + 1$. Since the Condition (4) was not satisfied by $r^* + 1$, we deduce that $\frac{1}{4}c\alpha(W) - \beta(W^*) < |\delta_{W_{(r^*+1)}}(S')|$. Combining these inequalities and using that $\beta(W^*) \leq \beta(W)$ yields $\beta(W) + |\delta_{W^*}(S^*)| \geq \frac{1}{8}c\alpha(W) - 1$. Since $\alpha(W) \geq 16/c$, the claim follows. □

Furthermore, we have the following lemma.

Lemma 2. $\alpha(S^*)$ and $\alpha(W^* \setminus S^*)$ are both at least $\frac{1}{4}\alpha(W)$.

Proof. Let $\rho = r^* - t_{\min}(W) + 1$. Observe that $\rho \cdot \alpha(S^*) \geq |S^*| \geq (\kappa - 8c)|W^*| \geq (\frac{1}{2} - 9c)|W^*|$. These inequalities follow by the definition of ρ and $\alpha(\cdot)$, Inequality (2), and the choice of κ and c, respectively. Since $|W^*|$ is at least ρ times the number of permanent items in W, $\alpha(S^*) \geq (\frac{1}{2} - 9c)(\alpha(W) - \beta(W)) \geq (\frac{1}{2} - 9c)(1 - c)\alpha(W) \geq \frac{1}{4}\alpha(W)$, where the last inequality follows by our choice of c. A symmetric argument holds for $W^* \setminus S^*$. □

Since the number of temporary items in S^* is at most the number of temporary items in W^* plus the size of the cut $\delta_{W^*}(S^*)$, (4) yields that $\beta(S^*) \leq \frac{1}{4}c\alpha(W)$, and so $\beta(S^*) \leq c\alpha(S^*)$ by the lemma. Similarly, $\beta(W^* \setminus S) \leq c\alpha(W^* \setminus S^*)$. This shows that the algorithm may recurse in S^* and $W^*\backslash S^*$. Eventually this recursion yields labeled families $\mathcal{L}_{\text{left}}$ and $\mathcal{L}_{\text{right}}$ on S^* and $W^*\backslash S^*$, respectively. If $W\backslash W^* \neq \emptyset$, the algorithm also iterates on $W\backslash W^*$, obtaining $\mathcal{L}_{\text{rest}}$. The resulting labeled laminar family $\mathcal{L}$ is $\mathcal{L}_{\text{left}} \cup \mathcal{L}_{\text{right}} \cup \mathcal{L}_{\text{rest}} \cup \{S^*, W^*\backslash S^*\}$, where S^* is labeled left and $W^*\backslash S^*$ is labeled right. We call S^* and $W^*\backslash S^*$ *siblings*. Note that W may have many direct children in $\mathcal{L}$, but each layer intersects precisely one sibling pair. Also note that since $|S^*| \in \left[(\frac{1}{2} - 9c)|W^*|, (\frac{1}{2} + 9c)|W^*|\right]$, $\mathcal{L}$ has logarithmic depth.

Once $\mathcal{L}$ has been constructed, the ordering $\prec_t$ of the algorithm's list A_t at time t is determined in the following manner. If there exists a set in $\mathcal{L}$ containing (x,t) but not (y,t), and the maximal such set S is labeled left, then x is to the left of y in A_t, that is, $x \prec_t y$; if S is labeled right, x is to the right of y, that is, $y \prec_t x$. If there is no set in $\mathcal{L}$ containing (x,t) but not (y,t), we let $x \prec_t y$ if and only if $x < y$, i.e., we order x and y according to the initial ordering. In the latter case, we say $x \prec_t y$ *by default*. Note that this rule yields the correct ordering for A_0, since for any $x < y$, at the moment where they are separated, we ensure that $(x,0) \in S^*$ and $(y,0) \notin S^*$.

2.3   Algorithm Analysis

Definition 3. *Given two orderings A, B of the items, and any two distinct items x, y, we say that (x, y) is a* discordant pair *(for A and B) if x and y have a different relative order in A and in B.*

Note that the number of discordant pairs for A and B is precisely the permutation distance between A and B, i.e., the minimum number of swaps of adjacent items required to obtain order B starting from order A.

Definition 4. *For $S \in \mathcal{L}$, define $\text{cost}(S)$ to be $\alpha(S) \cdot |\delta(S)|$ and $\text{cost}(\mathcal{L})$ to be $\sum_{S \in \mathcal{L}} \text{cost}(S)$. Let $\text{parent}(S)$ be the parent of S, meaning the minimal set $S' \in \mathcal{L} \cup \{V\}$ with $S' \supsetneq S$ (which is unique by the laminarity of $\mathcal{L}$). (By including V here, we ensure that every set in $\mathcal{L}$ has a parent.) Let $\text{pair}(S)$ be the union of S and its sibling (i.e., the other child of $\text{parent}(S)$ in $\mathcal{L}$ which covers the same layers as S). Note that $S \subsetneq \text{pair}(S) \subseteq \text{parent}(S)$.*

Lemma 3. *The cost of the algorithm is at most $8 \cdot \text{cost}(\mathcal{L}) + O(\text{OPT})$ (irrespective of the random choices made in the algorithm).*

Proof. We argue separately for each time t. First consider the swap costs, so let $t \geq 1$, and define $E_t^c = \{\{(x, t-1), (x, t)\} : x \in [n]\}$. The proof strategy for this part is to consider discordant pairs for A_{t-1} and A_t, assigning them to certain sets $S \in \mathcal{L}$, and later counting the number of discordant pairs assigned to each set $S \in \mathcal{L}$. Hence suppose $x \prec_{t-1} y$ and $y \prec_t x$ (so (x, y) is a discordant pair for A_{t-1} and A_t).

First consider the case that $x \prec_{t-1} y$ by default; the case where $y \prec_t x$ by default is analogous. As $x \prec_{t-1} y$ by default, there is a set T minimal in $\mathcal{L}$ with $(x, t-1), (y, t-1) \in T$. Since (x, y) is a discordant pair, however $y \prec_t x$ not by default. Hence, the construction of A_t from $\mathcal{L}$ tells us that there is a unique left-labeled set $U \in \mathcal{L}$ with $(y, t) \in U$, $(x, t) \in \mathrm{parent}(U)\backslash U$. We say that U *certifies* that $y \prec_t x$. Note that $U \cap T = \emptyset$ or $(\mathrm{parent}(U)\backslash U) \cap T = \emptyset$ (or both). Assign discordant pair (x, y) to the corresponding set out of $U, \mathrm{parent}(U)\backslash U$.

Now consider the case that neither $x \prec_{t-1} y$ by default nor $y \prec_t x$ by default. Again we know that there is a left-labeled set $U \in \mathcal{L}$ certifying $y \prec_t x$. Similarly, there is a left-labeled set T certifying that $x \prec_{t-1} y$. We prove the following claim.

Claim. $(x, t) \notin T$ or $(y, t-1) \notin U$ (or both).

Proof (Claim). Suppose not. Then T contains $(x, t-1)$ and (x, t), and U contains $(y, t-1)$ and (y, t). Moreover T is a maximal set in $\mathcal{L}$ containing $(x, t-1)$ and not $(y, t-1)$, and U is a maximal set containing (y, t) and not (x, t); we deduce that $T' := \mathrm{parent}(T) = \mathrm{parent}(U)$. But this contradicts the assumption that T and U are both labeled left; within the subproblem induced by T', the algorithm produces only one set labeled left containing nodes at time t. □

So in this case assign discordant pair (x, y) to T if $(x, t) \notin T$, and to U if $(y, t-1) \notin U$ (if both occur, the assignment can be arbitrary). Now consider the assignment of discordant pairs (in both cases) from the perspective of a set $S \in \mathcal{L}$. If (x, y) is assigned to S, then for $(z, \bar{z}) \in \{(x, y), (y, x)\}$ we have

- $(z, t-1) \in S$, $(\bar{z}, t-1) \in \mathrm{parent}(S)\backslash S$, and $(z, t) \notin S$,
- or $(z, t) \in S$, $(\bar{z}, t) \in \mathrm{parent}(S)\backslash S$, and $(z, t-1) \notin S$.

Thus we can bound the total number of discordant pairs assigned to S by

$$2 \cdot (|\{z : (z, t-1) \in S, (z, t) \notin S\}|$$
$$+ |\{z : (z, t) \in S, (z, t-1) \notin S\}|) \cdot \alpha(\mathrm{parent}(S))$$
$$= 2 \cdot |\delta(S) \cap E_t^c| \cdot \alpha(\mathrm{parent}(S))$$
$$\leq 8 \cdot |\delta(S) \cap E_t^c| \cdot \alpha(S),$$

where the last inequality follows by Lemma 2. So the cost to optimally reorder A_{t-1} to A_t, being exactly the number of discordant pairs, is at most $8 \cdot \sum_{S \in \mathcal{L}} \alpha(S) \cdot |\delta(S) \cap E_t^c|$.

We now consider the service cost at time t. The request pair is an edge $e_t = \{(x, t), (y, t)\}$ in G; assume that $x \prec_t y$, otherwise relabel x and y. The algorithm pays the distance between x and y in A_t. If $x \prec_t y$ by default, the distance between x and y in A_t is at most $16/c$ by construction, and the cost is taken care of by the $O(\mathrm{OPT})$ term in the cost bound, because OPT pays at least one for the considered request. Otherwise consider the set S certifying that $x \prec_t y$. Then $S' := \mathrm{parent}(S)$ contains both (x, t) and (y, t), and $\alpha(S')$ clearly bounds the distance between x and y in A_t (because any item outside of S' is either to the left of both x and y or to the right of both). Since $\alpha(S') \leq 4 \cdot \alpha(S)$ (again Lemma 2), we conclude that the service cost at time t is at most $4 \cdot \sum_{S \in \mathcal{L}} \alpha(S) \cdot 1_{e_t \in \delta(S)}$. Combining the swap and service costs at all times, we obtain the lemma. □

Before we state the next lemma, we note that for any $S \in \mathcal{L}$, pair(S) is exactly the set within which S was a good balanced cut, and $|\delta_{\text{pair}(S)}(S)|$ is exactly the cost of S in this balanced cut problem.

Definition 5. *We define* $\text{cost}_{\text{core}}(S)$ *to be* $\alpha(S) \cdot |\delta_{\text{pair}(S)}(S)|$, *and* $\text{cost}_{\text{core}}(\mathcal{L})$ *to be* $\sum_{S \in \mathcal{L}} \text{cost}_{\text{core}}(S)$.

Lemma 4. *Irrespective of the random choices made by the algorithm, we have* $\text{cost}_{\text{core}}(\mathcal{L}) = \Omega(\text{cost}(\mathcal{L}))$.

Proof. Begin by assigning to each pair (e, S), where $S \in \mathcal{L}$ and $e \in \delta(S)$, a charge of $\alpha(S)$. Our goal is to redistribute this charge to pairs (e, S) where $e \in \delta_{\text{pair}(S)}(S)$, and where each such pair gets a total charge of $O(\alpha(S))$. This clearly implies the lemma.

For a set $S \in \mathcal{L}$, we call an edge of the form $\{(x, t_{\max}(S)), (x, t_{\max}(S) + 1)\}$ that crosses S a *top shuffle edge for S*, and similarly an edge of the form $\{(x, t_{\min}(S) - 1), (x, t_{\min}(S))\}$ a *bottom shuffle edge for S*. Let $Q(S)$ denote the set of shuffle edges (either top or bottom) for S. Note that $Q(S) \subseteq \delta(S) \setminus \delta_{\text{pair}(S)}(S)$. Now notice that we have the following downward-closed property.

Claim. If e is a shuffle edge for some $S \in \mathcal{L}$, and $e \in \delta(T)$ for some $T \in \mathcal{L}$ with $T \subset S$, then e is a shuffle edge for T as well.

We reassign the charge in stages. In the first stage, we reassign all the charges involving an edge e to a maximal $S \in \mathcal{L}$ with $e \in \delta(S)$. Note that there are two possible choices for S. If $e \in E^c$, choose the S containing the earlier endpoint of e and not the later one; if $e \in E^r$, make any choice. Now consider any (e, S). It may receive charge from multiple consistency edges, whose initial charges are geometrically decreasing starting from $\alpha(S)$, and a single request edge with initial charge $\alpha(S)$. So (e, S) has charge $O(\alpha(S))$ after this reassignment. Moreover, by the above choice of S, no bottom shuffle edges have any charge remaining.

For the next stage, we prove the following statement.

Claim. For any $S \in \mathcal{L}$, the total charge in $Q(S)$ is $O(\alpha(S) \cdot |\delta(S) \setminus Q(S)|)$.

Proof (Claim). Let $W = \text{parent}(S)$ and $U = \text{pair}(S)$. There are two cases. The first case is if $t_{\max}(S) = t_{\max}(W)$. In this case, all top shuffle edges for S cross W as well, and so have no charge. The second case is if $t_{\max}(S) < t_{\max}(W)$. In this case, since each shuffle edge of S currently has charge at most $O(\alpha(S))$, it suffices to show that $|Q(S)| = O(|\delta(S) \setminus Q(S)|)$. Notice that $|\delta(S) \setminus Q(S)| \geq |\delta_U(S)| + \beta(W)$, because every non-shuffle edge crossing S is either contained in U, or crosses W, meaning it was a temporary node in W. But since $t_{\max}(S) < t_{\max}(W)$, we know from Property 1 that $|\delta_U(S)| + \beta(W) = \Omega(\alpha(W))$. The claim follows since $|Q(S)| \leq \alpha(S)$ and $\alpha(S) \leq 4 \cdot \alpha(W)$, by Lemma 2. $\square$

It follows that, in the next stage, we can now redistribute all charge on the shuffle edges of a set $S \in \mathcal{L}$ to other edges, maintaining that no edge of $\delta(S)$ has a charge more than $O(\alpha(S))$.

In the final stage, we again reassign all charge of an edge to a maximal set that it crosses; each pair (e, S) still has a charge $O(\alpha(S))$. It remains true that no pair (e, S) with $e \in Q(S)$ gets any charge, because of Claim 2.3. So all charge for a set S is on edges that are not shuffle edges, and which do not cross parent(S); these are precisely the edges of $\delta_{\text{pair}(S)}(S)$. This completes the proof of Lemma 4. $\qquad\qquad\qquad\qquad\qquad\qquad\qquad\qquad\qquad\qquad\qquad\qquad\qquad\qquad\qquad\quad \square$

Lemma 5. $\mathbb{E}[\text{cost}_{\text{core}}(\mathcal{L})] = O(\log^2 n) \cdot \text{OPT}.$

Proof. Let A_t^* denote the ordering in the optimum solution after responding to request t. Let $S_1, \ldots S_k$ be the left elements of $\mathcal{L}$ that are of some depth d in this laminar family. Let $W_i = \text{parent}(S_i)$ and $U_i = \text{pair}(S_i)$. (Note that the U_i's are disjoint, but many of the W_i's may be the same.) We fix the random choices made by the algorithm above level d (thus we may consider W_i to be deterministic for each i, although U_i is random). Define, for any $i \in [k]$ and $S \subseteq \text{pair}(S_i)$, $\text{cost}_{\text{core}}(S) = \alpha(S) \cdot |\delta_{\text{pair}(S_i)}(S)|$. Then for each S_i, we will show how to derive from OPT a (random) balanced cut C_i of $G[U_i]$, such that $\text{cost}_{\text{core}}(S_i) = O(\log n) \cdot \text{cost}_{\text{core}}(C_i)$ and

$$\sum_{i=1}^{k} \mathbb{E}[\text{cost}_{\text{core}}(C_i)] = O(\text{OPT}). \tag{5}$$

The expectation is over the random choices made by the algorithm at layer d. The result then follows, since $\mathcal{L}$ has depth $O(\log n)$.

Now fix some $i \in [k]$. Note that we can assume $\alpha(W_i) \geq 16/c$. We define C_i as follows. Let κ_i denote the random choice made by the algorithm for the subproblem U_i. and let $m_i = \lfloor \kappa_i \alpha(W_i) \rfloor$. Now consider the permanent items of W_i as they appear in A_t^*, and let $p_{i,t}$ be the position of the m_i'th such item, for any $i \in [k]$ and $t_{\min}(U_i) \leq t \leq t_{\max}(U_i)$. Note that the probability that $p_{i,t}$ takes any specific value is $O(1/\alpha(W_i))$. Then define

$$C_i = \{(x, t) \in U_i : x \text{ is at or to the left of position } p_{i,t} \text{ in } A_t^*\}.$$

Note that the C_i's are obviously disjoint sets, since the U_i's are disjoint.

We first prove that $\text{cost}_{\text{core}}(S_i) = O(\log n) \cdot \text{cost}_{\text{core}}(C_i)$ (irrespective of the random choice of κ_i). To see this, consider some layer of U_i, and observe that the number of nodes in it is in $[\alpha(W_i) - \beta(W_i), \alpha(W_i)] \subseteq [(1 - c) \cdot \alpha(W_i), \alpha(W_i)]$ (using (1)). On the other hand, the number of nodes contained in any layer of C_i is in $[\lfloor \kappa_i \cdot \alpha(W_i) \rfloor, \lfloor \kappa_i \cdot \alpha(W_i) \rfloor + \beta(W_i)] \subseteq [\kappa_i \cdot \alpha(W_i) - 1, (\kappa_i + c) \cdot \alpha(W_i)]$ (again using (1)). Putting the two observations together and using $\alpha(W_i) \geq 16/c$ yields that any layer of C_i contains a fraction in $[\kappa_i - c, \kappa_i + 3c]$ of the nodes of the corresponding layer of U_i. Summing over all layers shows that the balance of C_i in U_i is in $[\kappa_i - c, \kappa_i + 3c] \subseteq [\frac{1}{2} - 2c, \frac{1}{2} + 4c]$. Therefore, $\alpha(C_i) = \Theta(\alpha(S_i))$. Applying this to $|\delta_{U_i}(S_i)| = O(\log n) \cdot |\delta_{U_i}(C_i)|$, which follows from the fact that C_i fulfills (3) in the definition of the algorithm, yields $\text{cost}_{\text{core}}(S_i) = O(\log n) \cdot \text{cost}_{\text{core}}(C_i)$.

Next, we show (5). Look at any request edge e_t. If it is not contained within U_i for some i, then it does not contribute to the left hand side of (5), so suppose it is contained in U_i. Let $q_1 < q_2$ be the positions of the endpoints of e_t in A_t^*; so OPT pays $q_2 - q_1$. Then the probability that e_t crosses C_i is $\mathbb{P}(q_1 \leq p_{i,t} < q_2)$, which is $O((q_2 - q_1)/\alpha(W_i))$. If e_t does cross C_i, it contributes $\alpha(C_i) = \Omega(\alpha(S_i))$, and so its expected contribution is $O(q_2 - q_1)$.

Now consider the swap cost. The swap cost in the optimal solution at time t is the number of discordant pairs for A_{t-1}^* and A_t^*. So assign the swap cost to an item x at time t to be equal to the number of such discordant pairs that include x. The sum of the costs assigned to all items over all times is then exactly twice the swap cost of OPT.

So fix some item x and time t. Suppose $\{(x, t-1), (x, t)\} \in E[U_i]$ for some i (otherwise again, it does not contribute to (5)). Let q_1 and q_2 be the number of permanent items in W_i to the left of x in A_{t-1}^* and A_t^*, respectively. Once again, the probability that $\{(x, t-1), (x, t)\} \in \delta_{U_i}(C_i)$ is at most $O(|q_2 - q_1|/\alpha(W_i))$, yielding an expected contribution to (5) of $O(|q_2 - q_1|)$. But $|q_2 - q_1|$ is at most the number of discordant pairs involving x. Summing over all x and t completes the proof. □

Combining Lemmas 3, 4 and 5 yields that the cost of the algorithm is $O(\log^2 n \cdot \text{OPT})$, as desired.

3   Lower Bound for ILU

We now prove that there is no randomized $o(\log n)$-competitive online algorithm for ILU.

Proof (Theorem 3). We apply Yao's principle: For a particular input distribution, we show that the expected cost for every deterministic online algorithm is a $\Omega(\log n)$ factor more than the expected optimal cost [8]. Conceptually, underlying the lower bound construction is a complete binary tree T, of depth $q = \Theta(\log n)$, with n leaves. We think of each internal node of T as having a left and right subtree; thus the leaves of T can be associated with the positions on the track. We begin by choosing an uniformly random initial assignment δ of items to leaves. The adversary initially orders the items in the track to match this assignment, and will not move the items after this.

The sequence of requests consists of q rounds. Each round i, $0 \leq i \leq q - 1$, the request sequence π_i is a permutation of $[n]$. We define a *depth-d subtree* to be a subtree rooted at a node of depth d in T. Here we assume the root of T has depth 0. To obtain π_i, the depth-$(q - i)$ subtrees, of which there are 2^{q-i}, are first ordered uniformly at random (and independent of all other random choices). Then, while maintaining the order of the subtrees, the 2^i leaves within each of the depth-$(q - i)$ subtrees are uniformly randomly ordered (again independent of all other random choices). To make this more precise, let $v_{\sigma(1)}, \ldots, v_{\sigma(2^{q-i})}$ be a random permutation of the vertices of depth $q - i$ in T. For each $1 \leq j \leq 2^{q-i}$, let $v_{\rho_j(1)}, \ldots, v_{\rho_j(2^i)}$ be a random permutation of the leaves of the subtree of T

rooted at $v_{\sigma(j)}$. Then $v_{\rho_j(k)}$ precedes $v_{\rho_{j'}(k')}$ in π_i if and only if j occurs before j' in σ, or $j = j'$ and k occurs before k' in ρ_j.

We now bound the costs for the optimum. The only swap cost is incurred initially and is no more than n^2. During π_i, movement between two items in the same depth-$(q - i)$ subtree costs at most 2^i. Thus the total movement between items in the same depth-$(q - i)$ subtrees costs at most $n2^i$. Movement between two such items costs at most n.

The movement cost for the first item is at most n. Also, movement between two items in different depth-$(q - i)$ subtrees costs at most n. There are exactly $2^{q-i} - 1$ consecutive accesses to items in different depth-$(q - i)$ subtrees. Thus the total movement cost between items in different depth-$(q - i)$ subtrees is at most $n2^{q-i}$. Summing over i, we get that the adversary's cost is at most $\sum_{i=1}^{q} n2^i + n2^{q-i} = O(n^2)$.

We now bound the expected cost for the online algorithm. We can generously assume that the online algorithm knows the adversary's strategy for constructing the sequences π_i and that it sees π_i just before round i. Before seeing π_i, the online player knows the depth-$(q-i+1)$ subtrees of T, but it has no information at all on how depth-$(q - i + 1)$ subtrees are paired to form the depth-$(q - i)$ subtrees. So an alternative equivalent way to randomly generate δ would be to at this time randomly pair the depth-$(q - i + 1)$ subtrees.

For the moment assume that the online algorithm does not reorder the items during round i. Then consider the 2^i consecutive requests to the leaves in some depth-$(q - i)$ subtree in π_i. In expectation, at least a constant fraction of these consecutive accesses will be to items in different depth-$(q-i+1)$ subtrees. As the depth-$(q-i+1)$ subtrees are paired up randomly, any online algorithm will have to move in expectation $\Omega(n)$ positions in response to the requests in the different depth-$(q - i + 1)$ subtrees. Hence, the expected cost for the algorithm for each round is $\Omega(n^2)$ if the online algorithm makes no swaps after seeing permutation π_i. But note that any swap made after seeing π_i can reduce the movement cost in round i by at most 2 since each item is requested only once in π_i. Hence, the cost for the online algorithm is $\Omega(n^2)$ per round, and $\Omega(n^2 \log n)$ in total.

Note that this construction can be repeated to rule out the possibility of a $o(\log n)$-competitive algorithm with an additive error term.

Acknowledgements. We acknowledge Suzanne Den Hertog (née van der Ster) for many helpful discussions. Part of this work was done while several of the authors were participating in the Hausdorff Trimester on Discrete Mathematics in Fall 2015.

References

1. Albers, S., von Stengel, B., Werchner, R.: A combined BIT and TIMESTAMP algorithm for the list update problem. Inf. Process. Lett. **56**(3), 135–139 (1995)
2. Albers, S., Westbrook, J.: Self-organizing data structures. In: Fiat, A., Woeginger, G.J. (eds.) Online Algorithms. LNCS, vol. 1442, pp. 13–51. Springer, Heidelberg (1998). https://doi.org/10.1007/BFb0029563

3. Ambühl, C.: Offline list update is NP-hard. In: Paterson, M.S. (ed.) ESA 2000. LNCS, vol. 1879, pp. 42–51. Springer, Heidelberg (2000). https://doi.org/10.1007/3-540-45253-2_5

4. Ambühl, C.: SIGACT news online algorithms column 31. SIGACT News **48**(3), 68–82 (2017)

5. Arora, S., Rao, S., Vazirani, U.V.: Expander flows, geometric embeddings and graph partitioning. J. ACM **56**(2), 5:1–5:37 (2009)

6. Avissar, O., Rajeev, R.B., Stewart, D.: An optimal memory allocation scheme for scratch-pad-based embedded systems. ACM Trans. Embed. Comput. Syst. **1**(1), 6–26 (2002)

7. Banakar, R., Steinke, S., Lee, B.S., Balakrishnan, M., Marwedel, P.: Scratchpad memory: design alternative for cache on-chip memory in embedded systems. In: Symposium on Hardware/Software Codesign, pp. 73–78. ACM, New York (2002)

8. Borodin, A., El-Yaniv, R.: On randomization in online computation. In: IEEE Conference on Computational Complexity (CCC), pp. 226–238 (1997)

9. Feige, U., Lee, J.R.: An improved approximation ratio for the minimum linear arrangement problem. Inf. Process. Lett. **101**(1), 26–29 (2007)

10. Gu, S., Sha, E., Zhuge, Q., Chen, Y., Hu, J.: Area and performance co-optimization for domain wall memory in application-specific embedded systems. In: Proceedings of the 52nd Annual Design Automation Conference, pp. 20:1–20:6. ACM, New York (2015)

11. Hansen, M.D.: Approximation algorithms for geometric embeddings in the plane with applications to parallel processing problems. In: IEEE Symposium on Foundations of Computer Science (FOCS), pp. 604–609 (1989)

12. Kamali, S., López-Ortiz, A.: A survey of algorithms and models for list update. In: Brodnik, A., López-Ortiz, A., Raman, V., Viola, A. (eds.) Space-Efficient Data Structures, Streams, and Algorithms. LNCS, vol. 8066, pp. 251–266. Springer, Heidelberg (2013). https://doi.org/10.1007/978-3-642-40273-9_17

13. Kandemir, M., Ramanujam, J., Choudhary, A.: Exploiting shared scratch pad memory space in embedded multiprocessor systems. In: Design Automation Conference, pp. 219–224 (2002)

14. Kandemir, M., Ramanujam, J., Irwin, J., Vijaykrishnan, N., Kadayif, I., Parikh, A.: Dynamic management of scratch-pad memory space. In: Proceedings of the 38th Annual Design Automation Conference, pp. 690–695 (2001)

15. Leighton, F.T., Rao, S.: Multicommodity max-flow min-cut theorems and their use in designing approximation algorithms. J. ACM **46**(6), 787–832 (1999)

16. Panda, P.R., Dutt, N.D., Nicolau, A.: Efficient utilization of scratch-pad memory in embedded processor applications. In: European Design and Test Conference (1997)

17. Rao, S., Richa, A.W.: New approximation techniques for some linear ordering problems. SIAM J. Comput. **34**(2), 388–404 (2004)

18. Reingold, N., Westbrook, J.: Off-line algorithms for the list update problem. Inf. Process. Lett. **60**(2), 75–80 (1996)

19. Sleator, D., Tarjan, R.E.: Amortized efficiency of list update and paging rules. CACM **28**(2), 202–208 (1985)

20. Teia, B.: A lower bound for randomized list update algorithms. Inf. Process. Lett. **47**, 5–9 (1993)

21. Williamson, D.P., Shmoys, D.B.: The Design of Approximation Algorithms. Cambridge University Press, Cambridge (2011)

The Price of Fixed Assignments in Stochastic Extensible Bin Packing

Guillaume Sagnol[1,2](✉), Daniel Schmidt genannt Waldschmidt[1], and Alexander Tesch[2]

[1] Institut für Mathematik, Technische Universität Berlin, Straße des 17. Juni 136, 10623 Berlin, Germany
{sagnol,dschmidt}@math.tu-berlin.de
[2] Zuse Institute Berlin, Takustr. 7, 14195 Berlin, Germany
tesch@zib.de

Abstract. We consider the *stochastic extensible bin packing problem* (SEBP) in which n items of stochastic size are packed into m bins of unit capacity. In contrast to the classical bin packing problem, the number of bins is fixed and they can be extended at extra cost. This problem plays an important role in stochastic environments such as in surgery scheduling: Patients must be assigned to operating rooms beforehand, such that the regular capacity is fully utilized while the amount of overtime is as small as possible.

This paper focuses on essential ratios between different classes of policies: First, we consider the price of non-splittability, in which we compare the optimal non-anticipatory policy against the optimal fractional assignment policy. We show that this ratio has a tight upper bound of 2. Moreover, we develop an analysis of a fixed assignment variant of the LEPT rule yielding a tight approximation ratio of $(1 + e^{-1}) \approx 1.368$ under a reasonable assumption on the distributions of job durations.

Furthermore, we prove that the price of fixed assignments, related to the benefit of adaptivity, which describes the loss when restricting to fixed assignment policies, is within the same factor. This shows that in some sense, LEPT is the best fixed assignment policy we can hope for.

Keywords: Approximation algorithms · Stochastic scheduling · extensible bin packing

1 Stochastic Extensible Bin Packing

In the *extensible bin packing problem* (EBP), we must put n items of size $(p_1, \ldots, p_n)$ in m bins, where the bins can be extended to hold more than the regular unit capacity. The cost of a bin is its regular capacity together with its

The research of the first two authors is carried out in the framework of MATHEON supported by Einstein Foundation Berlin.

L. Epstein and T. Erlebach (Eds.): WAOA 2018, LNCS 11312, pp. 327–347, 2018.
https://doi.org/10.1007/978-3-030-04693-4_20

extension costs: Specifically, a bin holding the items $I \subseteq \{1, \ldots, n\}$ has a cost of $\max\left(\sum_{i \in I} p_i, 1\right)$. The goal is to minimize the total cost of the m bins.

The model of extensible bin packing naturally arises in scheduling problems with machines available for some amount of time at a fixed cost, and an additional cost for extra-time. So we stick to the scheduling terminology in this article (bins are machines, items are jobs, and item sizes are processing times). Recently, the model of EBP was adopted to handle surgery scheduling problems [3,11,23]: here, the machines are operating rooms, and the jobs are operations to be performed on elective patients. The extension of the regular working time of a machine corresponds to overtime for the medical staff. This application to surgery scheduling motivates the present paper: in practice, the duration of a surgical operation on a given patient is not known with certainty. Therefore, we want to study the stochastic counterpart of the extensible bin packing problem, in which the processing durations p_j's are only known probabilistically, and the expected cost of the machines is to be minimized.

Related Work. EBP is closely related to another scheduling problem, where each job j has a due date d_j and the goal is to minimize the *total tardiness* $\sum_j T_j$, where T_j is the positive part of the difference of its completion time and its due date. This problem can not be approximated within any constant factor in polynomial time, unless $P = NP$ [17]. It relies on the fact that an approximation algorithm could differentiate YES and NO instances of PARTITION, since for YES instances the objective is equal to 0. Therefore, several articles studied approximation algorithms for a modified tardiness criterion, $\sum T_j + d_j$; see [16, 19]. The situation is very similar for extensible bin packing: the problem of minimizing the amount by which bins have to be extended is not approximable, and the criterion of EBP is obtained by adding the constant m to the objective.

The (deterministic version of) EBP was introduced by [9], who showed that the problem is strongly NP-hard, by reducing from 3–PARTITION; cf. [13]. Moreover, they prove that the *longest processing time first* (LPT) algorithm –which considers the jobs sorted in nonincreasing order of their processing time and assigns them sequentially to the machine with the largest remaining capacity– is a $\frac{13}{12}$–approximation algorithm. For equal bins, LPT can also be interpreted as iteratively assigning the jobs to the machine with the currently smallest load. In [10] the LPT algorithm was shown to be a $2(2 - \sqrt{2}) \simeq 1.1716$–approximation algorithm for the case of unequal bin sizes. In a more general framework, Alon et. al. present a polynomial time approximation scheme [1].

The online version of the problem also attracted attention. Here, the jobs arrive one at a time and they must be assigned to a machine irrevocably. The list scheduling algorithm LS that assigns an incoming job to the machine with the largest remaining capacity was shown to have a competitive ratio of $\frac{5}{4}$ for equal bin sizes in [10] and was generalized in [27] for the case with unequal bin sizes. Furthermore, it was proven that no algorithm can achieve a performance of $\frac{7}{6}$ or smaller compared to the offline optimum. An improved online algorithm with a competitive ratio of 1.228 was also presented in [27].

In the context of surgery scheduling, a slightly more general framework has been introduced in [11]: the decision maker also chooses the number of bins of size S to open, at a fixed cost c^f, and there is a variable cost c^v for each minute of overtime. It is observed in [3] that every $(1 + \rho)$−approximation algorithm for EBP yields a $(1 + \rho \frac{Sc^v}{c^f})$-approximation algorithm in this more general setting. They also consider a two-stage stochastic variant of the problem, in which emergency patients should be allocated to operating rooms with pre-allocated elective patients. For this problem (in the case $S = c^v = c^f = 1$), a particular fixed assignment policy was shown to be a $\frac{5\theta}{4}$-approximation algorithm, when each job has a duration with bounded support $P_j \in [0, p_j^{\max}]$ such that $p_j^{\max} \leq \theta \mathbb{E}[P_j]$. To the best of our knowledge, this has been the only attempt to consider stochastic jobs in the literature on EBP.

When considering stochastic optimization problems adaptive and non-adaptive policies are the solution concepts of matter. Especially, the greatest ratio between the cost of an optimal non-adaptive and the cost of an optimal adaptive policy over all instances is a quantity of interest. This so-called *benefit of adaptivity* or *adaptivity gap* has drawn attention dating back to the work in [8] and is getting popular, see e.g. [2,7,14]. In this work, we will work with another slightly different ratio closely related to it, since in the field of stochastic scheduling we are concerned with non-anticipatory policies that can make time-dependent decisions, such as idling. This can make a difference in the setting of parallel machines.

In the remaining of this section, we introduce the *stochastic extensible bin packing problem* (SEBP). Throughout, we consider the (offline) problem of scheduling n stochastic jobs on m parallel identical machines non-preemptively. We will assume that the distribution of the processing times are given beforehand and that their expectation is finite and computable[1]. The set of machines and jobs are denoted by $\mathcal{M} = \{1, \ldots, m\}$ and $\mathcal{J} = \{1, \ldots, n\}$, respectively.

Stochastic Scheduling. Now, we want to give the intuition and main ideas of the required background in the field of stochastic scheduling. Precise definitions are given in [22]. The processing times are represented by a vector $P = (P_1, \ldots, P_n)$ of random variables. We denote by $\boldsymbol{p} = (p_1, \ldots, p_n) \in \mathbb{R}_{\geq 0}^n$ a particular realization of P. We assume that the P_j's are mutually independent, and that each processing time has a finite expected value. Unlike the deterministic case, a scheduling strategy can take more general forms than just an allocation of jobs to machines, as information is gained during the execution of the schedule. Indeed, job durations become known upon completion, and adaptive policies can react to the processing times observed so far.

We define a *schedule* as a pair $S = (\boldsymbol{s}, \boldsymbol{a}) \in \mathbb{R}_{\geq 0}^n \times \mathcal{M}^n$, where $s_j \geq 0$ is the starting time of job j and $a_j \in \mathcal{M}$ is the machine to which job j is assigned. A

---

[1] We do not specify how the processing time distributions should be represented in the input of the problem, as the policies we study only require the expected value of the processing times. In fact, we could even assume a setting in which the input consists only of the mean processing times $\mu_j = \mathbb{E}[P_j]$ $(\forall j \in \mathcal{J})$, and an adversary chooses some distributions of the P_j's matching the vector $\boldsymbol{\mu}$ of first moments.

schedule S is said to be *feasible* for the realization $\boldsymbol{p}$ if each machine processes at most one job at a time:

$$\forall i \in \mathcal{M}, \ \forall t \geq 0, \quad \left| \{ j \in \mathcal{J} : \ a_j = i, \ s_j \leq t < s_j + p_j \} \right| \leq 1.$$

We denote by $\mathcal{S}(\boldsymbol{p})$ the set of all feasible schedules for the realization $\boldsymbol{p}$. A *planning rule* is a function Π that maps a vector $\boldsymbol{p} \in \mathbb{R}^n_{\geq 0}$ of processing times to a schedule $S \in \mathcal{S}(\boldsymbol{p})$. A planning rule is called a *scheduling policy* if it is *non-anticipatory*, which intuitively means that decisions taken at time t (if any) may only depend on the observed durations of jobs completed before t, and the probability distribution of the other processing times (conditioned by the knowledge that ongoing jobs have not completed before t).

Stochastic Extensible Bin Packing (SEBP). For a scheduling policy Π, we denote by S_j^Π and A_j^Π the random variables for the starting time of job j, and the machine to which j is assigned, respectively. The completion time of job j is $C_j^\Pi = S_j^\Pi + P_j$. We further introduce the random variable W_i^Π for the completion time of machine i, which is defined as the latest completion time of a job on machine i:

$$W_i^\Pi := \max\{ C_j^\Pi \mid j \in \mathcal{J}, \ A_j^\Pi = i \}.$$

It is easy to see that when Π is *non-idling*, i.e., if the starting time of any job is either 0 or equal to the completion time of the previous job assigned to the same machine, then

$$W_i^\Pi = \sum_{\{j \in \mathcal{J} : \ A_j^\Pi = i\}} P_j.$$

The realizations of the random vectors S^Π, A^Π, C^Π and W^Π for a vector of processing times $\boldsymbol{p}$ are denoted by appending $\boldsymbol{p}$ as an argument. For example, the workload of machine i for a non-idling policy Π in the scenario $\boldsymbol{p} \in \mathbb{R}^n_{\geq 0}$ is

$$W_i^\Pi(\boldsymbol{p}) = \sum_{j \xrightarrow{\Pi(\boldsymbol{p})} i} p_j,$$

where $j \xrightarrow{\Pi(\boldsymbol{p})} i$ means that $\Pi(\boldsymbol{p})$ assigns job j to machine i, i.e., we sum over indices $\{ j \in \mathcal{J} : A_j^\Pi(\boldsymbol{p}) = i \}$.

Remark 1. We want to point out that other authors (e.g., in [1]) use the notation C_i for the machine completion times. We prefer to use the symbol W_i (which stands for *workload* in the non-idling case) to avoid the risk of confusion with the job completion times C_j.

We assume that jobs are scheduled on machines with an extendable working time, each machine having a unit regular working time. The cost incurred on machine i is equal to $\max(W_i^\Pi, 1)$, which accounts for the fixed costs, plus the amount by which the regular working time has to be extended. We are interested in strategies that minimize the expected value of the sum of costs over all machines:

$$\Phi(\Pi) := \mathbb{E}\Big[\sum_{i \in \mathcal{M}} \max(W_i^{\Pi}, 1) \Big].$$

The criterion can also be defined realization-wise: we define $\phi(\Pi, \boldsymbol{p}) := \sum_{i \in \mathcal{M}} \max(W_i^{\Pi}(\boldsymbol{p}), 1)$, so that $\Phi(\Pi) := \mathbb{E}_P[\phi(\Pi, P)]$.

Classes of Scheduling Policies. We define the following classes of scheduling policies:

- $\mathcal{P}$ denotes the class of all scheduling policies (non-anticipatory planning rules).
- $\mathcal{F}$ denotes the set of all *non-idling fixed-assignment policies*. Such policies are characterized by a vector of job-to-machine assignments $\boldsymbol{a} \in \mathcal{M}^n$, so that $A^{\Pi}(\boldsymbol{p}) = \boldsymbol{a}$ does not depend on the realization of processing times. For such a policy Π, it holds

$$\Phi(\Pi) = \sum_{i \in \mathcal{M}} \mathbb{E}\Big[\max \big(\sum_{j \xrightarrow{\Pi} i} P_j, 1 \big) \Big],$$

where the sum indexed by " $j \xrightarrow{\Pi} i$ " goes over all jobs j such that $A_j^{\Pi} = i$.

The distinction between fixed assignment policies and other, more sophisticated adaptive policies plays a central role in this article. Indeed, in the context of surgery scheduling, committing to a fixed assignment policy is a common practice [3,11,23], because fixed assignments yield simple schedules, that are easier to apprehend for both the medical staff and the patients. Hence, they cause less stress and are better suited to handle the human resources of an operating theatre [12]. Nonetheless, there is currently active research on the use of reactive policies for operating room scheduling [29]. As *"fully adaptive scheduling models and policies are infeasible in operating room scheduling practice"*, the focus is now on hybrid scheduling policies with a large amount of static decisions, and a limited amount of adaptivity [28]. While more flexible policies could arguably lead to an important gain of efficiency over static policies, there are still many obstacles for their introduction in the operating theatre. In particular, it must be ensured that adaptive policies do not harm the quality of health care [30], and computer-assisted scheduling techniques need to gain acceptance among practitioners [15]. In this context, one goal of the present paper is to study the gap between fixed assignment and adaptive policies from a theoretical perspective.

In addition, we define the following class of fractional policies, which is related to scheduling problems concerning moldable work preserving tasks (see [18]). It cannot be considered as non-anticipatory planning rules, but will be useful to derive bounds:

- $\mathcal{R}$ denotes the class of fractional assignment policies, in which a fraction $a_{ij} \in [0, 1]$ of job j is to be executed on machine i, with $\sum_{i \in \mathcal{M}} a_{ij} = 1$, for all

$j \in \mathcal{J}$. For a "policy" $\Pi \in \mathcal{R}$, the different fractions of a job can be executed simultaneously on different machines, so

$$\Phi(\Pi) := \sum_{i \in \mathcal{M}} \mathbb{E}\Big[\max \big(\sum_{j \in \mathcal{J}} a_{ij}^{\Pi} P_j, 1 \big) \Big].$$

LEPT Policies. There is no unique way to generalize the LPT algorithm used in the deterministic case. We distinguish two variants of the "longest expected processing time first" (LEPT) policy. The policy LEPT$_{\mathcal{F}}$ is the fixed assignment policy that results in the same assignments as the LPT algorithm for the deterministic processing times $p_j = \mathbb{E}[P_j]$. In other words, job to machine assignments are precomputed offline, as follows: jobs are considered in decreasing order of $\mathbb{E}[P_j]$, and sequentially assigned to the least loaded machine (in expectation). An example of LEPT$_{\mathcal{F}}$ is depicted in Fig. 1. The second policy, which we denote by LEPT$_{\mathcal{P}}$, is the priority list policy which considers jobs in the order of decreasing $\mathbb{E}[P_j]$'s, and start them (in this order) as early as possible. Unlike LEPT$_{\mathcal{F}}$, the job to machine assignments of the list policy LEPT$_{\mathcal{P}}$ depend on the realization $\boldsymbol{p}$ of the processing times. By [27] it immediately follows that LEPT$_{\mathcal{P}}$ is a $\frac{5}{4}$-approximation with respect to $OPT_{\mathcal{P}}$, since in every realization the schedule produced by LEPT$_{\mathcal{P}}$ is obtained by list scheduling.

As discussed earlier, given the prominence of fixed assignment policies in the context of surgery scheduling, we focus on the policy LEPT$_{\mathcal{F}}$ in the remaining of this article.

Performance Ratios. For a given instance $I = (P, m)$ of the SEBP, we denote the optimum value in the class $\mathcal{C}$ of scheduling policies by

$$OPT_{\mathcal{C}}(I) = \inf_{\Pi \in \mathcal{C}} \Phi(\Pi).$$

Whenever the instance is clear from the context, or when $I = (P, m)$ is an arbitrary instance, we will drop I from the argument, so we simply write $OPT_{\mathcal{C}}$. We also denote by $OPT(\boldsymbol{p})$ the optimal value of the criterion for the deterministic problem with processing times $\boldsymbol{p}$. In this case, it is clear that we can restrict our attention to fixed assignment policies $\Pi \in \mathcal{F}$:

$$OPT(\boldsymbol{p}) = \inf_{\Pi \in \mathcal{F}} \phi(\Pi, \boldsymbol{p}).$$

We now define various performance ratios. We say that $\Pi \in \mathcal{C}$ is an α-*approximation in the class* $\mathcal{C}$ if the inequality $\Phi(\Pi) \leq \alpha OPT_{\mathcal{C}}$ holds for all instances of SEBP. The *price of fixed assignments* and the *price of non-splittability* are respectively defined by

$$\mathrm{PoFA} = \sup_I \frac{OPT_{\mathcal{F}}(I)}{OPT_{\mathcal{P}}(I)} \quad \text{and} \quad \mathrm{PoNS} = \sup_I \frac{OPT_{\mathcal{P}}(I)}{OPT_{\mathcal{R}}(I)},$$

where the suprema go over all instances $I = (P, m)$ of SEBP.

The first ratio (PoFA) describes the loss if we restrict our attention to fixed assignment policies. In other words, it is a measure of what can be gained by

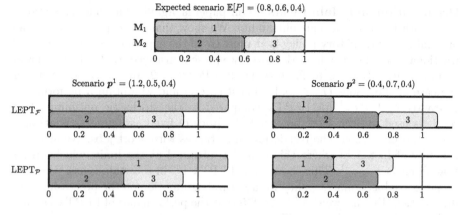

Fig. 1. Example of a fixed assignment policy: assume machines $\mathcal{M} = \{1, 2\}$ and jobs $\mathcal{J} = \{1, 2, 3\}$ with processing time distributions $p_1 \in \{0.4, 1.2\}$, $p_2 \in \{0.5, 0.7\}$, $p_3 = 0.4$ where the duration of each stochastic job is attained with probability $\frac{1}{2}$. Since $\mathbb{E}[P_1] = 0.8 \geq \mathbb{E}[P_2] = 0.6 \geq p_3 = 0.4$, LEPT$_\mathcal{F}$ assigns the jobs in order $1 \to 2 \to 3$ to the machines before their realization is known. The figure on the top depicts the resulting job to machine assignments with the average durations. For the realization $p_1 = (1.2, 0.5, 0.4)$ (lower left), LEPT$_\mathcal{F}$ is optimal with cost 2.2. For the realization $p_2 = (0.4, 0.7, 0.4)$ (lower right), LEPT$_\mathcal{F}$ yields cost 2.1. In contrast, note that LEPT$_\mathcal{P}$ would have started job 3 on the first machine after completion of job 1, giving a cost of 2.

allowing the use of more flexible, adaptive policies. This quantity gained attention in classical scheduling problems, e.g., in [21] and [25], whereby the latter shows that it can be arbitrarily large for the objective of minimizing the expected sum of completion times on parallel identical machines as the coefficient of variation grows.

The second ratio (PoNS) is related to the power of preemption, see e.g. [5, 6, 24, 26], but should not be mixed up with it, because the class $\mathcal{R}$ allows different parts of a job to be processed simultaneously on several machines for fractional assignment policies. However, this quantity has a simple interpretation in the context of surgery scheduling. Consider a hospital that assigns patients to a particular day until the total expected duration of the booked surgeries exceeds a certain threshold, but ignores the actual allocation of patients to operating rooms. The precise assignment of patients to operating rooms is deferred to a later stage, typically one week to one day before the day of surgery, when the set of all elective patients will be known. In fact, this simplification amounts to assuming that jobs of a particular day are placed in a single bin of size m (rather than in m bins of unit size). We will see in Proposition 1 that this can be interpreted as splitting the patient durations arbitrarily, and hence, evaluating the costs within this simplified one-bin model can yield a multiplicative error of up to PoNS.

Organization and Main Results. Our paper is organized as follows. Section 2 deals with the price of non-splittability. We show that the expected cost of an optimal non-anticipatory policy is at most twice the expected cost of an optimal fractional assignment policy. Moreover, we present instances that achieve a lower bound arbitrarily close to 2, showing that PoNS = 2. In Sect. 3, we consider the case of short jobs ($P_j \in [0,1]$ almost surely) and we obtain a performance guarantee of $1 + e^{-1}$ for LEPT$_\mathcal{F}$ compared to the stochastic optimum. This result is used in Sect. 4 to show that the price of fixed assignments is at most $1 + e^{-1}$, even without the restriction to instances with short jobs. We also give a family of instances where this bound is attained at the limit, which proves that PoFA = $1 + e^{-1}$. This shows that LEPT$_\mathcal{F}$ is –in a certain sense– the best possible fixed assignment policy for a natural assumption on the processing time distribution. Finally, we show in Sect. 5 that the performance of LEPT$_\mathcal{F}$ can not be better than $\frac{4}{3}$ in the class $\mathcal{F}$.

2   The Price of Non-splittability

Proposition 1. *Let (P, m) be an instance of SEBP and let $\rho := \frac{1}{m} \sum_{j \in \mathcal{J}} \mathbb{E}[P_j]$ be the expected workload averaged over all machines. Then the following holds:*

$$OPT_\mathcal{F} \geq OPT_\mathcal{P} \geq \mathbb{E}_P[OPT(P)] \geq OPT_\mathcal{R} = \mathbb{E}\left[\max(\sum_{j \in \mathcal{J}} P_j, m)\right] \geq m\max(\rho, 1).$$

Proof. The first inequality follows immediately since $\mathcal{F} \subseteq \mathcal{P}$.

Next, for all policies $\Pi \in \mathcal{P}$ and all realizations $\boldsymbol{p}$ it holds $\phi(\Pi, \boldsymbol{p}) \geq OPT(\boldsymbol{p})$, by definition of an optimal policy for the deterministic processing times $\boldsymbol{p}$. Taking the expectation on both sides yields the second inequality.

Before we go on to the next inequality, we first show that $OPT_\mathcal{R} = \mathbb{E}\left[\max(\sum_{j \in \mathcal{J}} P_j, m)\right]$. To do so we show that for any realization $\boldsymbol{p}$ an optimal fractional assignment policy assigns all jobs uniformly to all machines. More precisely, we show that $a_{ij} = \frac{1}{m}$ for all $i \in \mathcal{M}$ and $j \in \mathcal{J}$ solves the following problem of finding the optimal fractional assignment:

$$\underset{0 \leq a_{ij} \leq 1}{\textbf{minimize}} \quad \sum_{i \in \mathcal{M}} \max(\sum_{j \in \mathcal{J}} a_{ij}p_j, 1), \quad \text{such that} \quad \sum_{i \in \mathcal{M}} a_{ij} = 1, \quad \forall j \in \mathcal{J}. \tag{1}$$

A trivial lower bound on the optimal value of Problem (1) is $\max(\sum_{j \in \mathcal{J}} p_j, m)$. This is true since for any feasible fractional assignment $(a_{ij})_{i \in \mathcal{M}, j \in \mathcal{J}}$, $\sum_{i \in \mathcal{M}} \max(\sum_{j \in \mathcal{J}} a_{ij}p_j, 1) \geq \sum_{i \in \mathcal{M}} \sum_{j \in \mathcal{J}} a_{ij}p_j = \sum_{j \in \mathcal{J}} p_j$, and similarly, $\sum_{i \in \mathcal{M}} \max(\sum_{j \in \mathcal{J}} a_{ij}p_j, 1) \geq \sum_{i \in \mathcal{M}} 1 = m$. Choosing all fractions to be $\frac{1}{m}$ we obtain $\sum_{i \in \mathcal{M}} \max(\sum_{j \in \mathcal{J}} \frac{1}{m}p_j, 1) = m \cdot \max(\sum_{j \in \mathcal{J}} \frac{1}{m}p_j, 1) = \max(\sum_{j \in \mathcal{J}} p_j, m)$ which exactly matches the lower bound and hence, it must be optimal. Since this holds for any realization we can take the expected value resulting into the desired identity.

In order to show $\mathbb{E}_P[OPT(P)] \geq OPT_\mathcal{R}$, we observe that for any realization $\boldsymbol{p}$, Problem (1) is the continuous relaxation of the problem with binary variables for finding the optimal assignments for the deterministic problem with processing times p. Hence, by again taking expectations this yields the inequality.

Finally, the last inequality is Jensen's inequality applied to the convex function $x \mapsto \max(x, m)$.

In the next proposition, which we prove in the appendix, we show the intuitive fact that among non-idling policies, the worst case is to assign all jobs to the same machine.

Proposition 2. *Let $\Pi \in \mathcal{P}$ be non-idling and let Π_1 be the fixed assignment policy that schedules all jobs on machine 1. Then, $\Phi(\Pi) \leq \Phi(\Pi_1)$.*

We show that any non-idling policy is a 2-approximation in the class of non-anticipatory policies (and hence in the class of fixed-assignment policies).

Proposition 3. *Let Π be any non-idling policy. Then,*

$$\Phi(\Pi) \leq 2\,OPT_{\mathcal{R}}.$$

Proof. Let Π be a non-idling policy and Π_1 be the naive fixed assignment policy in which all jobs are scheduled on one machine without idle time. Proposition 2 yields that $\Phi(\Pi) \leq \Phi(\Pi_1)$, and we have

$$\Phi(\Pi_1) = \mathbb{E}[\max(\sum_{j \in \mathcal{J}} P_j, 1)] + (m-1) \leq \mathbb{E}[\max(\sum_{j \in \mathcal{J}} P_j, m)] + m - 1.$$

We know that $m \leq \mathbb{E}[\max(\sum_{j \in \mathcal{J}} P_j, m)] = OPT_{\mathcal{R}}$ from Proposition 1, so we have

$$\Phi(\Pi) \leq \Phi(\Pi_1) \leq 2\,OPT_{\mathcal{R}} - 1 \leq 2\,OPT_{\mathcal{R}}.$$

Consequently, we are only interested in finding $\alpha-$approximation algorithms for $\alpha < 2$, since a $2-$approximation algorithm performs no better (in the worst case) than the naive policy that puts all jobs on a single machine.

The last proposition also shows that the price of non-splittability is upper bounded by 2. In fact, this bound is tight:

Theorem 1. *The price of non-splittability of SEBP is PoNS $= 2$.*

The proof relies on a technical lemma which is proved in the appendix:

Lemma 1. *Let $Y \sim Poisson(\lambda)$ for some $\lambda \in \mathbb{N}$. Then,*

$$\frac{1}{\lambda}\mathbb{E}\Big[\max(Y, \lambda)\Big] = 1 + \frac{e^{-\lambda}\lambda^\lambda}{\lambda!}.$$

Proof (of Theorem 1). It follows from Propositions 1 and 3 that $OPT_{\mathcal{P}} \leq OPT_{\mathcal{F}} \leq 2OPT_{\mathcal{R}}$.

Let $\lambda \in \mathbb{N}$ and consider the instance I with $n = m \geq \lambda$ independent and identically distributed jobs in which the processing time of each job j takes the value $\frac{m}{\lambda}$ with probability $\frac{\lambda}{m}$ and 0 otherwise. In other words, for all $j \in \mathcal{J}$

we have $P_j \sim \frac{m}{\lambda}$ Bernoulli($\frac{\lambda}{m}$). As $n = m$, an optimal non-idling policy clearly assigns each job to a different machine. This yields

$$OPT_{\mathcal{P}}(I) = m \cdot \mathbb{E}[\max(P_1, 1)] = m \cdot \left((1 - \frac{\lambda}{m}) \cdot 1 + \frac{\lambda}{m} \cdot \frac{m}{\lambda} \right) = 2m - \lambda.$$

For the objective value of an optimal fractional assignment policy we can use Proposition 1. We will also use the fact that the sum of i.i.d. Bernoulli random variables is binomially distributed, i.e., $X := \frac{\lambda}{m} \cdot \sum_{j \in \mathcal{J}} P_j \sim$ Binomial($m, \frac{\lambda}{m}$). Moreover, it is folklore that X converges in distribution to $Y \sim$ Poisson(λ) as $m \to \infty$.

Therefore, we have $\frac{\lambda}{m} OPT_{\mathcal{R}}(I) = \frac{\lambda}{m} \mathbb{E}\left[\max\left(\sum_{j \in \mathcal{J}} P_j, m \right) \right] = \mathbb{E}[\max(X, \lambda)]$, which converges in distribution to $1 + \frac{e^{-\lambda} \lambda^\lambda}{\lambda!}$ as $m \to \infty$ by Lemma 1. Putting all together, the ratio $OPT_{\mathcal{P}}(I)/OPT_{\mathcal{R}}(I)$ converges to $2(1 + \frac{e^{-\lambda} \lambda^\lambda}{\lambda!})^{-1}$ as $m \to \infty$, and this quantity can be made arbitrarily close to 2 by choosing λ large enough.

3   Approximation Ratio of LEPT: The Case of Short Jobs

In this section, we show that $LEPT_{\mathcal{F}}$ is an $(1 + e^{-1})$-approximation algorithm when the instance only contains *short jobs*.

Definition 1. *We say job j is* short *if its processing time P_j is less than or equal to 1 almost surely, i.e.,*

$$\mathbb{P}[0 \leq P_j \leq 1] = 1.$$

It is reasonable to assume that jobs are short: In real world applications, such as in surgery scheduling, the duration of a single operation rarely exceeds the regular capacity of an operating room. Moreover, this assumption is not uncommon; cf. [10,27]. The proof of the performance guarantee of $LEPT_{\mathcal{F}}$ relies on three lemmas which we prove in the appendix. The first lemma gives a tight bound on the expected cost incurred on one machine.

Lemma 2. *Let k be some positive integer and let all jobs $j \in [k]$ be short. Then,*

$$\mathbb{E}\left[\max\left(\sum_{j=1}^{k} P_j, 1 \right) \right] \leq \sum_{j=1}^{k} \mathbb{E}[P_j] + \prod_{j=1}^{k} (1 - \mathbb{E}[P_j]).$$

Moreover, this bound is tight, and attained for the two point distributions $P_j^* \sim$ Bernoulli($\mathbb{E}[P_j]$).

The second lemma gives bounds on the expected workload of any machine in an LEPT$_{\mathcal{F}}$ schedule. Interestingly, the gap between the lower and upper bounds becomes smaller when the number of jobs scheduled on a machine grows.

Lemma 3. *Let x_i denote the expected load of machine $i \in \mathcal{M}$ produced by $LEPT_{\mathcal{F}}$, i.e., $x_i := \mathbb{E}[W_i^{LEPT_{\mathcal{F}}}] = \sum\limits_{j \xrightarrow{LEPT_{\mathcal{F}}} i} \mathbb{E}[P_j]$. Then, there exists $\ell \geq 0$*

such that for all $i \in \mathcal{M}$,

$$\ell \leq x_i \leq \frac{n_i}{n_i - 1}\ell,$$

where $n_i := |\{j \in \mathcal{J} : j \xrightarrow{LEPT_{\mathcal{F}}} i\}|$ denotes the number of jobs assigned to machine i, and we use the convention $\frac{n_i}{n_i-1} = \frac{1}{0} := +\infty$ whenever $n_i = 1$.

We need a third lemma with a technical result:

Lemma 4. *Let $\ell \geq 0$ and $\rho \geq \ell$. We define the function $h : [0,1] \to \mathbb{R}$, $y \mapsto (1 - y)^{1+\frac{\ell}{y}}$, which is defined by continuity at $y = 0$ with $h(0) = e^{-\ell}$. Let $\boldsymbol{y} \in [0,1]^m$ be any vector satisfying the equality $\sum_{i \in \mathcal{M}} y_i = m(\rho - \ell)$. Then,*

$$\sum_{i \in \mathcal{M}} h(y_i) \leq me^{-\rho}.$$

We are now ready to prove the main result of this section:

Theorem 2. *Consider an instance (P, m) with only short jobs. Let $\rho := \frac{1}{m}\sum_{j \in \mathcal{J}} \mathbb{E}[P_j]$ denote the expected workload averaged over all machines. Then it holds*

$$\frac{\Phi(LEPT_{\mathcal{F}})}{m\max(\rho, 1)} \leq \frac{\rho + e^{-\rho}}{\max(\rho, 1)} \leq 1 + e^{-1}.$$

Proof. Let J_i denote the subset of jobs that $LEPT_{\mathcal{F}}$ assigns to machine $i \in \mathcal{M}$ and let $n_i := |J_i|$. As in Lemma 3, let $LEPT_{\mathcal{F}}$ produce an expected workload of $x_i = \sum_{j \in J_i} \mathbb{E}[P_j]$ on machine i. Then, by Lemma 2 we can bound the expected cost incurred on machine i as

$$\mathbb{E}[\max(W_i^{LEPT_{\mathcal{F}}}, 1)] \leq \sum_{j \in J_i} \mathbb{E}[P_j] + \prod_{j \in J_i}(1 - \mathbb{E}[P_j]) \leq x_i + \left(1 - \frac{x_i}{n_i}\right)^{n_i} \quad (2)$$

where the last inequality follows from the Schur-concavity of $\boldsymbol{\mu} \mapsto \prod_{j \in J_i}(1 - \mu_j)$ over $[0,1]^{n_i}$; cf. [20, Proposition 3.E.1]. Next, we apply Lemma 3, so there exists an $\ell \geq 0$ such that $\ell \leq x_i \leq \frac{n_i}{n_i-1}\ell$. Let $y_i := x_i - \ell \geq 0$. The second inequality can be rewritten as

$$n_i \leq \frac{x_i}{x_i - \ell} = 1 + \frac{\ell}{y_i}, \quad (3)$$

which remains valid for $y_i = 0$ if we define $\ell/0 := +\infty$. We know that $P_j \in [0,1]$ almost surely, in particular $\mathbb{E}[P_j] \leq 1$, and hence, $x_i \leq n_i$. For this reason, the above inequality implies $x_i \leq \frac{x_i}{x_i-\ell}$ and therefore, $y_i \leq 1$. By combining (2) and (3), and using the fact that $(1 - \frac{x_i}{n_i})^{n_i}$ is a nondecreasing function of n_i, we obtain

$$\mathbb{E}[\max(W_i^{LEPT_{\mathcal{F}}}, 1)] \leq x_i + (1 - (x_i - \ell))^{\frac{x_i}{x_i-\ell}} = \ell + y_i + h(y_i), \quad (4)$$

where h is the function defined in Lemma 4, and the y_i's satisfy $y_i \in [0, 1]$. Moreover, we have $\sum_{i \in \mathcal{M}} y_i = m(\rho - \ell) \iff \sum_{i \in \mathcal{M}} (\ell + y_i) = \sum_{i \in \mathcal{M}} x_i = \rho m$. Summing up the inequalities (4) over all $i \in \mathcal{M}$ and using Lemma 4 yields

$$\Phi(\text{LEPT}_{\mathcal{F}}) = \sum_{i \in \mathcal{M}} \mathbb{E}[\max(W_i^{\text{LEPT}_{\mathcal{F}}}, 1)] \leq \rho m + \sum_{i \in \mathcal{M}} h(y_i) \leq m(\rho + e^{-\rho}).$$

As a consequence, we obtain

$$\frac{\Phi(\text{LEPT}_{\mathcal{F}})}{m \max(\rho, 1)} \leq \frac{\rho + e^{-\rho}}{\max(\rho, 1)}.$$

Finally, the second inequality of the theorem follows from the fact that the above ratio is maximized for $\rho = 1$. This is true because $\frac{\rho + e^{-\rho}}{\max(\rho, 1)} = \rho + e^{-\rho}$ on $[0, 1]$, hence increasing, and $\frac{\rho + e^{-\rho}}{\max(\rho, 1)} = 1 + \frac{e^{-\rho}}{\rho}$ on $[1, +\infty]$, hence decreasing.

Combining this result with the inequality $OPT_{\mathcal{P}} \geq m \max(1, \rho)$ from Proposition 1 yields the following

Corollary 1. *The LEPT$_{\mathcal{F}}$ policy is an $(1 + e^{-1})$-approximation algorithm in the class $\mathcal{P}$, over the set of instances with short jobs only.*

As we will see in the next section, our analysis of LEPT$_{\mathcal{F}}$ is tight.

4   The Price of Fixed Assignments

In this section, we are going to show that the price of fixed assignments is equal to $1 + e^{-1}$. To do this, we require a lemma that will allow us to focus on instances with short jobs. Our analysis relies on a parameter $\alpha \geq 0$ which quantifies the *length excess* of jobs (for an instance with only short jobs, it holds $\alpha = 0$).

Lemma 5. *Let $I = (P, m)$ be an instance of SEBP, and let $I' = (P', m)$ denote the instance in which the processing time P_j of all jobs is replaced by $P_j' = \min(P_j, 1)$. Let $\alpha = \sum_{j \in \mathcal{J}} \alpha_j$, where we define $\alpha_j := \mathbb{E}[\max(P_j - 1, 0)] \geq 0$. The new P_j's are short jobs, and we have*

$$OPT_{\mathcal{F}}(I') = OPT_{\mathcal{F}}(I) - \alpha \quad \text{and} \quad \mathbb{E}_{P'}[OPT(P')] = \mathbb{E}_P[OPT(P)] - \alpha.$$

Proof. Let J_i and J_i' denote the subsets of jobs assigned to machine i in an optimal fixed assignment policy Π for instance I, and in an optimal fixed assignment policy Π' for instance I', respectively. Let $\boldsymbol{p}$ be a realization of the processing times for instance I, and let $\boldsymbol{p}'$ denote the vector with elements $p_j' = \min(p_j, 1)$. We compute the difference between the costs incurred by $\Pi(\boldsymbol{p})$ and $\Pi(\boldsymbol{p}')$ on machine i:

$$\max(W_i^{\Pi}(\boldsymbol{p}), 1) - \max(W_i^{\Pi}(\boldsymbol{p}'), 1) = \max\left(\sum_{j \in J_i} p_j, 1\right) - \max\left(\sum_{j \in J_i} \min(p_j, 1), 1\right). \tag{5}$$

It is easy to see that $\sum_{j \in J_i} p_j \leq 1 \iff \sum_{j \in J_i} \min(p_j, 1) \leq 1$. Hence, we distinguish two cases. If $\sum_{j \in J_i} p_j \leq 1$, then the right hand side of (5) vanishes. Otherwise, the right hand side of (5) becomes $\sum_{j \in J_i} p_j - \min(p_j, 1) = \sum_{j \in J_i} \max(p_j - 1, 0)$. In both cases, it holds $\max(W_i^{\Pi}(\boldsymbol{p}), 1) - \max(W_i^{\Pi}(\boldsymbol{p}'), 1) = \sum_{j \in J_i} \max(p_j - 1, 0)$. Taking the expectation and summing up over all machines yields

$$\Phi_I(\Pi) - \Phi_{I'}(\Pi) = \sum_{i \in \mathcal{M}} \sum_{j \in J_i} \alpha_j = \sum_{j \in \mathcal{J}} \alpha_j = \alpha,$$

where the symbol $\Phi_I(\Pi)$ emphasizes that the expected value in the criterion is taken with respect to the processing time distributions of instance I.

Since Π is optimal in the class $\mathcal{F}$ for instance I, we have $\Phi_I(\Pi) = OPT_{\mathcal{F}}(I)$ and $\Phi_{I'}(\Pi) \geq OPT_{\mathcal{F}}(I')$. Hence,

$$\Phi_{I'}(\Pi) = OPT_{\mathcal{F}}(I) - \alpha \geq OPT_{\mathcal{F}}(I'). \tag{6}$$

Similarly, the comparison of the costs incurred by $\Pi'(\boldsymbol{p})$ and $\Pi'(\boldsymbol{p}')$ on machine i yields $\max(W_i^{\Pi'}(\boldsymbol{p}), 1) - \max(W_i^{\Pi'}(\boldsymbol{p}'), 1) = \sum_{j \in J_i'} \max(p_j - 1, 0)$. Again, by taking the expectation and summing over all machines we obtain $\Phi_I(\Pi') - \Phi_{I'}(\Pi') = \sum_{j \in \mathcal{J}} \alpha_j = \alpha$. Now, we observe that $\Phi_{I'}(\Pi') = OPT_{\mathcal{F}}(I')$ and $\Phi_I(\Pi') \geq OPT_{\mathcal{F}}(I)$, so we have

$$\Phi_I(\Pi') = OPT_{\mathcal{F}}(I') + \alpha \geq OPT_{\mathcal{F}}(I). \tag{7}$$

Finally, by combining (6) and (7) we obtain $OPT_{\mathcal{F}}(I) - \alpha \geq OPT_{\mathcal{F}}(I') \geq OPT_{\mathcal{F}}(I) - \alpha$, which shows the desired equality:

$$OPT_{\mathcal{F}}(I) - \alpha = OPT_{\mathcal{F}}(I').$$

The proof of the equality $\mathbb{E}_{P'}[OPT(P')] = \mathbb{E}_P[OPT(P)] - \alpha$ works in a similar manner, but we must take sums over a different subset of jobs $J_i(\boldsymbol{p})$ for each scenario $\boldsymbol{p}$, corresponding to the jobs that an optimal policy assigns to machine i for the deterministic problem with processing times $\boldsymbol{p}$.

We can now prove the main result of this section:

Theorem 3. *The price of fixed assignments for SEBP is equal to* $(1 + e^{-1})$:

$$\mathrm{PoFA} = 1 + e^{-1}.$$

Proof. Let $I = (P, m)$ denote an instance of SEBP and $I' = (P', m)$ the reduced instance as in Lemma 5. We have:

$$\frac{OPT_{\mathcal{F}}(I)}{OPT_{\mathcal{P}}(I)} \leq \frac{OPT_{\mathcal{F}}(I)}{\mathbb{E}_P[OPT(P)]} = \frac{OPT_{\mathcal{F}}(I') + \alpha}{\mathbb{E}_{P'}[OPT(P')] + \alpha} \leq \frac{OPT_{\mathcal{F}}(I')}{\mathbb{E}_{P'}[OPT(P')]} \leq 1 + e^{-1},$$

where the first inequality follows from Proposition 1, the equality is a consequence of Lemma 5, the second inequality follows from $\alpha \geq 0$, and the last inequality results from Proposition 1 and Theorem 2. Therefore, it remains to show that for all $\epsilon > 0$ there exists an instance I in which we have

$$\frac{OPT_{\mathcal{F}}(I)}{OPT_{\mathcal{P}}(I)} \geq 1 + e^{-1} - \epsilon.$$

For this purpose, we consider an instance $I = (P, m)$ in which we have $n = km$ jobs for some $k \in \mathbb{N}$, where $P_j \sim \text{Bernoulli}(\frac{1}{k})$ for all $j \in \mathcal{J}$. An optimal fixed assignment policy assigns each machine the same number of jobs, in this case k. The cost on one machine is hence the expected value of $\max(Z, 1)$, where $Z := \sum_{j=1}^{k} P_j \sim \text{Binomial}(k, \frac{1}{k})$. So,

$$OPT_{\mathcal{F}}(I) = m \cdot \mathbb{E}[\max(Z, 1)] = m \cdot \Big(\mathbb{E}[Z|Z \geq 1]\ \mathbb{P}[Z \geq 1] + \mathbb{E}[1|Z < 1]\ \mathbb{P}[Z < 1]\Big)$$

$$= m \cdot (\mathbb{E}[Z] + \mathbb{P}[Z = 0]) = m \cdot \left(1 + \left(1 - \frac{1}{k}\right)^{k}\right),$$

which converges to $m(1 + e^{-1})$ as $k \to \infty$. On the other hand, an optimal policy in $\mathcal{P}$ lets a job run whenever a machine becomes idle. The cost of an optimal policy is hence m whenever less than m jobs have duration 1, and is equal to $\sum_{j=1}^{km} p_j$ otherwise. This shows that $OPT_{\mathcal{P}}(I) = \mathbb{E}[\max(U, m)]$, where $U := \sum_{j=1}^{km} P_j \sim \text{Binomial}\left(km, \frac{1}{k}\right)$. Now, we can argue as in Theorem 2 that U converges in distribution to $Y \sim \text{Poisson}(m)$ as $k \to \infty$. So, by Lemma 1, we have

$$OPT_{\mathcal{P}}(I) \to m \cdot \left(1 + \frac{e^{-m} m^m}{m!}\right) \quad \text{as} \quad k \to \infty.$$

Finally, we have shown that the ratio of $OPT_{\mathcal{F}}(I)$ to $OPT_{\mathcal{P}}(I)$ can be made arbitrarily close to $(1 + e^{-1}) \cdot \left(1 + \frac{e^{-m} m^m}{m!}\right)^{-1}$ by choosing k large enough. We conclude by observing that $\lim_{m \to \infty} \frac{m^m e^{-m}}{m!} = 0$, so this ratio can be arbitrarily close to $1 + e^{-1}$.

This proves that our analysis of $LEPT_{\mathcal{F}}$ is tight. It even shows that $LEPT_{\mathcal{F}}$ is the best fixed assignment policy in the following sense: Since there exists instances for which the ratio of an optimal fixed assignment policy to an optimal non-anticipatory policy is arbitrarily close to $1 + e^{-1}$ and the fact that $LEPT_{\mathcal{F}}$ is a $1 + e^{-1}$-approximation (for short jobs), we cannot hope to find a policy $\Pi \in \mathcal{F}$ with a better approximation guarantee in the class $\mathcal{P}$.

5   Performance of LEPT in the Class of Fixed Assignment Policies

It would also be interesting to characterize the approximation guarantee of $LEPT_{\mathcal{F}}$ in the class of fixed assignment policies. The next proposition gives a lower bound:

Proposition 4. *For all $\epsilon > 0$, there exists an instance I of SEBP such that* $\frac{\Phi(LEPT_\mathcal{F})}{OPT_\mathcal{F}(I)} = \frac{4-\epsilon}{3}$.

Proof. We construct an instance with $m = 2$ machines and $n = 3$ jobs. The first two jobs are deterministic and have duration $P_1 = P_2 = 1$. The distribution of the third job is $P_3 = \frac{1}{\epsilon}X$, where $X \sim$ Bernoulli(ϵ), so $\mathbb{E}[P_3] = 1$. We assume that the LEPT$_\mathcal{F}$ policy assigns both deterministic jobs to the first machine and the stochastic job to the other machine, which gives $\Phi(LEPT_\mathcal{F}) = 2 + (1 - \epsilon) + \frac{\epsilon}{\epsilon} = 4 - \epsilon$. In contrast, for any policy Π^* which assigns the two deterministic jobs on different machines, we have $\Phi(\Pi^*) = 1 + (1 - \epsilon) + (1 + \frac{1}{\epsilon})\epsilon = 3$. The policy Π^* reaches the lower bound $m \max(\rho, 1)$ of Proposition 1, hence it is optimal.

As will be discussed below, we believe that the performance guarantee of $1 + e^{-1}$ of LEPT$_\mathcal{F}$ can be extended for all instances (even when some jobs are not short). In this case, this would show that the best approximation factor for LEPT$_\mathcal{F}$ in the class of fixed assignment policies lies between $\frac{4}{3} \approx 1.333$ and $1 + e^{-1} \approx 1.368$.

6   Conclusion and Future Work

We showed that LEPT$_\mathcal{F}$ is, in some sense, the best algorithm among the class of fixed assignment policies we can hope for. This result might inspire future work to consider the same or similar and related ratios for other scheduling problems, in which we compare within or against several subclasses of policies, in order to obtain more interesting and precise results on the performance of algorithms.

We believe that the $(1 + e^{-1})$-approximation guarantee of LEPT$_\mathcal{F}$ can be extended for instances containing long jobs, i.e., jobs whose duration may exceed 1. It can be shown –using a similar approach as in Theorem 2– that $\frac{\Phi(LEPT_\mathcal{F})}{OPT_\mathcal{R}} \leq 1 + e^{-\frac{1}{d_{max}}}$ for instances where each job satisfies $P_j \in [0, d_{max}]$ almost surely for some $d_{max} \geq 1$, and that this bound is tight. Letting $d_{max} \to \infty$ just gives the trivial approximation guarantee of 2, so we have to use a better lower bound on $OPT_\mathcal{P}$ in order to prove that LEPT$_\mathcal{F}$ is a $(1 + e^{-1})$-approximation algorithm. Our next candidate is the bound $OPT_\mathcal{P} \geq \mathbb{E}_P[OPT(P)]$, cf. Proposition 1. We think that an analysis relying on the parameters $\rho = \frac{1}{m}\sum_j \mathbb{E}[P_j]$ and $\alpha = \sum_j \mathbb{E}[\max(0, P_j - 1)]$ introduced in Lemma 5 could lead to the desired result. So far, we obtained encouraging intermediate results that support our claim, but we did not obtain an analytical proof.

An interesting direction for future work on SEBP is the study of the case of unequal bins, which is relevant for the application to surgery scheduling, where operating rooms may have different opening hours. Since the class of fixed assignment policies is relevant for surgery scheduling, another interesting open question is whether there exists a policy $\Pi \in \mathcal{F}$ with a performance guarantee $< \frac{4}{3}$ in the class $\mathcal{F}$. A good candidate could be the variant of LEPT that considers more than just first moment information on the P_j's, and inserts sequentially the job j on the machine minimizing $\mathbb{E}[\max(X_i + P_j, 1)]$, where X_i is the random variable for the load already assigned to machine i. We also observe that the coefficient of variation of the jobs tend to infinity in all our tight examples, so it is natural to ask if we can obtain better bounds when these coefficients are upper bounded by a constant Δ. Last but not least, a two-stage stochastic online extension of the EBP could yield a better understanding of policies for the surgery scheduling problem with add-on cases (emergencies).

A   Proofs of Intermediate Results

Proof (of Proposition 2). To prove this result, we examine the change in the objective value of Π when we move one job to the machine with highest load in Π, for a realization p of the processing times. W.l.o.g. let machine 1 be the one with highest workload in $\Pi(p)$. Consider another machine $i \in \mathcal{M} \setminus \{1\}$ on which at least one job is scheduled. Let k be the last job on machine i, i.e., $C_k^\Pi(p) = W_i^\Pi(p)$. For the sake of simplicity, we define $A := \{j \in \mathcal{J} | j \xrightarrow{\Pi(p)} i\} \setminus \{k\}$ and $B := \{j \in \mathcal{J} | j \xrightarrow{\Pi(p)} 1\}$. We consider another schedule $\Pi'(p)$ which coincides with $\Pi(p)$ except that job k is scheduled on machine 1 right after all jobs in B. We obtain

$$
\begin{aligned}
&\phi(\Pi, p) - \phi(\Pi', p) \\
=\ & \max\Big(\sum_{j \in A} p_j + p_k, 1\Big) + \max\Big(\sum_{j \in B} p_j, 1\Big) - \Big(\max\Big(\sum_{j \in A} p_j, 1\Big) + \max\Big(\sum_{j \in B} p_j + p_k, 1\Big)\Big) \\
=\ & \begin{cases} 1 + \max\Big(\displaystyle\sum_{j \in B} p_j, 1\Big) - \Big(1 + \max\Big(\displaystyle\sum_{j \in B} p_j + p_k, 1\Big)\Big) & \text{if } \displaystyle\sum_{j \in A} p_j + p_k \leq 1 \\ \displaystyle\sum_{j \in A} p_j + p_k + \sum_{j \in B} p_j - \Big(\max\Big(\displaystyle\sum_{j \in A} p_j, 1\Big) + \sum_{j \in B} p_j + p_k\Big) & \text{otherwise} \end{cases} \\
\leq\ & 0.
\end{aligned}
$$

Hence, iteratively moving some job k to the fullest machine yields $\phi(\Pi, p) \leq \phi(\Pi_1, p)$. Finally, the result follows by taking the expectation.

Proof (of Lemma 1). The proof simply works by exploiting the analytical form of Poisson probabilities:

$$
\begin{aligned}
\frac{1}{\lambda} \mathbb{E}\Big[\max(Y, \lambda)\Big] &= \frac{1}{\lambda} \sum_{k=0}^{\infty} \max(k, \lambda) \cdot \frac{e^{-\lambda} \lambda^k}{k!} \\
&= \frac{1}{\lambda} \sum_{k=0}^{\infty} k \cdot \frac{e^{-\lambda} \lambda^k}{k!} + \frac{1}{\lambda} \sum_{k=0}^{\infty} \max(0, \lambda - k) \cdot \frac{e^{-\lambda} \lambda^k}{k!} \\
&= 1 + \sum_{k=0}^{\lambda} \Big(1 - \frac{k}{\lambda}\Big) \cdot \frac{e^{-\lambda} \lambda^k}{k!}
\end{aligned}
$$

$$= 1 + e^{-\lambda} \cdot \left(\sum_{k=0}^{\lambda} \frac{\lambda^k}{k!} - \sum_{k=1}^{\lambda} \frac{\lambda^{k-1}}{(k-1)!} \right)$$

$$= 1 + \frac{e^{-\lambda} \lambda^\lambda}{\lambda!},$$

where the last step follows from the property of a telescoping sum.

Proof (of Lemma 2).
Let X and Y be random variables with $\mathbb{P}[0 \leq X \leq 1] = 1$. Observe that $0 \leq \mathbb{E}[X] \leq 1$. We are going to show that $\mathbb{E}[\max(X+Y, 1)]$ can be bounded from above by choosing the two point distribution $X^* \sim \text{Bernoulli}(\mathbb{E}[X])$, such that $\mathbb{P}[X^* = 0] = (1 - \mathbb{E}[X])$ and $\mathbb{P}[X^* = 1] = \mathbb{E}[X]$. To do so, we define the function $g : [0, 1] \to \mathbb{R}, x \mapsto \mathbb{E}_Y[\max(x + Y, 1)]$. This function is convex, since it is the expectation of a pointwise maximum of two affine functions [4]. Therefore, for all $x \in [0, 1]$ we have $g(x) \leq g(0) + x(g(1) - g(0))$. Then, by definition of g,

$$\mathbb{E}[\max(X + Y, 1)] = \mathbb{E}_X[g(X)] \leq g(0) + \mathbb{E}_X[X] \cdot (g(1) - g(0))$$
$$= \mathbb{E}_{X^*}[g(X^*)] = \mathbb{E}[\max(X^* + Y, 1)].$$

Using this bound for all $j \in [k]$, we obtain $\mathbb{E}\left[\max\left(\sum_{j=1}^k P_j, 1\right)\right] \leq \mathbb{E}\left[\max\left(\sum_{j=1}^k P_j^*, 1\right)\right]$, where $P_j^* \sim \text{Bernoulli}(\mathbb{E}[P_j])$. Then, by the law of total expectation, we have:

$$\mathbb{E}\left[\max\left(\sum_{j=1}^k P_j^*, 1\right)\right] = \mathbb{E}\left[\sum_{j=1}^k P_j^* \Big| \sum_{j=1}^k P_j^* \geq 1\right] \mathbb{P}[\sum_{j=1}^k P_j^* \geq 1] + \mathbb{E}[1] \, \mathbb{P}[\sum_{j=1}^k P_j^* < 1].$$

Since the random variable $\sum_{j=1}^k P_j^*$ is a nonnegative integer, it cannot lie in the interval $(0, 1)$, so the first term in the above sum is equal to $\mathbb{E}\left[\sum_{j=1}^k P_j^*\right] = \sum_{j=1}^k \mathbb{E}\left[P_j\right]$, and the second term is equal to $\mathbb{P}[P_1^* = \ldots = P_k^* = 0] = \prod_{j=1}^k (1 - \mathbb{E}[P_j])$.

Proof (of Lemma 3). We set $\ell := \min\{x_i : i \in \mathcal{M}\}$. Then, the first inequality follows immediately. Next, we will show that in each step that $\text{LEPT}_{\mathcal{F}}$ assigns a job to a machine the second inequality is fulfilled. Let j denote the job which is put on machine i in the current step. Furthermore, let ℓ' and ℓ denote the minimum expected load among all machines before and after the allocation, respectively. Trivially, $\ell' \leq \ell$ is true. Moreover, let x_i' and x_i denote the expected workload of i before and after assigning j to it, respectively. Clearly, we have

$$x_i = x_i' + \mathbb{E}[P_j].$$

Observe, that $\ell' = x_i'$, because $\text{LEPT}_{\mathcal{F}}$ assigns j to the machine with the smallest expected load. In addition, let n_i denote the number of jobs running on machine

i after the insertion of j. Since LEPT$_{\mathcal{F}}$ sorts jobs in decreasing order of their expected processing times, it holds

$$\mathbb{E}[P_j] \leq \frac{x_i'}{n_i - 1} = \frac{\ell'}{n_i - 1}.$$

Consider a machine other than i. If the inequality of the statement was fulfilled in an earlier step, then by setting the new ℓ it still is true. In the beginning, when we have no job at all, the inequality is true, so we only have to take care of machine i.

Finally, we obtain on machine i

$$\frac{x_i}{\ell} = \frac{x_i' + \mathbb{E}[P_j]}{\ell} \leq \frac{x_i' + \mathbb{E}[P_j]}{\ell'} \leq 1 + \frac{\mathbb{E}[P_j]}{\ell'} \leq 1 + \frac{\ell'}{\ell'(n_i - 1)} = \frac{n_i}{(n_i - 1)}.$$

Proof (of Lemma 4). First, we argue that $h : y \mapsto (1-y)^{1+\frac{\ell}{y}}$ is convex over $[0,1]$. To see this, we compute its second derivative:

$$h''(y) = \frac{\ell(1-y)^{\frac{\ell}{y}-1}}{y^4} h_2(y),$$

where $h_2(y) := y^2(\ell - y + 2) + \ell(y-1)^2 \log^2(1-y) - 2(\ell+1)(y-1)y \log(1-y)$. Now, we use the fact that $\log(1-y) = -\sum_{k=1}^{\infty} \frac{y^k}{k}$ for all $y \in [0,1)$. Hence, $\log^2(1-y) = \sum_{k=2}^{\infty} \gamma_k y^k$, where $\gamma_k := \sum_{i=1}^{k-1} \frac{1}{i(k-i)}$. After some calculus, the terms of order 2 and 3 vanish and we obtain the following series representation of h_2 over $[0,1)$:

$$h_2(y) = \left(\frac{\ell}{4} + \frac{1}{3}\right)y^4 + \sum_{k=5}^{\infty}\left(\frac{2(\ell+1)}{(k-1)(k-2)} + \ell(\gamma_k + \gamma_{k-2} - 2\gamma_{k-1})\right)y^k.$$

We are going to show that $\gamma_k + \gamma_{k-2} - 2\gamma_{k-1} \geq 0$ for $k \geq 5$ implying that $h''(y) \geq 0$ for all $y \in [0,1)$. To do so, we rewrite the sums using the partial fraction decomposition. As a consequence, we obtain

$$
\begin{aligned}
\gamma_k + \gamma_{k-2} - 2\gamma_{k-1} &= \frac{2}{k}\sum_{i=1}^{k-1}\frac{1}{i} + \frac{2}{k-2}\sum_{i=1}^{k-3}\frac{1}{i} - \frac{4}{k-1}\sum_{i=1}^{k-2}\frac{1}{i} \\
&= \frac{2}{k}\left(\frac{1}{k-2} + \frac{1}{k-1}\right) - \frac{4}{(k-1)(k-2)} + \left(\frac{2}{k} + \frac{2}{k-2} - \frac{4}{k-1}\right)\sum_{i=1}^{k-3}\frac{1}{i} \\
&= -\frac{6}{k(k-1)(k-2)} + \frac{4}{k(k-1)(k-2)}\sum_{i=1}^{k-3}\frac{1}{i} \\
&\geq 0.
\end{aligned}
$$

The last inequality results from the fact that for all $k \geq 5$ we have $4\sum_{i=1}^{k-3}\frac{1}{i} \geq 6$. Hence, h is convex on $[0,1)$, and even on $[0,1]$ by continuity. Now, let $v^*(\rho, \ell)$ denote the optimal value of the problem

$$\underset{y \in \mathbb{R}^m}{\text{maximize}} \quad \sum_{i \in \mathcal{M}} h(y_i) \tag{8a}$$

$$s.t. \quad \sum_{i \in \mathcal{M}} y_i = m(\rho - \ell) \tag{8b}$$

$$0 \le y_i \le 1, \quad (\forall i \in \mathcal{M}). \tag{8c}$$

As h is convex, a maximizer of the optimization problem above is an extreme point of the polytope induced by the constraints (8b) and (8c). Let $k := \lfloor m(\rho - \ell) \rfloor$ and $u := m(\rho - \ell) - k$, where $\lfloor . \rfloor$ denotes the floor function, that is, $\lfloor x \rfloor$ is the largest integer less than or equal to x. By construction, it holds $0 \le u \le 1$, and $u + k = m(\rho - \ell)$. At an extreme point, at least $m - 1$ inequalities of (8c) must be tight. Hence, one coordinate of y must be u, k coordinates must be 1 and the remaining $(m - k - 1)$ coordinates must be 0.

It follows that $v^*(\rho, \ell) = (m - k - 1)h(0) + h(u) = (m - k - 1)e^{-\ell} + (1 - u)^{1 + \ell/u}$. Now, we observe that $(1 - u)^{\ell/u} \le e^{-\ell}$, so

$$(1 - u)^{1 + \ell/u} \le (1 - u)e^{-\ell}$$

$$\Longleftrightarrow \quad (1 - u)^{1 + \ell/u} \le (1 + k - m(\rho - \ell))e^{-\ell}$$

$$\Longleftrightarrow \quad \underbrace{(m - k - 1)e^{-\ell} + (1 - u)^{1 + \ell/u}}_{v^*(\rho, \ell)} \le m(1 - \rho + \ell)e^{-\ell},$$

where the first equivalence is due to the decomposition $m(\rho - \ell) = k + u$.

Finally, the inequality of the proposition follows from the fact that $(1 + \ell - \rho)e^{-\ell}$ is a nondecreasing function of ℓ over $[0, \rho]$.

References

1. Alon, N., Azar, Y., Woeginger, G., Yadid, T.: Approximation schemes for scheduling on parallel machines. J. Sched. **1**(1), 55–66 (1998)
2. Bansal, N., Nagarajan, V.: On the adaptivity gap of stochastic orienteering. Math. Program. **154**(1–2), 145–172 (2015)
3. Berg, B., Denton, B.: Fast approximation methods for online scheduling of outpatient procedure centers. INFORMS J. Comput. **29**(4), 631–644 (2017)
4. Boyd, S., Vandenberghe, L.: Convex Optimization. Cambridge University Press, Cambridge (2004)
5. Canetti, R., Irani, S.: Bounding the power of preemption in randomized scheduling. SIAM J. Comput. **27**(4), 993–1015 (1998)
6. Correa, J.R., Skutella, M., Verschae, J.: The power of preemption on unrelated machines and applications to scheduling orders. Math. Oper. Res. **37**(2), 379–398 (2012)
7. Dean, B.C., Goemans, M.X., Vondrák, J.: Adaptivity and approximation for stochastic packing problems. In: Proceedings of the Sixteenth Annual ACM-SIAM Symposium on Discrete Algorithms, pp. 395–404. Society for Industrial and Applied Mathematics (2005)
8. Dean, B.C., Goemans, M.X., Vondrák, J.: Approximating the stochastic knapsack problem: the benefit of adaptivity. Math. Oper. Res. **33**(4), 945–964 (2008)

9. Dell'Olmo, P., Kellerer, H., Speranza, M., Tuza, Z.: A 13/12 approximation algorithm for bin packing with extendable bins. Inf. Process. Lett. **65**(5), 229–233 (1998)
10. Dell'Olmo, P., Speranza, M.: Approximation algorithms for partitioning small items in unequal bins to minimize the total size. Discret. Appl. Math. **94**(1–3), 181–191 (1999)
11. Denton, B., Miller, A., Balasubramanian, H., Huschka, T.: Optimal allocation of surgery blocks to operating rooms under uncertainty. Oper. Res. **58**(4–part-1), 802–816 (2010)
12. Dexter, F., Traub, R.: How to schedule elective surgical cases into specific operating rooms to maximize the efficiency of use of operating room time. Anesth. Analg. **94**(4), 933–942 (2002)
13. Garey, M.R., Johnson, D.S.: Computers and Intractability: A Guide to NP-Completeness. WH Freeman and Company, San Francisco (1979)
14. Gupta, A., Nagarajan, V., Singla, S.: Algorithms and adaptivity gaps for stochastic probing. In: Proceedings of the Twenty-seventh Annual ACM-SIAM Symposium on Discrete Algorithms, pp. 1731–1747. SIAM (2016)
15. Isern, D., Sánchez, D., Moreno, A.: Agents applied in health care: a review. Int. J. Med. Inform. **79**(3), 145–166 (2010)
16. Kolliopoulos, S., Steiner, G.: Approximation algorithms for scheduling problems with a modified total weighted tardiness objective. Oper. Res. Lett. **35**(5), 685–692 (2007)
17. Kovalyov, M.Y., Werner, F.: Approximation schemes for scheduling jobs with common due date on parallel machines to minimize total tardiness. J. Heuristics **8**(4), 415–428 (2002)
18. Leung, J.Y.T.: Handbook of Scheduling: Algorithms, Models, and Performance Analysis. CRC Press, Boca Raton (2004)
19. Liu, M., Xu, Y., Chu, C., Zheng, F.: Online scheduling to minimize modified total tardiness with an availability constraint. Theor. Comput. Sci. **410**(47–49), 5039–5046 (2009)
20. Marshall, A., Olkin, I., Arnold, B.: Inequalities: Theory of Majorization and its Applications. Elsevier, Amsterdam (1979)
21. Megow, N., Uetz, M., Vredeveld, T.: Models and algorithms for stochastic online scheduling. Math. Oper. Res. **31**(3), 513–525 (2006)
22. Möhring, R., Radermacher, F., Weiss, G.: Stochastic scheduling problems I-general strategies. Z. für Oper. Res. **28**(7), 193–260 (1984)
23. Sagnol, G., et al.: Robust allocation of operating rooms: a cutting plane approach to handle lognormal case durations. Eur. J. Oper. Res. (2018). https://doi.org/10.1016/j.ejor.2018.05.022, e-pub ahead of print
24. Schulz, A.S., Skutella, M.: Scheduling unrelated machines by randomized rounding. SIAM J. Discret. Math. **15**(4), 450–469 (2002)
25. Skutella, M., Sviridenko, M., Uetz, M.: Unrelated machine scheduling with stochastic processing times. Math. Oper. Res. **41**(3), 851–864 (2016)
26. Soper, A.J., Strusevich, V.A.: Power of preemption on uniform parallel machines. In: 17th International Workshop on Approximation Algorithms for Combinatorial Optimization Problems (APPROX 2014), pp. 392–402 (2014)
27. Speranza, M., Tuza, Z.: On-line approximation algorithms for scheduling tasks on identical machines with extendable working time. Ann. Oper. Res. **86**, 491–506 (1999)

28. Xiao, G., van Jaarsveld, W., Dong, M., van de Klundert, J.: Models, algorithms and performance analysis for adaptive operating room scheduling. Int. J. Prod. Res. **56**(4), 1389–1413 (2018)
29. Zhang, Z., Xie, X., Geng, N.: Dynamic surgery assignment of multiple operating rooms with planned surgeon arrival times. IEEE Trans. Autom. Sci. Eng. **11**(3), 680–691 (2014)
30. Zhu, M., et al.: Managerial decision-making for daily case allocation scheduling and the impact on perioperative quality assurance. Transl. Perioper. Pain Med. **1**(4), 20 (2016)

Author Index

Azar, Yossi   21

Bensmail, Julien   36
Bienkowski, Marcin   51
Böckenhauer, Hans-Joachim   102, 118
Borodin, Allan   69
Bousquet, Nicolas   295
Boyar, Joan   69
Byrka, Jarosław   87

Calinescu, Gruia   134
Chang, Minjun   149
Christodoulou, George   165
Czumaj, Artur   181

Dosa, Gyorgy   204
Driemel, Anne   218

Fairstein, Yaron   238
Fuchs, Janosch   102

Geissmann, Barbara   259

Hochbaum, Dorit S.   149

Ikeda, Motoki   277

Jacob Fanani, Amit   21

Kellerer, Hans   204
Komm, Dennis   118
Kortsarz, Guy   134
Kraska, Artur   51
Krivošija, Amer   218

Larsen, Kim S.   69
Lewandowski, Mateusz   87
Liu, Hsiang-Hsuan   51

Mansour, Yishay   181
Mary, Arnaud   295
Mazauric, Dorian   36
Mc Inerney, Fionn   36
Melissourgos, Themistoklis   165

Naor, Seffi (Joseph)   238
Nisse, Nicolas   36
Nutov, Zeev   134

Olver, Neil   310

Pankratov, Denis   69
Pérennes, Stéphane   36
Pruhs, Kirk   310

Raz, Danny   238

Sagnol, Guillaume   327
Schewior, Kevin   310
Schmidt, Paweł   51
Schmidt genannt Waldschmidt, Daniel   327
Sitters, René   310
Spaen, Quico   149
Spirakis, Paul G.   165
Spoerhase, Joachim   87
Stougie, Leen   310

Tanigawa, Shin-ichi   277
Tesch, Alexander   327
Tuza, Zsolt   204

Unger, Walter   102

Vardi, Shai   181
Velednitsky, Mark   149

Wegner, Raphael   118
Woeginger, Gerhard J.   3

Printed in the United States
By Bookmasters